INSTRUCTOR'S MANUAL AND RESOURCE GUIDE

TO ACCOMPANY

THE NATURE OF LIFE

INSTRUCTOR'S MANUAL AND RESOURCE GUIDE

TO ACCOMPANY

JOHN H. POSTLETHWAIT & JANET L. HOPSON

Prepared by

Dennis Todd

University of Oregon

McGRAW-HILL, INC.

NEW YORK ST. LOUIS SAN FRANCISCO AUCKLAND BOGOTÁ CARACAS HAMBURG LISBON LONDON MADRID MEXICO MILAN MONTREAL NEW DELHI PARIS SAN JUAN SÃO PAULO SINGAPORE SYDNEY TOKYO TORONTO

Instructor's Manual and Resource Guide to accompany
The Nature of Life, Second Edition

1 2 3 4 5 6 7 8 9 0 MAL MAL 9 0 9 8 7 6 5 4 3 2

ISBN 0-07-050651-5

The editors were Mary M. Eshelman and Eileen Burke.
The production supervisor was Pattie Myers.
Editorial assistance was provided by Susan D. Borowitz and Charissa Hogeboom.
Malloy Lithographing, Inc., was printer and binder.

CONTENTS

TO THE INSTRUCTOR

This manual is written to assist instructors using *The Nature of Life*, Second Edition, by Postlethwait and Hopson. Biology teachers face daunting tasks: they must keep up with rapid changes in diverse specialized fields and they must present old and new material to their students clearly and convincingly. It is my hope that this manual aids them in these tasks.

Section I, the largest part of this manual, is a chapter-by-chapter treatment of *The Nature of Life*. Each chapter in this manual includes:

- **Perspective**: Puts the text chapter into context by listing other chapters to which it is related.
- **Themes and concepts:** Defines some important subjects and facts that appear in the chapter.
- **Chapter objectives:** A summary of what the student should gain from the chapter.
- **Lecture outline:** Includes the major concepts, facts, and arguments of the text chapter, formatted for use as lecture notes or for overhead projection.
- **Instructional aids:** Videos, films, and references that may be used as classroom materials, additional reading, or background information.
- **Key terms:** An alphabetical list of important terms and the pages on which they appear in the text.
- **Fun facts:** A page of background or collateral reading that students might enjoy. May be photocopied for class distribution.
- **Messages:** A list of key points that can serve as a framework for lectures. May be photocopied for class distribution or for overhead projection.

Section II, General Resources, includes lists of video and film distributors, computer software distributors, and sources of biological supplies and teaching materials.

Section III, Answers to Text Study Questions, includes essay questions from the text and their answers.

I would like to thank John Postlethwait of the University of Oregon for his exemplary teaching, Mary Eshelman of McGraw-Hill for her editorial skills and encouragement, and the Honors College at the University of Oregon for its support.

Dennis Todd
October 1991
Eugene, Oregon

KEY TO NOTATIONS USED IN CHAPTER OUTLINES

(msg 7)	Number of message that applies to the lecture topic.
(p 777)	Text page on which this topic is discussed.
(tpq 7)	Thought-provoking question that may be asked of students during a lecture. (Questions are at the bottom of the page.)
(tr 77)	Transparency number.

ALTERNATE TEACHING PLANS

Instructors can rearrange chapters and sets of chapters for *The Nature of Life*, taking into account the order of subject matter they prefer, the amount of reading their students can do in a week, and the number of weeks in their school's semester, quarter, or academic year. The following sample teaching plans are designed to accommodate different types of course emphasis. Of course, these too may be rearranged to meet individual instructors' needs and preferences.

One-Year Course

OPTION 1: Hierarchy of Life

Semester 1:

Text chapter	Subject area
Ch. 1-19	Cells, Genetics, Diversity

Semester 2:

Text chapter	Subject area
Ch. 20-38	Plant and Animal Biology, Evolution and Ecology

OPTION 2: Ecology/Evolution Emphasis

Semester 1:

Text chapter	Subject area
Ch. 1	Organizing Concepts
Ch. 14-19	Life's Origin and Diversity
Ch. 7, 8	Mendelian Genetics
Ch. 33	Evolution and Population Genetics
Ch. 34-37	Ecology
Ch. 38	Behavior

Semester 2:

Text chapter	Subject area
Ch. 2-6	Cells and Energy
Ch. 9-12	DNA and Genetics
Ch. 13, 14	Development
Ch. 20-29	Animal Biology
Ch. 30-32	Plant Biology

OPTION 3: Zoology

Semester 1:

Ch. 1	Organizing Concepts
Ch. 2-5	Cells and Energy
Ch. 7-14	Genetics and Development

Semester 2:

Ch. 20-29	Animal Physiology and Anatomy
Ch. 33	Evolution and Population Genetics
Ch. 34-37	Ecology
Ch. 38	Behavior

One-Semester Course

OPTION 1: Plant Science Emphasis

Ch. 1	Organizing Concepts
Ch. 2-6	Cells, Energy, and Photosynthesis
Ch. 7-11	Genetics
Ch. 15-17	Life's Origin and Plant Diversity
Ch. 30-32	Plant Anatomy and Physiology
Ch. 33-37	Ecology and Evolution

OPTION 2: Cell Biology Emphasis

Ch. 1	Organizing Concepts
Ch. 2-6	Cells and Energy
Ch. 7-12	Genetics
Ch. 13, 14	Developmental Biology
Ch. 15	Life's Origins
Ch. 20	Overview of Physiology
Ch. 22, 26, 27	Immune, Endocrine, and Nervous Systems
Ch. 31	Plant Development

OPTION 3: General Biology

Ch. 1	Organizing Concepts
Ch. 2-6	Cells and Energy
Ch. 7-12	Genetics
Ch. 13, 14	Developmental Biology
Ch. 15	Life's Origins
Ch. 17	Plants and Fungi
Ch. 18, 19	Animal Diversity
Ch. 20, 21, 26, 27	Animal Physiology
Ch. 30, 32	Plant Anatomy and Physiology
Ch. 33-37	Evolution and Ecology

SUPPLEMENTS

A comprehensive and integrated package of supplementary materials accompanies *The Nature of Life,* Second Edition. In addition to this *Instructor's Manual,* these supplements include:

For the student

- **Biology Write Now!** A brief text intended to incorporate the process of writing into the biology curriculum; includes essays that promote students' ability to integrate the processes and principles of biology through writing exercises.

- **BioPartner.** Offers computerized testing and tutoring of text material; allows students to take practice tests and/or review text material with immediate feedback. Available in IBM and Macintosh versions.

- **Critical Thinking Workbook.** Allows students to review material and concepts presented in the text, to test retention, and to develop abilities of active learning and logical thought.

For the instructor

- **Computerized Instructor's Manual.** A computerized version of this manual. Allows teachers to assemble customized outlines and transparency masters. Includes lecture outlines, chapter perspectives, and other materials; in IBM, Macintosh, and Apple.

- **Test Bank.** Includes VisiQuizzes based on text art and more than 2000 multiple-choice, true/false, completion, and essay questions and answers; tests ability to recall, to integrate concepts, and to analyze principles.

- **Computerized Test Bank.** Allows teachers to assemble customized examinations by incorporating their own questions into any selection of questions from the *Test Bank.* Available for IBM, Macintosh, and Apple.

For lecture/lab

- **Video Series.** Brief videos include nature footage, graphics from the text, and narrative that capture students' attention and make the principles of biology come alive.

- **Biology Slides and Acetate Package.** Colorful, instructive visual aids that include reproductions of text art; for overhead projection or slide projection.

- **Biology Encyclopedia Videodisk.** Contains animations, still images, and moving pictures that bring biology to life; may be accessed by disk player's controls or through the use of the accompanying Hypercard software program.

- **Hands-On Biology** and accompanying **Preparator's Guide**. Presents 20 active laboratory exercises using minimal equipment that offer students the opportunity to work with real organisms and to use the scientific method.

- **Laboratory Manual.** Presents 27 laboratory topics in the same sequence in which they appear in the text.

For further information regarding the supplements, please contact your McGraw-Hill sales representative.

OVERHEAD TRANSPARENCIES OF TEXT ART

OVERHEAD TRANSPARENCIES OF LIGHT AND ELECTRON MICROGRAPHS

SECTION I
INSTRUCTION MANUAL

1

THE NATURE OF LIFE: AN INTRODUCTION

PERSPECTIVE

This chapter introduces the five most important themes in biology: the role of energy in life; the perpetuation of life through reproduction and development; evolution, or biological change over time; the interaction of organisms and their environment; and the process of science as a way of learning about the natural world.

There is a hierarchy of levels of organization ranging from the atomic level to the whole biosphere. These levels provide a general outline for the book: Section I discusses the nature of atoms, molecules, and cells; Section II covers heredity and development; Section III describes the range of living organisms; Sections IV and V explain how organisms function and survive; and Sections VI and VII cover evolution and ecology.

The themes and concepts of this chapter include:

- Biologists have compiled a list of characteristics common to all things.
- Living things exhibit a high degree of order in a hierarchy of organization.
- Genes are inheritable units of information.
- Evolution accounts for the unity and diversity of life.
- Natural selection is the major mechanism leading to evolutionary change.
- Each organism may be classified into genus and species as well as a number of increasingly inclusive groups.
- The process of science includes the principles of causality and uniformity, inductive and deductive reasoning, and the testing of hypotheses.

CHAPTER OBJECTIVES

Students who master the material in this chapter will understand the five main themes in biology: the use of energy to maintain order; reproduction to overcome the death of individuals; the evolution of new traits; the interaction of organisms with the environment; and the utility of the scientific process in learning about the natural world. They will understand that beneath the diversity of life all organisms are united by descent from a common ancestor. They will appreciate the organization of life and see that order appears on many levels. Students should understand the principles of evolution by natural selection and the characteristics of the scientific method.

LECTURE OUTLINE

(p 3) (tr 1)

I. Strawberry frogs provide an introduction to the complexity of life. Their mating habits, parental care, communication, and poisonous skin introduce five themes:

- **A. The role of energy in life.**
- **B. The perpetuation of life through reproduction and development.**
- **C. Biological change over time, or evolution.**
- **D. The interaction of organisms and their environment.**
- **E. The process of science as a way of learning about the natural world.**

(p 4)

II. Life is defined through a set of observable characteristics.

(msg 1)

- **A. Traits related to the use of energy and the maintenance of organization include:**

1. Order refers to the precise arrangement of structural units and activities.
2. Metabolism is the chemical processing of energy compounds.
3. Motility describes the tendency of an organism to sense and react to its surroundings.
4. Responsiveness is the tendency of a living thing to sense and react to its surroundings.

B. Characteristics related to the perpetuation of life include:

1. Reproduction is the process that gives rise to offspring.
2. Development is the orderly sequence of physical and behavioral changes during an organism's life cycle.
3. Genes are units of inheritance that control many daily functions.

C. Adaptations are structures and activities that allow an organism to make better use of its environment.

D. Populations of living things evolve, or adjust to environmental variations through biological changes over time.

(p 4) **III. Living things combat disorder by taking energy and materials from their surroundings and employing them for maintenance, growth, and other survival activities.**

(msg 2)

(msg 3) **A. Living organisms possess a degree of order greater than anything in the nonliving world, as in the hierarchy of life.**

1. Levels of organization of the biosphere include ecosystems made up of communities of organisms and their physical environments. Communities are made up of populations of interbreeding organisms.

2. Each organism is made up of organ systems comprising organs made of tissues composed of cells, the least complicated units that are truly alive.

3. Cells include organelles, and are made of molecules, clusters of atoms composed of subatomic particles.

B. Metabolism includes the chemical steps that transform energy and materials in living organisms.

(tpq 1) **C. Motility, of cellular contents and in many cases of whole organisms or their parts, is a diagnostic feature of life.**

---

tpq 1. A tornado moves but a seed doesn't. Does that mean that the tornado is alive but the seed is not?

D. Responsiveness to the environment can be instantaneous or gradual.

(p 8) IV. Some life characteristics perpetuate a species over time.

A. Reproduction may be simple or complex, sexual or asexual.

B. Development is the process of growth of an individual in size and complexity from the fertilized egg to the adult.

C. Genes are the units of inheritance that direct the development and determine the traits of an organism.

1. Genes are parts of DNA molecules.

(tpq 2) 2. Mutations are genetic variations that arise occasionally. Many mutations are harmful but a few are advantageous. The gradual accumulation of such beneficial mutations is called evolution.

(p 10) V. Some life characteristics are related to evolution and environment.

A. Life forms evolve as time passes. All living things are descended from a common ancestor and arose as the result of genetic modification in species that lived before them.

---

tpq 2. Would it be an evolutionary advantage for a species to have perfect DNA replication and repair with no mutations?

B. Adaptations are the different ways of extracting energy and materials from the environment, of reproducing, of moving, of behaving and appearing, and other characteristics that increase an organism's chances of survival and reproduction.

(p 11) **VI. Evolution is biology's central theme.**

(msg 4) A. Biologists organize living things into groups according to their similarities.

1. Closely related individuals are grouped into species. A species includes structurally similar individuals that all descend from the same initial group and that have the potential to successfully breed with one another.
2. A genus includes similar species; related genera are grouped into families, families into orders, orders into classes, classes into phyla or divisions, and ultimately into kingdoms.

B. The simplest kingdom, Monera, includes bacteria.

C. Kingdom Protista comprises more complicated single-celled organisms.

D. Kingdom Fungi includes single-celled and multicellular organisms that absorb energy and materials after decomposing other living or dead organisms.

E. Plants are usually multicellular and generate their own food using the energy in sunlight.

F. Animals are multicellular and get their energy by ingesting other organisms.

G. Life is divided into three domains.

1. Bacteria includes all members of the kingdom Monera.
2. Archaea includes the archaebacteria.
3. Eucarya includes the protists, fungi, plants, and animals.

H. The genetic similarity among all organisms suggests that all living things on earth arose from a single group of ancestral cells present at the dawn of life, powerful evidence for the unity of life.

(msg 5) I. Natural selection, a concept developed by Charles Darwin and A. R. Wallace, is the main mechanism of evolution.

(tpq 3) 1. Lamarck suggested that evolution occurs by the inheritance of acquired characteristics, a theory not supported by observation and experiment.

(tr 2) 2. Darwin noted two facts:

a. Individuals in a population vary in many ways, and some of these variations are inheritable.

---

tpq 3. Sheep are born with long tails. Sheepherders cut the long tails short; are they likely to cause sheep to evolve short tails?

b. Populations have the inherent ability to produce many more offspring than the environment can support. As a consequence, individuals compete for limited resources.

3. Darwin concluded that individuals with traits that allow them to cope efficiently with the local environment leave more offspring than individuals with less-adaptive traits. As a result, certain heritable variations become more common in succeeding generations.
4. Darwin used the term natural selection to describe the greater reproductive success displayed by those individuals with more adaptive characteristics.

(msg 6) 5. An important feature is that natural selection chooses the best-adapted individuals to be parents for the next generation but does not cause the variations, which are products of gene mutations.

(p 16) VII. Science is an ongoing process for observing the world and gaining facts.

(msg 7) A. Causality and uniformity are fundamental principles.

(msg 8) B. Inductive reasoning involves generalizing from specific cases to arrive at broad principles.

C. Deductive reasoning involves analyzing specific cases on the basis of preestablished general principles.

(tpq 4) D. A hypothesis is a tentative generalization. A broad hypothesis that survives repeated attempts to disprove it becomes accepted as a theory.

E. The scientific method may be used to test generalizations. The method is summarized in a series of steps:

(msg 9)

1. Ask a question based on observation of the natural world.
2. Propose a hypothesis.
3. Predict the observations that will occur if the hypothesis is correct.
4. Test the prediction by performing an experiment. A control is included to check the experiment.

F. The scientific method does not apply to matters of religion, politics, culture, ethics, or art.

(p 19) VIII. Biological science can help solve problems like deforestation, loss of natural resources, excessive human reproduction, malnutrition, and disease. We live amidst a revolution in biology with the advent of genetic manipulation and new levels of knowledge about how organisms and ecosystems function.

---

tpq 4. Why must scientists try to disprove theories they believe?

INSTRUCTIONAL AIDS

VIDEOS, FILMS, SLIDES

Biology Set. Carolina Biological Supply Company, Burlington, NC. A starter set of transparencies covering ecology, zoology, histology, and botany.

Life Around Us. Time-Life Video, Paramus, NJ. A series of fourteen fun and educational films covering a wide range of scientific topics. Video, film.

Life on Earth. Carolina Biological Supply Company, Burlington, NC. A spectacular video encyclopedia of natural history, based on the television series hosted by David Attenborough. Spans the history of evolution.

REFERENCES

Case, Christine and Ted Johnson. 1984. *Laboratory Exercises in Microbiology*. Menlo Park, CA: Benjamin/Cummins. Includes 75 class-tested exercises, covering a wide range of topics in microbiology.

Hancock, Judith M. 1987. *Variety of Life: A Biology Teacher's Sourcebook*. Portland, ME: J. Weston Walch. A concise, handy reference on biological diversity for teachers.

Heidemann, Merle K. 1985. *Exercises in Biological Science*. Belmont, CA: Wadsworth. A laboratory manual for lower-division college courses, covering topics from biochemistry to evolution.

Poole, Robert M., ed. 1986. *The Incredible Machine*. Washington, DC: National Geographic Society. A beautifully illustrated, clearly written journey through the human body.

KEY TERMS

adaptation	page 10	motility	7
biology	3	natural selection	12
causality	16	order	5
control	19	prediction	19
deductive reasoning	17	reproduction	8
development	9	responsiveness	7
evolution	10	scientific method	18
gene	9	theory	18
hypothesis	17	uniformity	17
inductive reasoning	17		

FUN FACTS

A FROG PHARMACOPOEIA

The strawberry frog and its relatives in the family Dendrobatidae produce several different toxins. One poisons the signal-conducting mechanism in nerves. Another causes muscles to contract and stay contracted. A third toxin blocks the transmission of signals from nerves to muscles. Scientists have identified more than 200 different compounds secreted by dendrobatid frogs, including some of the most potent naturally occurring toxins. Biologists use the compounds to investigate the ways that nerves and muscles work.

See: Myers, C.W. and J.W. Daly. 1983. Dart-poison frogs. *Scientific American* 248 (2):120 – 132.

The African clawed frog also produces toxins in its skin. A snake that tries to eat one will experience uncontrollable fits of yawning and gaping and will be unable to close its mouth. Biologists consider the clawed frog to be evolutionarily primitive and point out that more highly evolved amphibians produce poisons that taste awful or make predators sick. The advantage of the sickening toxins over those that paralyze muscles is that one sickening or foul-tasting bite is enough to teach a predator not to repeat its folly, but a snake that tries to bite a frog and can't close its mouth will try again and again to capture the frog.

See: Weisburd, S. 1987. Frog defense: Make snakes yawn. *Science News* 132:215.

African clawed frogs, despite the bacteria, fungi, and parasites in ponds, don't often develop infections in their skin wounds, thanks to the skin's production of two protein molecules that inhibit disease organisms. Called magainins (after the Hebrew word for shield), these proteins are the first chemical defense system discovered in vertebrates (other than the immune system). Humans may produce similar substances, which may be the agents that prevent infection of the surface of the eyes and the lining of the lung.

See: Zasloff, Micheal. 1987. Magainins, a class of antimicrobial peptides from *Xenopus* skin. *Proceedings of the National Academy of Science* 84 (15):5449 – 5453.

CHAPTER 1 THE NATURE OF LIFE: AN INTRODUCTION

MESSAGES

1. Nine observable characteristics define life: order, metabolism, movement, responsiveness, reproduction, development, genes, evolution, and adaptations.

2. Organisms must expend energy to combat disorder.

3. The organization of life occurs at many levels, from subatomic particles to the biosphere.

4. Biologists classify living things in groups, from species to kingdoms.

5. Natural selection is the evolutionary mechanism that brings about the diversity of new species from ancestral lines.

6. Genes are inheritable units of information that determine many of an organism's characteristics. Mutations provide the variation on which natural selection acts.

7. Scientific principles include causality and uniformity.

8. Inductive reasoning involves generalizing from specific cases to arrive at broad principles. Deductive reasoning involves analyzing specific cases on the basis of preestablished general principles.

9. The scientific method involves asking a question, proposing a hypothesis, making a prediction, and testing the prediction.

2

ATOMS, MOLECULES, AND LIFE

PERSPECTIVE

This chapter presents the concept that all biological organization and processes depend ultimately on the physical and chemical properties of atoms and molecules. The basics of atomic theory and chemical bonds are discussed in the context of biological molecules and activities. These topics reappear in Chapter 5, which deals with energy use, and Chapter 6, which covers photosynthesis. The importance of water's physical and chemical properties are emphasized; Chapter 3, Cells, Part IV, How Animals Survive, and Part V, How Plants Survive, will discuss the roles of water. The four main classes of biological polymers are introduced, and will reappear throughout the remainder of the text.

The themes and concepts of this chapter include:

- All biological organization rests ultimately on atoms, and the properties of atoms in turn emerge from the precise arrangement of their parts.
- Atoms with unfilled outer energy levels tend to combine to form molecules.
- Water molecules are polar and are held to one another by hydrogen bonds, leading to properties that are biologically important.
- Living things contain four main classes of organic compounds: carbohydrates, lipids, proteins, and nucleic acids.
- A polymer's properties are determined by the nature of the monomers and the way those monomers are arranged.
- Proteins are keys to the diversity of life.
- Molecules comprising nucleic acids, including ATP, DNA, and RNA, are involved in the transfer of energy or the transmission of hereditary information.

CHAPTER OBJECTIVES

Students who master the material in this chapter will understand that all biological organization rests ultimately on atoms and their interactions. They will know that one element is different from another in the number of protons in the nucleus, that electrons in the outermost energy level form chemical bonds, and that polarity and hydrogen bonds affect the properties of water and other substances.

This chapter introduces four main classes of biological molecules: carbohydrates, proteins, lipids, and nucleic acids. Students will see that biological polymers are made of specific families of monomers and have specific functions. They will see begin to understand the importance of proteins and nucleic acids.

LECTURE OUTLINE

(p 24) **I. Australian aborigines' use of water introduces the topic of chemistry.**

- **A. All life depends on water.**
- **B. Living things are made up mostly of water molecules.**
- **C. The behavior of atoms and molecules underlies and explains the behavior of living cells.**
- **D. The physical structure of atoms and molecules determines their chemical properties.**
- **E. <u>Emergent properties</u> of the whole emerge not just from the properties of the individual parts, but from the way those parts are arranged and interact.**

(p 26) **II. <u>Elements</u> are pure substances; <u>compounds</u> are combinations of elements.**

- **A. Earth and life have different combinations of elements.**

1. Eight elements (O, Si, Al, Na, Ca, Fe) make up 98 percent of the earth's surface layer.
2. Six elements (H, O, C, N, P, S) make up 99 percent of a typical organism.

B. Atomic theory states that atoms are the smallest particles into which an element can be divided that still display the properties of that element.

(msg 1)

1. Atoms are composed of protons, neutrons, and electrons.
2. An atom's nucleus includes a set number of protons and a variable number of neutrons.
   a. The atomic number is the number of protons.
   b. The atomic mass is the number of protons plus neutrons.
   c. Isotopes are atoms of the same element with different numbers of neutrons.
   d. Radioactive isotopes are unstable and may emit energy as they change to stable forms.

(msg 2)

3. A set number of electrons, equal to the number of protons, orbits the nucleus.
   a. Ions are atoms with a positive or negative charge due to a gain or loss of electrons.
   b. Electrons are attracted by the nucleus but repelled by each other.
   c. Electrons occur in discrete energy levels. The energy levels nearest the nucleus (the levels with the lowest energy) are filled first.

(tpq 1)

4. An atom's chemical properties are largely a result of the number of electrons in the outermost energy level.
   a. An atom with a full outer level is inert.
   b. Atoms with unfilled outer levels will react with other atoms to fill or empty those levels.

C. Molecules are groups of atoms held together by molecular bonds, which are links of pure energy based on shared or donated electrons.

(tr 3)

1. Covalent bonds form when atoms share electrons.
   a. A single bond results from the sharing of one pair of electrons.
   b. Double bonds result from the sharing of two pairs of electrons.
2. Nonpolar covalent bonds occur between two atoms that share electrons equally.
3. Polar covalent bonds occur when the shared electrons spend more time around one of the nuclei. A slight negative charge results around that atom and a slight positive charge results around the other atom.

---

tpq 1. In what way are the properties of the elements *emergent properties*?

---

4. Hydrogen bonds are weak bonds that involve the attraction between a polar molecule that has a negative pole and a polar molecule that has a hydrogen atom bearing a slight positive charge.

(tr 4) 5. Ionic bonds occur between positively and negatively charged ions.

(p 33) (msg 3) **III. Hydrogen bonding between water molecules and other substances makes life possible.**

(tpq 2) **A. Hydrogen bonds affect the physical properties of water related to temperature.**

1. Water is slow to heat. Heat energy breaks and stretches hydrogen bonds between molecules before raising the temperature.
2. An unusual amount of heat is required to evaporate water. Hydrogen bonds hold liquid water molecules together, and a molecule must have a high velocity to escape as a gas. By evaporating, water removes heat from its surroundings.

B. Water's hydrogen bonds affect its mechanical properties.

---

tpq 2. Methane (CH_4) is a molecule similar to water in size and weight. Why is it a gas at room temperature while water is a liquid?

(tpq 3)

1. Cohesion is the tendency of like molecules to cling to each other. Surface tension is the tendency of water molecules at the surface of liquid water to stick to each other but not to the molecules of air above them.
2. Adhesion is the tendency of unlike molecules to cling together. Capillarity is the tendency of a liquid such as water to move upward through a narrow space.
3. Ice is less dense than water and floats on it.

C. Water has been called the universal solvent, a substance that can dissolve solutes including polar molecules and ions.

1. Hydrophilic substances dissolve readily in water.
2. Hydrophobic substances do not dissolve readily in water.

D. Water molecules tend to ionize, or dissociate, into a hydrogen ion, H^+, and a hydroxide ion, OH^-

1. An acid is any substance that gives off hydrogen ions when dissolved in water.
2. A base is any substance that accepts hydrogen ions when dissolved in water. Such substances are called alkaline.

---

tpq 3. Could trees grow as tall (or would they grow taller) if water molecules didn't form hydrogen bonds?

(tr 5) 3. The <u>pH scale</u> measures the concentration of H^+. The scale is logarithmic.

4. <u>Buffers</u> moderate changes in pH.

(p 38) **IV. Compounds that contain carbon are the stuff of life.**

(msg 4) A. Carbon-containing molecules are called <u>organic</u>; all other substances are <u>inorganic</u>.

B. Bonding versatility is carbon's key characteristic.

C. Carbon atoms have four electrons in their outermost energy level and thus are four electrons short of a filled outer level.

D. Biological molecules are composed of chains or rings of carbon atoms.

(tr 6) 1. <u>Functional groups</u> are small clusters of atoms that impart specific chemical properties to the molecules to which they are attached.

2. <u>Polymers</u> are made of <u>monomers</u> linked by covalent bonds. <u>Carbohydrates</u>, <u>lipids</u>, <u>proteins</u>, and <u>nucleic acids</u> are the four main classes of biological molecules.

(msg 5) 3. Carbohydrates, the most abundant carbon compounds in organisms, are made of C, H, and O in a 1:2:1 ratio.

a. The monomers that make up carbohydrates are sugar molecules called <u>monsaccharides</u>; they include <u>glucose</u>, <u>fructose</u>, and <u>galactose</u>.

b. Monosaccharides are linked to form disaccharides, such as sucrose, lactose, and maltose; and to form polysaccharides, such as starch, glycogen, cellulose, and chitin.

(tpq 4) c. The different three-dimensional structures of the molecules impart different chemical properties.

(msg 6) 4. Lipids can serve as energy storage molecules or as waterproof coverings around cells.

a. Triglycerides include fats and oils. Each molecule contains a glycerol molecule bonded to three fatty acids.

b. All the carbon atoms of saturated fats are bonded to at least two hydrogen atoms.

c. Unsaturated fats have one or more double bonds between carbon atoms.

d. Waxes have different structures from triglycerides.

(tr 7) e. Phospholipids are like lipids but have a phosphate group in place of one of the fatty acid chains, making the molecule hydrophilic. They are components of cell membranes.

---

tpq 4. How can different sugar molecules have the same formula, $C_6H_{12}O_6$?

f. Steroids include sex hormones and cholesterol. They can pass through the hydrophobic molecules that make up cell membranes.

(msg 7)

5. Proteins come in a wide variety of forms and functions.

   a. Proteins include structural proteins, regulatory proteins, hormones, transport proteins, antibodies, and enzymes.

   b. The specialized shapes and functions of different cell types depend on proteins.

   c. Proteins are polypeptides, long chains of amino acid subunits, folded into characteristic three-dimensional shapes.

   d. There are 20 common amino acids.

(tr 8)

(msg 8)

   e. Proteins have complex shapes based on four levels of structure.

      i. A protein's unique linear sequence of amino acids is its primary structure.

      ii. A protein chain will assume a folding pattern, the secondary structure, that allows the maximum number of hydrogen bonds between amino acids: either an alpha helix, a beta sheet, or a disordered loop.

      iii. The tertiary structure is the three-dimensional shape of the protein molecule.

iv. Quaternary structure is the way two or more folded polypeptide chains fit together.

v. Cofactors are add-on substances that help proteins function.

(msg 9) 6. Nucleic acids are information-carrying molecules or energy-carrying molecules.

a. Nucleic acids are made of monomers called nucleotides.

(tr 9) b. DNA is a double helix, RNA a single-stranded molecule.

c. Adenosine phosphates are not polymers. Adenosine phosphates include ATP, the universal energy molecule, and cAMP, which carries chemical signals.

d. Nucleotides called coenzymes are transport compounds necessary for energy harvest and building new cellular structures.

INSTRUCTIONAL AIDS

VIDEOS, FILMS, SLIDES

Chemistry in Today's World. Educational Images Ltd., Elmira, NY. A series of slide presentations covering carbon, nitrogen, and other chemical topics.

Dust of Life. Films Incorporated, Chicago, IL. An exploration of carbon in its many guises. Video, film.

Inquiring Into Life. Coronet/MTI, Deerfield, IL. Traces the levels of organization from molecules to complex living systems. Video, film.

REFERENCES

Holum, John R. 1986. *Organic and Biological Chemistry*, Second Edition. New York: John Wiley and Sons. A well-written textbook on the molecular basis of life.

Laing, Michael. 1987. No rabbit ears on water. *Journal of Chemical Education* 64:124 – 128. A concise discussion of water's molecular shape and structure.

Piel, Jonathan, ed. 1985. *The Molecules of Life: Issue of Scientific American.* New York: W. H. Freeman. October, 1985. A collection of articles discussing the molecular basis of life. Excellent photographs and graphics.

KEY TERMS

FUN FACTS

NAMES OF THE ELEMENTS

The names of the elements give us a fascinating glimpse into the history of chemistry. "Carbon" comes from the Latin word for charcoal. "Phosphorus" comes from Greek roots that mean "light bringer." Gold, silver, and other metals were known to the classical Romans and their contemporaries, and the names they gave these elements are preserved in the chemist's shorthand. Gold (*Au*) was *aurum*, "shining dawn." Silver, *Ag* in shorthand, comes from *argentum*, a Greek word that meant "white." Mercury, called *hydrargyrum*, or "watery silver," is *Hg* to chemists. Lead, *Pb* in the periodic table, was *plumbum*, meaning "heavy." Copper in ancient times came from mines on the island of Cyprus, giving this metal the name *cuprum*, abbreviated *Cu*. Iron's symbol, *Fe*, comes from a Sanskrit word that meant "firmness." *Natrium*, "soda" in Latin, is our sodium (*Na*). Some elements' names end with *-gen*, which means "giving rise to." *Hydrogen* gives rise to water (*hydro-*) when burned, *nitrogen* forms soda, and *oxygen* produces acid (in Greek, *oxys* literally meant "sharp," and by extension, "acid").

See: Moeller, Therald, et al. 1984. *Chemistry*, Second edition. Academic Press, Orlando.

BUCKMINSTERFULLERENE

Carbon's celebrated ability to form bonds underlies its ability to form intriguing molecules. Recently discovered hollow-cage structures, formed primarily or exclusively of carbon atoms, open new avenues of investigation. One exceptionally stable hollow-cage molecule has been named "buckminster-fullerene" to honor famed architect Buckminster Fuller, who developed the geodesic dome. The molecule is composed of 60 carbon atoms bonded to each other to form a soccer-ball-like structure (inspiring its other name, "footballene") with hexagonal and pentagonal faces. C_{60} has been detected in soot and in cosmic dust clouds. Other stable fullerenes may be composed of hundreds of carbon atoms.

See: Kroto, Harold. 1988. Space, stars, C_{60}, and soot. *Science* 242:1139 – 1145.

CHAPERONE PROTEINS

The folding of proteins into functional three-dimensional shapes has long been thought a spontaneous process, but recent evidence indicates that in many cases these events are aided by other proteins, termed "chaperone proteins" because they stabilize and protect newly formed polypeptides. Although chaperones were originally thought to be enzymes, several features puzzled scientists. They act much slower than enzymes, they cannot catalyze the formation of covalent bonds, and many can interact with a wide variety of target proteins that seem to have little in common. Chaperones instead seem to stabilize segments of polypeptides that would bind to other segments inappropriately. Thus, chaperones allow the polypeptide to fold into its proper three-dimensional configuration.

See: Sambrook, Joe, and Mary-Jane Gething. 1989. Chaperones, paperones. *Nature* 342:224 – 225.

CHAPTER 2 ATOMS, MOLECULES, AND LIFE

MESSAGES

1. Atoms are made of protons, neutrons, and electrons. All atoms of an element have a specific number of protons.

2. The number of electrons in the outer energy level determines to a large extent an atom's chemical properties. It is these electrons that form chemical bonds.

3. The polarity of water molecules and their ability to form hydrogen bonds affects many of water's physical and chemical properties.

4. Carbon-based molecules are the basic components of life. Biological macromolecules are polymers made of monomers of a particular family.

5. Carbohydrates are energy reserves and structural components of cells.

6. Lipids serve as energy reserves or waterproof coverings around cells.

7. Proteins are made of some 20 different amino acids, which have identical backbones but different side groups.

8. Protein chains fold into specific three-dimensional shapes. This shape is essential to their function.

9. Nucleic acids include polymers such as DNA and RNA, which are information-carrying molecules, and monomers such as ATP, an energy-carrying molecule.

3

CELLS: THE BASIC UNITS OF LIFE

PERSPECTIVE

This chapter outlines the similarities and differences found among cells of different types. Cells must take in and process raw materials and energy; these activities are discussed in the remaining chapters of this section and in the sections on plant and animal physiology. The role of DNA and the production of proteins are introduced and will be covered in more detail in Chapters 7 through 11 in the genetics section.

The themes and concepts of this chapter include:

- All cells have certain basic functions and structures.
- Life is an emergent property based on the organization of cells and their integrated functions.
- All organisms are made of cells, which are the basic units of life.
- All cells arise from preexisting cells.
- Prokaryotes are structurally simple; eukaryotes are structurally complex and include intracellular membrane systems.
- Cells must be separated from yet exchange materials with their environment and obtain and process energy.
- Information molecules from the nucleus direct the cell's activities.
- Mitochondria produce ATP by breaking down biological molecules.
- Chloroplasts carry out photosynthesis.
- Cells are often surrounded by walls or matrices and may be attached to other cells.

CHAPTER OBJECTIVES

Students who master the material in this chapter will see that all cells share certain basic functions and physical structures and that a cell's specialization depends on its size, shape, and organelles. They will understand that life is an emergent property based on the activities of the cell's parts. They will be able to describe the differences between prokaryotic and eukaryotic cells.

This chapter discusses the features and functions of cell membranes, the nucleus, ribosomes, the cytoskeleton, the endoplasmic reticulum, mitochondria, chloroplasts, flagella, and cilia; students should understand the functions and interactions of these organelles.

LECTURE OUTLINE

(p 52) **I. Euglena, a one-celled aquatic creature with both plant and animal characteristics, demonstrates unifying themes that relate to all cells.**

(msg 1) **A. All cells share basic functions performed by physical structures. The universality of these functions and structures is a strong argument for the unity of life.**

(tpq 1) **B. Each cell's specialization depends on its shape, size, and organelles.**

(msg 2) **C. The life of a cell depends on the arrangement of its parts and their integrated activities. The living state is an emergent property.**

(p 54) **II. Microscopes were developed in sixteenth and seventeenth century Europe.**

tpq 1. How does *Euglena* demonstrate that the divisions "plant" and "animal" are not useful when discussing protists?

(msg 3)

A. Anton van Leeuwenhoek observed animalcules in drops of water.

B. Robert Hooke discovered cells in cork.

C. Other scientists derived the cell theory.

1. All living things are made of cells.
2. Cells are the basic living units.
3. Living cells arise only from preexisting cells. (tpq 2)

(p 55) III. Organisms may be prokaryotes or eukaryotes.

(tr 10) A. The domains Bacteria and Archaea include all prokaryotes.

(msg 4)

1. Prokaryotes lack a membrane-bound nucleus; the DNA is a naked strand in a region called the nucleoid.
2. Prokaryotes do not have other membrane-bound organelles.
3. Many prokaryotic cells are surrounded by cell walls.
4. Bacteria are only distantly related to the members of the domain Archaea.
5. Their biochemical diversity allows prokaryotes to exploit many environments.

(msg 5) B. The domain Eucarya includes the kingdoms Protista, Fungi, Plantae, and Animalia, which are eukaryotes.

---

tpq 2. Is it possible to make a living cell by combining isolated cellular organelles?

1. **Like prokaryotes, eukaryotic cells are surrounded by a cell membrane.**

(tpq 3) 2. **Eukaryotic cells have internal membranes that surround the nucleus and other internal structures.**

3. **Eukaryotes have a highly organized cytoplasm with an internal latticework, the cytoskeleton, that contributes to structure and movement within the cell.**

4. **In multicellular eukaryotes, cells may specialize for given tasks in a division of labor.**

(tr 11) 5. **Common features of plant and animal cells include a nucleus, mitochondria, ribosomes, endoplasmic reticulum, Golgi apparatus, cytoskeleton, and a plasma membrane.**

   a. **Plant cells may be surrounded by a cell wall and may be largely filled by a storage organelle, the vacuole.**

   b. **Plant cells may contain chloroplasts, which carry out photosynthesis.**

6. **Cellular specialization leads to specialization of tissues, organs, and organisms.**

C. **Cells vary in size and number in organisms.**

---

tpq 3. How might the internal organization of a eukaryotic cell allow it to be larger than a prokaryotic cell?

---

1. **Most cells are small. Prokaryotes are the smallest cells. The largest animal cells include ostrich eggs and giraffe nerve cells.**

(tpq 4) 2. **The surface-to-volume ratio determines a cell's ability to exchange materials with its environment. A cell's shape affects its surface-to-volume ratio.**

(p 61) **IV. All cells must perform certain tasks.**

(msg 6) **A. The cytoplasm must be separated from its environment so that appropriate internal conditions can be maintained.**

B. The cell must take in raw materials and expel wastes through the barrier that separates it from its environment.

C. It must take in energy and convert it to a form useful for powering the cellular machinery.

D. It must synthesize molecules and cell parts for repair, growth, and reproduction.

E. It must coordinate and regulate its activities.

F. All eukaryotic cells have certain structures.

(tr 12) 1. **The plasma membrane separates the cell from its environment and controls the movement of substances into and out of the cell.**

---

tpq 4. If you doubled the diameter of a cell without changing its shape, how would you affect its ability to exchange materials with the environment?

---

a. The membrane is a lipid bilayer composed of phospholipid molecules with hydrophilic heads and hydrophobic tails. It is selectively permeable.

b. The membrane is studded with proteins that stabilize the lipid bilayer and act as gates, pumps, markers, or signal receptors, a concept known as the fluid mosic model.

(tpq 5) c. The membrane may fold inward to import material or outward to expel material from the cell.

(tr 13) 2. The nucleus is the largest organelle and contains the genetic material, DNA.

(msg 7) a. The genetic information passes from DNA to RNA to proteins, which carry out the work of the cell.

b. The nucleus is surrounded by a double-layer membrane, the nuclear envelope, which is perforated by nuclear pores. Each pore is a cluster of proteins that form a channel.

c. Nucleoli make ribosomal RNA, which is necessary for the production of proteins.

3. Most of the cell's mass is cytoplasm.

(tr 14) a. Ribosomes make protein molecules.

---

tpq 5. How might a cell take in or expel something too large to pass through a membrane protein channel?

---

b. **The cytoskeleton is a three-dimensional latticework that maintains the cell's shape and moves materials within the cell.**

4. **A system of internal membranes is involved in the manufacture, storage, transport, and export of proteins and raw materials.**

(tr 15)

a. **The endoplasmic reticulum (ER) is a series of membrane channels that may be studded on the inside with ribosomes (rough ER) for protein synthesis or that may be without ribosomes (smooth ER) and involved in the synthesis of non-proteins such as lipids.**

b. **At the ends of the ER channels, membrane sacs (vesicles) pinch off and carry the products to their destination, which may be another membrane system, the Golgi apparatus, which modifies and packages proteins, lipids, and other substances and exports most of them from the cell.**

c. **Lysosomes contain digestive enzymes that break down ingested food or, if broken open, digest the cellular components.**

d. **Microbodies are vesicles such as peroxisomes.**

(tr 16) (msg 8)

5. **<u>Mitochondria</u> provide chemical fuel for cellular processes by converting the energy in carbon-containing molecules into the energy of ATP molecules. This process is called <u>aerobic respiration</u>.**
   a. **A mitochondrion has a smooth <u>outer membrane</u> and an <u>inner membrane</u> thrown into folds called <u>cristae</u>, where ATP production takes place.**
   b. **Mitochondria are the most abundant in cells with the highest activity level.**
   c. **Mitochondria have their own DNA, ribosomes, and reproduction cycle, indicating that they are the descendants of bacteria.**

   (tpq 6)

   d. **Mitochondria are passed to an animal only from its mother.**

(p 73)

V. Some cells have specialized structures and functions.

A. **<u>Plastids</u> harvest solar energy and produce and store food.**

(tr 17)

1. **<u>Chloroplasts</u> are the organelles of photosynthesis, converting solar energy into chemical energy in the form of carbon compounds.**

---

tpq 6. Why is it that an inherited disease caused by a mitochondrial genetic defect can never be passed from a father to his children?

---

a. Chloroplasts contain chlorophyll and other light-absorbing pigments.

b. Chloroplasts contain their own DNA and may have evolved from bacteria that came to inhabit larger cells.

2. Chromoplasts are plastids that store yellow, orange, or red energy-trapping pigments and give color to fruits and flowers.

3. Leucoplasts are colorless storage organelles.

B. Many plant cells have a central vacuole that contains water and various storage products.

1. This reduces the volume of cytoplasm and thus increases the cell's surface-to-volume ratio.

2. The pressure of the water in the vacuole keeps the cell inflated. A plant wilts when the vacuole pressure drops because of drought.

C. Many protists that live in fresh water have a contractile vacuole that collects and pumps out water that could otherwise build up and burst the cell.

D. Organelles of movement include cilia and flagella.

E. The plasma membranes of most cells are surrounded by cell coverings that protect the delicate membrane.

(msg 9)

1. An extracellular matrix, a meshwork of secreted molecules, protects many cells that live within multicellular organisms.

2. Cell walls made largely of cellulose surround plant cells.
   a. First the cell lays down a primary cell wall outside the plasma membrane.
   b. Once plant cells stop growing, many lay down a secondary cell wall.
3. Virtually all animal cells secrete a meshwork of molecules that surrounds them. These molecules are mostly fibrous proteins.
   a. Collagen is the most common of these fibrous proteins, making up tendons and other connective tissue.
   b. This extracellular matrix cushions the cell, strengthens the tissue, and maintains cell shape.

(tr 18)

4. Cells in multicellular organisms are attached to neighboring cells. The junctions provide impermeability, adherence, and communication.

INSTRUCTIONAL AIDS

VIDEOS, FILMS, SLIDES

Cell Structure and Function. Educational Images Ltd., Elmira, NY. Compares prokaryotic and eukaryotic cells and presents the structure and function of organelles. Transparencies, video.

Cell Motility and Microtubules. Films for the Humanities and Sciences, Princeton. A study of ciliary and flagella movements in various organisms. Video.

The Cell: Its Structure. Carolina Biological Supply Company, Burlington, NC. A set of transparencies and a narrative cassette presenting the cell as a dynamic unit of life.

REFERENCES

Bubel, A. 1989. *Microstructure and Function of Cells: Electron Micrographs of Cell Ultrastructure.* New York: John Wiley & Sons. Photomicrographs emphasizing the diversity and structural variation of cell components in eukaryotes.

Carroll, Mark. 1989. *Organelles*. New York: Guilford Press. A comprehensive treatment of cell organelles and their interactions. Intended for advanced under-graduates.

de Duve, C. 1984. *A Guided Tour of the Living Cell*. New York: Scientific American Books. In the *Scientific American* tradition, well illustrated and packed with information.

Fulton, A. B. 1984. *The Cytoskeleton: Cellular Architecture and Choreography*. New York: Chapman and Hall. This slim text is a quite readable overview.

Robertson, R. N. 1983. *The Lively Membranes*. Cambridge: Cambridge University Press. A lively, readable essay that emphasizes visualization and minimizes mathematics.

KEY TERMS

FUN FACTS

ENDOSYMBIOSIS AND EUKARYOTES

Eukaryotic cells arose more than two billion years after prokaryotes appeared, and may have arisen by the cooperative amalgamation of different prokaryotes. Mitochondria and chloroplasts seem to be endosymbiotic descendants of formerly free-living bacteria. They have their own DNA, which resembles prokaryotic DNA more than the DNA of the host cell's nucleus, their own protein-making apparatus, and their own life cycle.

The incorporation of photosynthetic cells into nonphotosynthetic hosts may have occurred independently several times in evolution. Red algae have chloroplasts that are probable descendants of cyanobacteria. The ancestors of green algae may have incorporated different photosynthetic bacteria, known as prochlorophytes, ultimately leading to land plants. Eukaryotic algae are endosymbionts in a number of different organisms, including protists, cnidarians, and mollusks.

The degree of a chloroplast's dependence on the nucleus indicates the antiquity of the symbiosis. Organelles incorporated recently tend to be more independent and can often be cultured in isolation. But plastids that have long been symbionts have lost their independence as genes for chloroplast proteins have been transferred to the nucleus. Recent research indicates that "promiscuous" genes jump between chloroplasts, mitochondria, and nucleus.

See: Kite, G. 1986. Evolution by endosymbiosis: the inside story. *New Scientist* 111:50–52. July 3, 1986.

MEMBRANE PUMPS AND CANCER

A persistent problem in cancer therapy has been tumor resistance to chemotherapy. During the first treatment stages, chemotherapy seems effective; the tumor shrinks and apparently vanishes, only to reappear later. Subsequent drug treatments show progressively reduced effectiveness as drug-resistant tumor cells proliferate.

Many cancer cells show cross-resistance to several unrelated drugs, due to membrane protein pumps that actively transport drugs out of the cells. The pumps, called P-glycoproteins, seem conserved in evolution and appear in several mammalian species, from hamsters to humans. In normal cells, they may function to remove toxins from the cell interior. They are normally expressed in the kidneys, adrenal glands, liver, and parts of the gastrointestinal tract. Research suggests that inhibiting P-glycoprotein function may be essential to countering multidrug-resistant cancers.

See: Kartner, Norbert, and Victor Ling. 1989. Multidrug resistance in cancer. *Scientific American* 260(3):44–51.

CHAPTER 3 CELLS: THE BASIC UNITS OF LIFE

MESSAGES

1. **All cells share basic functions and structures that keep them alive and allow them to reproduce. Each cell's specialization depends on its size, shape, and organelles.**
2. **The state of being alive is an emergent property dependent on the arrangement of the cell's parts and on their integrated activities.**
3. **Modern cell theory says that all living things are made of one or more cells, that cells are the basic living units within organisms, and that all cells arise from preexisting cells.**
4. **Prokaryotes are structurally simple and biochemically diverse. They do not have membrane-bound organelles.**
5. **Eukaryotes are structurally complex, with intracellular membrane systems and membrane-bound organelles.**
6. **All cells must be separated from their environment, must exchange materials with their environment, must absorb and use energy, must produce biological molecules, and must regulate and control their activities.**
7. **Information molecules from the nucleus direct a eukaryotic cell's activities.**
8. **Mitochondria are the organelles that produce ATP by breaking down biological molecules. Chloroplasts are the organelles of photosynthesis. Both are endosymbiotic descendants of prokaryotes.**
9. **Cells are often surrounded by walls or matrices and may be attached to other cells.**

4

THE DYNAMIC CELL

PERSPECTIVE

This chapter introduces metabolism and presents the laws of thermodynamics and their implications to life. The vocabulary and concepts of Chapters 2 and 3 are integrated into a discussion of the flow of energy that powers and sustains cells. Life counteracts disorder by capturing and processing energy that emanates from the sun. The principles of exergonic and endergonic chemical reactions apply to the biochemical activities that occur in a cell, where reactions are often coupled by ATP. Enzymes act to speed up reactions, diffusion and osmosis move substances into and out of cells, and membrane pumps actively transport materials through membranes.

Chapter 5 discusses how living things harvest energy from nutrient molecules. Chapter 6 presents photosynthesis, how energy that powers life's processes is originally captured. Concepts presented in Chapter 4 will be important in Part IV, How Animals Survive, and Part V, How Plants Survive.

The themes and concepts of this chapter include:

- The ability to carry out organized biological work depends on a continual flow of energy from the environment.
- Energy exists as potential energy and kinetic energy. Losses accompany all energy conversions as heat is produced.
- Cells expend energy to construct and maintain cell parts and processes.
- Chemical reactions underlie virtually all cellular energy conversions. Exergonic reactions are coupled with endergonic reactions.
- Enzymes accelerate chemical reactions by reducing the activation energy necessary to achieve the transition state between reactants and products.
- By diffusion, osmosis, and active and passive transport, materials move through membranes.
- Cells expend energy to import or export solutes to help maintain an internal solute concentration close enough to the concentration in the extracellular fluid that the cells will neither burst nor shrivel.

CHAPTER OBJECTIVES

This chapter introduces the reader to the fundamental laws that govern the flow of energy in cells and how metabolism harnesses this energy to maintain the cell against entropy. The student who masters this chapter will understand that the molecular activities of cells conform to the same rules that govern other chemical reactions, that energy-liberating reactions can be coupled to energy-consuming reactions, and that enzymes accelerate reaction rates by lowering activation energy. Materials move into and out of cells by passive and active processes. Cells perform various mechanical tasks.

The student should understand that chemical and mechanical processes occur continuously in cells, that the orderliness of life is maintained by a continuous flow of energy that emanates from the sun, and that energy conversions are accompanied by an increase in entropy.

LECTURE OUTLINE

(p 80) **I. Red blood cells introduce cellular function.**

A. Red blood cells lose their nuclei, mitochondria, and ribosomes as they mature.

B. Cells carry out <u>chemical tasks</u>, <u>transport tasks</u>, and <u>mechanical tasks</u>.

(msg 1) **C. Cells depend on a continual flow of energy and materials to counteract the universal tendency toward disorder.**

(p 82) **II. The laws of thermodynamics describe the flow of energy.**

(msg 2) **A. <u>Potential energy</u> is energy that is stored and ready to do work. It can be converted to <u>kinetic energy</u>, or energy of motion.**

B. The <u>first law of thermodynamics</u> says that energy can be converted from one form to another but can neither be created nor destroyed.

C. The <u>second law of thermodynamics</u> states that systems tend toward greater states of disorder.

1. Energy conversions are never 100 percent efficient. Some energy is lost as heat.
2. <u>Heat</u> is the random motion of atoms and molecules.
3. <u>Entropy</u> is the measure of disorder in a system.

(tpq 1) (msg 3) D. A living cell is a temporary repository of order purchased at the cost of a constant flow of energy to power repair and replacement of membranes and organelles.

(p 84) III. Chemical energy fuels the maintenance of cellular order through <u>chemical reactions</u>.

(msg 4) A. <u>Reactants</u> interact to form <u>products</u>.

B. <u>Exergonic</u> reactions proceed spontaneously, releasing energy.

1. The reactants have higher energy than the products.
2. The entropy of the system increases, satisfying the second law of thermodynamics.

---

tpq 1. Does life violate the second law of thermodynamics by reversing disorder?

C. Endergonic reactions do not proceed spontaneously.

1. Energy must be added to drive the reaction.
2. The products have higher energy than the reactants.
3. The energy for the cell's endergonic reactions comes from the cell's exergonic reactions. The reactions are coupled.

D. Metabolic pathways are energy-converting chains of coupled reactions.

1. Catabolism includes the reactions that break down complex molecules to simpler products.
2. Anabolism includes the reactions that combine simple precursors into complex molecules. Anabolic reactions usually require energy input.
3. Catabolism and anabolism constitute metabolism: the energy and material changes withing a living thing that arise from interrelated chemical reactions and allow a cell to utilize energy from its environment and survive.

(tpq 2)
(tr 19)

E. ATP and other energy carriers act as intermediates between metabolic energy exchanges.

---

tpq 2. What might happen to a cell if it were exposed to a poison that prevented the formation of ATP?

(tpq 3) **F.** **In a productive collision, reactants collide with enough energy (the activation energy) to achieve the transition state. Such activated molecules can proceed to form products.**

(msg 5) **G.** **Enzymes speed up reactions without being used up or changed.**

(tr 20) **1.** **Enzymes lower the activation energy needed for biochemical reactions to take place.**

2. **Enzyme molecules have active sites that fit specific substrates.**

3. **An enzyme binds reactants in an enzyme-substrate complex, orienting and changing the shape of the reactants, straining the bonds and helping them reach the transition state.**

H. **Metabolism harvests energy to fuel work and build cell parts. Cell parts are constantly being replaced.**

(p 92) **IV.** **The flow of materials accompanies the flow of energy in a cell. The materials move in the intracellular fluid.**

A. **Cells of multicellular organisms are surrounded by extracellular fluid. Single-celled organisms are surrounded by water or the fluids of their host's body.**

---

tpq 3. Iron rusts (combines with oxygen) at room temperature. Does this mean that rusting does not require activation energy?

(msg 6) (tr 21)

B. Materials move through membranes, by passive transport down concentration gradients or by energy-consuming active transport against gradients.

C. Diffusion is the tendency of materials to move from areas of high concentration to areas of low concentration.

1. Most materials move by simple diffusion through the selectively permeable membrane surrounding the cell.

2. The membrane contains pores that allow materials to pass.

3. Carrier-facilitated diffusion involves carrier molecules that passively allow substances like glucose and urea to pass.

D. Osmosis is the movement of water through a semipermeable membrane from areas of high water concentration to areas of low water concentration.

(tpq 4)

E. In hypotonic solutions, water moves into a cell; in isotonic solutions, there is no net movement; in hypertonic solutions, water moves out of a cell. Plant cells have higher solute concentrations than the surrounding fluid, leading to turgor pressure that keeps the cells inflated.

---

tpq 4. What would happen to a red blood cell if you dropped it into distilled water? Into sea water?

(msg 7) **F. Membrane pumps use energy to move materials against gradients.**

1. **The <u>sodium-potassium pump</u> moves K^+ into the cell and Na^+ out.**
2. **Other <u>membrane transport proteins</u> move glucose, amino acids, and other raw materials into the cell.**

(p 96) (msg 8) **V. Cells must perform mechanical tasks, including movement, packaging and exporting materials, dividing, and reproducing.**

(msg 9) **A. Underlying the mechanical actions in the cell are the rapid assembly and disassembly of cytoskeletal filaments or the sliding of filaments past each other.**

B. These actions are accomplished by enzymes and fueled by ATP.

INSTRUCTIONAL AIDS

VIDEOS, FILMS, SLIDES

The Architecture of Cells. Human Relations Media, Pleasantville, NY. A three-part series that explores how different cells are specialized to carry out specific functions. Filmstrip, video.

The Cell: A Functioning Structure. Pennsylvania State University AV Services, University Park, PA. Two films that survey the structure and function of the living cell.

REFERENCES

Fenn, J. B. 1982. *Engines, Energy, and Entropy: A Thermodynamics Primer*. New York: W. H. Freeman. A concise and easy-to-understand discussion of thermodynamics.

Rowe, H. Alan, and Morris Brown. 1988. Practical enzyme kinetics. *Journal of Chemical Education* 65:548–549. A simple laboratory exercise that demonstrates the kinetics of the enzyme papain.

Williams, Roberta B. 1987. Understanding human energy requirements. *American Biology Teacher* 49:429–433. A lab exercise/discussion that deals with caloric intake, energy metabolism, and body composition.

KEY TERMS

activation energy	page 89	heat	83
active site	90	hypertonic	95
active transport	92	hypotonic	95
anabolism	86	intracellular fluid	92
ATP	87	isotonic	95
carrier-facilitated diffusion	93	kinetic energy	82
catabolism	86	metabolic pathway	86
catalyst	90	metabolism	86
chemical reaction	84	osmosis	94
diffusion	93	passive transport	92
endergonic	85	potential energy	82
energy	82	product	85
entropy	83	productive collision	89
enzyme	90	reactant	85
enzyme-substrate complex	90	second law of thermodynamics	82
exergonic	85	sodium-potassium pump	96
extracellular fluid	92	substrate	90
first law of thermodynamics	82	transition state	89
gradient	92	turgor pressure	95

FUN FACTS

NEURON CYTOSKELETON

The cytoskeleton of a neuron faces daunting tasks. It must stabilize the long axon to resist great mechanical stress and must supply the axon with proteins and other molecules produced by the nucleus and cell body. To meet these challenges, the cytoskeletal polymers are densely packed and unusually stable. These macromolecules are produced near the nucleus, then transported slowly (at rates of millimeters per day) toward the axon terminal. Experiments involving the extraction, purification, fluorescent labelling, and reintroduction of cytoskeletal subunits allow scientists to follow the migration and incorporation of the building blocks of the cytoskeleton.

Microtubules and microfilaments do not seem to move after being incorporated into the cytoskeleton. A section of the cytoskeleton may be marked by bleaching with bright light. If the cytoskeleton were slowly flowing from its source, the nuclear region, toward the axon terminus, one would expect to see the bleached spot moving toward the axon end. Observations indicate that once formed, the cytoskeleton stays in place but its subunits move around and gradually replace the bleached units. A mystery remains: How are molecules rapidly moved to the axon terminal if the cytoskeleton is fixed in place? The answer may lie in neurofilaments, which contain three different protein subunits.
See: Hollenbeck, Peter J. 1990. Cytoskeleton on the move. *Nature* 343:408–409.
February 1, 1990.

NUCLEAR PORES

To pass into or out of a nucleus, material must pass through nuclear pore complexes in the nuclear membrane. When photographed with electron microscopes, nuclear pores often seem to be plugged with what was thought to be macromolecules passing through. But it may be that the plug is a protein complex acting as a gate in the nuclear pore.

On the sides of a pore are coaxial rings, one facing the cytoplasm and the other, the nucleoplasm. The rings are made of eight spoke-like subunits hinged at the edge of the complex. These units act like the leaves of a camera iris, opening and closing the pore. The pore complex, estimated to contain up to 100 proteins, can bidirectionally and apparently simultaneously transport protein into the nucleus and RNA out, driven by ATP energy.
See: Dingwall, Colin. 1990. Plugging the nuclear pore. *Nature* 346:512–514.
August 9, 1990.

CHAPTER 4 THE DYNAMIC CELL

MESSAGES

1. The ability to carry out organized biological work depends on a continual flow of energy from the environment.

2. Energy exists as potential energy and as kinetic energy. Energy can be converted from one form to another but entropic losses accompany all conversions.

3. Cells expend energy to construct and maintain cell parts and processes.

4. Chemical reactions underlie virtually all cellular energy conversions. Exergonic reactions are coupled with endergonic reactions, usually through ATP.

5. Enzymes accelerate chemical reactions by reducing the activation energy necessary to achieve the transition state between reactants and products.

6. By diffusion, osmosis, and active and passive transport, materials move through membranes.

7. Cells expend energy to import or export solutes to help maintain an internal solute concentration close enough to the concentration in the extracellular fluid that the cells will neither burst nor shrivel.

8. The cell's mechanical tasks include movements of the entire cell, of external parts such as cilia, and of internal parts.

9. The rapid assembly and disassembly of microtubules and microfilaments, catalyzed by enzymes and powered by ATP, underlies intracellular movement.

5

HOW LIVING THINGS HARVEST ENERGY FROM NUTRIENT MOLECULES

PERSPECTIVE

This chapter reveals how cells process energy-rich molecules such as glucose, breaking them down into smaller molecules and capturing the energy stored in their chemical bonds. Anaerobic and aerobic catabolism are compared. The energy that is harvested is used to recharge ATP, which powers virtually all the cell's activities that are detailed throughout this text. Photosynthesis, the original source of high-energy molecules, is presented in Chapter 6. The sections on animal physiology and plants will describe how energy is used to build and maintain organisms.

The themes and concepts of this chapter include:

- Glycolysis is the first step in energy catabolism in all organisms; it uses a small portion of the energy in a glucose molecule to produce two ATPs.
- Fermentation follows glycolysis in anaerobic conditions and yields no additional ATPs.
- Aerobic respiration oxidizes the product of glycolysis to CO_2 and H_2O, producing 34 additional ATPs.
- ATP is an energy carrier that is involved in almost all exergonic and endergonic biological reactions.
- Oxidation-reduction reactions transfer electrons and energy through specialized pathways.
- Feedback systems control reaction rates and metabolic activity.

CHAPTER OBJECTIVES

Students who master this chapter will understand that all cells are powered by energy released during the breakdown of energy-rich carbon-based molecules. They will see that energy-releasing catabolism can take place in the absence or presence of molecular oxygen and that aerobic metabolism is much more efficient and productive than anaerobic metabolism. They will appreciate the universality of ATP as an energy link between exergonic and endergonic biological reactions and will learn of the efficiency of feedback systems that regulate metabolic activities.

LECTURE OUTLINE

(p 100) **I. A hike on a mountain trail introduces energy metabolism.**

A. Plants, animals, fungi, and protists use the same basic metabolic pathways for harvesting energy, indicating life's unity.

1. **Glycolysis is common to virtually all cells.**
2. **Fermentation follows glycolysis in anaerobic conditions. The harvest of ATP is low.**
3. **Aerobic respiration occurs when oxygen is present. The harvest of ATP is high.**

(tpq 1) (msg 1) **B. Metabolic pathways are gradual; they occur in small steps, improving the efficiency and controllability of the process.**

---

tpq 1. What disadvantage might there be to cells' performing catabolism in a few large steps?

(p 102) **II. ATP is an energy carrier that links energetic**
(msg 2) **reactions in cells.**

(tr 22) **A. The cleaving of one phosphate group from ATP yields 8 kilocalories of energy per mole and leaves ADP plus a phosphate ion.**

(tpq 2) **1. Energy is released when phosphate groups are cleaved.**

2. Energy is required to put phosphate groups back on ADP or AMP.

3. A phosphate group may be attached to another molecule, energizing the molecule in the process of <u>phosphorylation</u>.

(msg 3) **B. <u>Oxidation-reduction reactions</u> occur as electrons move from one atom or molecule to another.**

1. Oxidation is the removal of electrons, which increases the net charge on a molecule.

2. Reduction is the addition of electrons, which reduces the net charge.

3. In biological systems, special energy carriers transfer electrons. Hydrogen ions accompany the electrons.

a. <u>Coenzymes</u> include electron carriers like NAD^+ and FADH.

b. <u>Vitamins</u> include precursors to energy-carrying coenzymes.

---

tpq 2. Is the reaction ATP --> ADP exergonic or endergonic? Might it happen spontaneously?

---

4. The flow of electrons through carriers creates a current that is channeled into ATP formation.

(p 104) (tr 23) **III. All organisms share the same biochemical pathways for releasing energy stored in the molecular bonds of nutrient molecules.**

(tpq 3) **A. Heterotrophs must take in preformed nutrient molecules. Autotrophs generate their own nutrient molecules.**

(tr 24) (msg 4) **B. Glycolysis is the first step, breaking glucose down into two three-carbon pyruvate molecules.**

1. Glycolysis takes place in the cytoplasm, in the presence or absence of oxygen.
2. Two ATP molecules are formed for each glucose molecule.

(tr 25) (msg 5) **C. In the absence of oxygen, fermentation in the cytoplasm processes pyruvate, regenerating NAD^+ but not producing additional ATPs.**

(tr 26) (msg 6) **D. Aerobic respiration, which takes place in mitochondria, breaks pyruvate down to CO_2 and H_2O plus 34 molecules of ATP per molecule of glucose.**

1. Aerobic respiration occurs only in the presence of oxygen.

---

tpq 3. Why do photosynthesizers waste energy by going through all the conversions of metabolism instead of using the sun's energy directly to power their activities?

2. Aerobic respiration includes two phases, the <u>Krebs cycle</u> and the <u>electron transport chain</u>.

(p 105) **IV. Glycolysis occurs in nine sequential steps.**

(tpq 4) A. In the first three steps, a molecule of glucose is phosphorylated, representing an energy investment of 2 ATP.

B. In step 4, the phosphorylated molecule is split into two three-carbon molecules called G-3-P.

C. In step 5, the G-3-P is oxidized and energy is released. The energy and hydrogens are transferred to NAD^+, reducing it to NADH. Phosphates are added to G-3-P, forming DPGA.

D. In steps 6, 7, and 8, DPGA is shaped and reshaped into a series of compounds. The phosphates are transferred to ADP, forming four ATP molecules per initial molecule of glucose.

E. In the final step, pyruvate, the end product of glycolysis, is created.

F. For each glucose molecule, two ATPs are invested and four are formed, for a net gain of two ATPs, less than 5 percent of the energy stored in the bonds of the glucose molecule. One NADH is produced.

---

tpq 4. Why is energy added by phosphorylation to the glucose molecule, which already contains a great deal of energy in its chemical bonds?

(p 106) V. Fermentation is the major anaerobic metabolic pathway that processes pyruvate.

A. Fermentation regenerates NAD^+ but does not create any additional ATP. The NAD^+ is necessary for glycolysis.

B. Alcoholic fermentation yields the waste products carbon dioxide and ethyl alcohol, useful in baking and brewing.

C. Lactic acid fermentation reduces pyruvate to lactic acid, regenerating NAD^+ but no additional ATP. This process occurs in oxygen-starved animal muscle cells.

D. Anaerobic catabolism plays a key role in recycling organic materials in oxygen-poor environments.

(p 110) (msg 7) VI. Aerobic respiration oxidizes pyruvate to carbon dioxide and water, producing much more energy than does glycolysis alone.

(tr 27) A. The Krebs cycle strips the carbon atoms, releasing them as carbon dioxide, and transfers the energy by reducing carrier molecules.

1. The Krebs cycle is preceded by a reaction that oxidizes pyruvate, releasing one of its carbon atoms as carbon dioxide and attaching the remaining two-carbon molecule to coenzyme A. Energy is released and stored as NADH.

2. **In the Krebs cycle proper, acetyl-CoA is added to a four-carbon molecule.**

   a. **Two carbon atoms are stripped from the product in a series of steps. Two carbon dioxide molecules are produced.**

   b. **Energy is transferred, generating 2 ATP, 8 NADH, and 2 $FADH_2$ per molecule of glucose.**

   c. **The original four-carbon molecule is regenerated as the cycle completes one turn.**

(tr 28)

3. **Other energy-rich molecules, such as lipids and proteins, can be broken down into molecules that can enter the Krebs cycle for energy production.**

4. **Intermediates in the Krebs cycle may be withdrawn for use as building blocks for other molecules.**

(tpq 5)

B. **The electron transport chain converts the energy in NADH and $FADH_2$ into ATP energy.**

(tr 29)

1. **The electron transport chain includes proteins embedded in the inner mitochondrial membrane.**

   a. **The high-energy electrons stored in NADH and $FADH_2$ are passed from one protein to another in a series of oxidation-reduction reactions.**

---

tpq 5. Would the Krebs cycle be of any use to a cell without a functioning electron transport chain?

---

b. **Small amounts of energy are released during each reaction.**

c. **Oxygen is the final electron acceptor and combines with H^+ to form H_2O.**

(msg 8) 2. **Before they combine with oxygen, H^+ ions are concentrated in the outer compartment of the mitochondrion. As they migrate through membrane proteins, the energy released by their passage is used to phosphorylate ADP to ATP in a process known as chemiosmotic coupling.**

(p 117) **VII. Metabolism is controlled by feedback systems.**

(msg 9)
(tr 30)
(tpq 6)

A. **Feedback inhibition involves the buildup of a metabolic product and the product's inhibitory effect on the pathway that produced it.**

B. **An enzyme may have two binding sites: the catalytic site and the regulatory (allosteric) site.**

C. **When the regulatory site is occupied by a molecule (usually an intermediate or end product of the metabolic pathway), the catalytic site will not bind to its substrate.**

D. **Thus, in a feedback system, the activity of the enzyme is controlled by the presence of a product.**

---

tpq 6. What evolutionary advantage arises from controlling metabolic systems?

(p 118) **VIII. Energy for different forms of exercise comes from different parts of the metabolic pathway.**

A. The immediate system is instantly available for brief explosive action.

1. The small amount of ATP stored in muscle cells is immediately useful but quickly exhausted.

2. Creatine phosphate is an energy storage molecule that transfers phosphates to ADP, regenerating ATP. A cell's store of creatine phosphate may become depleted after a minute of strenuous work.

B. The glycolytic system fuels activities ranging from about 1 to 3 minutes. It produces lactic acid as a waste product.

C. The oxidative system supplies energy for activity of moderate intensity and long duration. It relies on aerobic respiration.

INSTRUCTIONAL AIDS

VIDEOS, FILMS, SLIDES

Cellular Respiration. Carolina Biological Supply, Burlington, NC. Analyzes the glycolytic system, citric acid cycle, and electron transport. Transparencies, video.

Cellular Respiration Set. Connecticut Falley Biological, Southampton, MA. A set of transparencies that demonstrate glycolysis, the citric acid cycle, and electron transport.

REFERENCES

Cavalier-Smith, T. 1987. Molecular evolution: eukaryotes with no mitochondria. *Nature* 326:332–333. A review of eukaryotes without mitochondria and a discussion of whether the absence of mitochondria is a primitive or secondary character.

Gould, J. L. and C. G. Gould, editors. 1989. *Life at the Edge*. W. H. Freeman, NY. Shows how life survives in extreme habitats, revealing the roles of energy in life cycles.

Grivell, Leslie A. 1983. Mitochondrial DNA. *Scientific American* 248:78–89. March, 1983. A review of the organization of mitochondrial genes, detailing the differences among mitochondria, bacteria, and the nucleus.

Holtzman, Eric, and A. B. Novikoff. 1984. *Cells and Organelles*, Third Edition. Philadelphia: Saunders. A readable college text with an integrated approach.

Lane, M. D., et. al. 1986. The mitochondrion updated. *Science* 234:526–527. A review of an international symposium on the mitochondrion and an overview of the history of mitochondrial research.

Reid, R. A., and R. M. Leech. 1980. *Biochemistry and Structure of Cell Organelles*. New York: John Wiley & Sons. A compact upper-division textbook that details the activities of mitochondria.

KEY TERMS

aerobic	page 101	fermentation	106
aerobic respiration	101	glycolysis	104
anaerobic	101	heterotroph	104
autotroph	104	Krebs cycle	110
chemiosmotic coupling	117	oxidation-reduction reaction	103
electron transport chain	110	phosphorylation	102
feedback inhibition	118		

FUN FACTS

FERMENTING STINKBIRDS

The hoatzin of Venezuela is commonly known as the stinkbird because it smells like fresh cow manure. It's no coincidence; both the hoatzin and cow are herbivores that rely on foregut fermentation for digestion.

The hoatzin, *Opisthocomus hoazin*, is one of the few birds that eat nothing but leaves. Its food is bulky, fibrous, low in nutrients, and rich in toxic compounds. The bird weighs less than a kilogram, and like all small endotherms, has a high energy demand. To satisfy its energy needs on such a low-quality diet, the hoatzin relies on microbial fermentation in its enlarged crop and esophagus. Its voluminous digestive apparatus takes up body space occupied by breastbone and flight muscles in other birds, making the hoazin a poor flier. The young have functional claws on the first and second digits of their wings; they use the claws when climbing trees.

Despite the energetic and physical limitations imposed by such a diet, there are advantages to flying folivory. Fliers can be selective about their food, reaching fresh new growth that is inaccessible to terrestrial herbivores. They can exploit patchy food resources. Scientists are puzzled by the apparent uniqueness of the hoatzin's digestive system among birds.

See: Grajal, Alejandro, et al. 1989. Foregut fermentation in the hoatzin, a neotropical leaf-eating bird. *Science* 245:1236–1238.

Grajal, Alejandro, and Stuart D. Strahl. 1991. A bird with the guts to eat leaves. *Natural History* 8/91:48–55. August, 1991.

CHAPTER 5 HOW LIVING THINGS HARVEST ENERGY FROM NUTRIENT MOLECULES

MESSAGES

1. Cells break down energy-rich molecules in gradual and controllable reactions and use the energy to power metabolism.

2. ATP links most energy exchanges within a cell.

3. Energy may be transferred as a flow of electrons in oxidation-reduction reactions.

4. The catabolism of glucose starts with glycolysis, which breaks glucose down into two three-carbon pyruvate molecules and yields two ATPs.

5. In anaerobic conditions, fermentation follows glycolysis, regenerating NAD^+ but producing no additional ATPs.

6. Aerobic respiration breaks glucose down into CO_2 and H_2O, producing 34 additional ATP molecules.

7. The Krebs cycle breaks down pyruvate into CO_2 and transfers the energy to electron carriers.

8. The electron transport chain converts the energy of flowing electrons into ATP energy through chemiosmotic coupling.

9. Metabolism is controlled by feedback systems.

6

PHOTOSYNTHESIS: TRAPPING SUNLIGHT TO BUILD NUTRIENTS

PERSPECTIVE

This chapter presents the processes that produce the high-energy organic molecules on which most organisms subsist, the raw materials whose breakdown was detailed in Chapter 5. This completes the survey of the chemistry of life, ending the section on cells and cellular chemistry.

Material presented in this chapter will reappear in Chapters 15 through 17, which cover life's origins, monerans, protists, plants, and fungi; it will be relevant to the plant section, which includes Chapters 30 through 32, and it will be discussed again in Chapters 35 through 37, which present evolution and ecology.

The themes and concepts of this chapter include:

- Photosynthesis is the process by which solar energy is trapped, converted to chemical energy, and stored in the bonds of organic molecules.
- Photosynthesis is nearly the reverse of aerobic respiration.
- During photosynthesis, water is oxidized and carbon dioxide is reduced.
- Chloroplasts are the sites of photosynthesis in eukaryotic autotrophs. Light-dependent and light-independent reactions take place in different parts of a chloroplast.
- Chlorophyll and other pigments absorb specific wavelengths of visible light, which is a small part of the electromagnetic spectrum.
- Noncyclic photophosphorylation involves two different reaction centers in tandem and yields more energy than cyclic photophosphorylation.
- The light-independent reactions fix carbon dioxide and construct glucose.
- C_4 plants can maintain a high ratio of carbon dioxide to oxygen under hot, dry conditions.
- The reciprocal processes of photosynthesis and respiration drive the global carbon cycle.

CHAPTER OBJECTIVES

Students who master the material presented in this chapter will see the connection between photosynthesis and energy metabolism and appreciate the symmetry of the carbon cycle. They will understand how pigments trap light energy, how that energy passes through electron transport systems to generate energy-carrying molecules, and how the energy in those carriers is used to fix carbon and build glucose molecules. They will have a new respect for corn, exemplifying C_4 plants, and its photosynthetic ability. They will begin to appreciate the dangers of the increase in greenhouse gases and of nuclear winter.

LECTURE OUTLINE

(p 122) **I. Corn, photosynthetic champion, introduces the topic of photosynthesis.**

(tpq 1) (msg 1) **A. Photosynthesis is the metabolic process by which solar energy is trapped, converted to chemical energy, and stored in the bonds of organic molecules.**

B. Photosynthetic pigments include chlorophyll, which is both widespread and vitally important to life.

C. Photosynthesis generates the nutrients that virtually all cells break down to power their activities.

(p 124) (msg 2) **II. Photosynthesis is nearly the reverse of aerobic respiration. Both operate through a series of reactions.**

---

tpq 1. How is it true that you are solar powered?

(tpq 2) **A. The general formula for photosynthesis is:**

(msg 3) $6\ CO_2 + 6\ H_2O$ + light energy --> $C_6H_{12}O_6 + 6\ O_2$

B. In respiration, chemical energy is released from glucose; in photosynthesis, solar energy is stored in the chemical bonds of glucose.

(tr 31) **C. During photosynthesis, light energy is used to remove electrons (oxidation) in a high-energy state from water and add them (reduction) to carbon dioxide.**

(msg 4) (tr 32) **1. During the light-dependent reactions, solar energy boosts electrons in a pigment to higher energy levels. The energy is released bit by bit and stored in ATP and NADPH.**

(tr 33) **2. During the light-independent (dark) reactions, the energy in these carriers is released and stored in the bonds of glucose molecules.**

D. Photosynthesis occurs in chloroplasts, which resemble mitochondria.

1. Chloroplasts have inner and outer membranes collectively enclosing a space called the stroma.

2. A third, inner membrane forms the thylakoids.

a. A stack of thylakoids is a granum.

b. Photosynthetic pigments and electron transport proteins are embedded in the thylakoid membrane.

---

tpq 2. How does the formula for aerobic respiration differ from the formula for photosynthesis?

---

3. **Light-independent reactions occur in the stroma and in the cytoplasm of the cell.**

(p 125) **III. Colored pigments in living cells trap light.**

A. Visible light is just a small part of the electromagnetic spectrum. Autotrophs use less than 1 percent of the visible light striking the earth.

1. **Radiation travels as photons.**
2. **The longer a photon's wavelength is, the lower its energy is.**
3. (tpq 3) **The range of light that a pigment absorbs is its absorption spectrum.**

B. Chlorophyll a, chlorphyll b, and carotenoids are the main pigments of photosynthesis. Together they absorb most frequencies of visible light.

(p 128) **IV. The light-dependent reactions involve three main stages.**

A. First, light excites electrons to higher energy levels.

B. Second, that energy is stored in stable energy carriers.

C. Third, water molecules are split, replacing the excited electrons with low-energy electrons and releasing oxygen.

---

tpq 3. Why does a leaf look green?

(msg 5) D. Antenna complexes are clusters of pigment molecules in thylakoid membranes that act together to capture light energy.

1. The pigment molecules are arranged around a central chlorophyll molecule, the reaction center, which receives energy from the other molecules.
2. There are two different photosystems that absorb different wavelengths of light.
3. Electrons in a chlorophyll molecule are raised to higher energy states when a photon is absorbed.

(msg 6) E. In noncyclic photophosphorylation, light energy captured by a photosystem is shunted through electron acceptors.

1. Electron acceptors store some of the energy in ATP molecules.
2. At the end of the electron transport chain, the electrons join a second photosystem and become boosted by light once more.
3. This path is the Z scheme of photosynthesis.
4. NADPH, which contains seven times the energy of ATP, is formed using the energy of the flowing electrons.

5. In an ancient type of photosynthesis, cyclic photophosphorylation, electrons are recycled to the same chlorophyll, producing a small amount of ATP.

(tpq 4) (msg 7) F. Oxygen gas is released as a waste product when a water molecule is split and electrons are removed. The electrons replace those ejected by chlorophyll.

(p 131) V. The light-independent reactions, called the Calvin-Benson cycle, use ATP and NADPH energy to create glucose.

A. The cycle includes three main stages.
   1. Carbon is fixed to a biological molecule.
   2. Carbohydrates are manufactured as stable energy storage molecules.
   3. The starting compound is regenerated.

B. Carbon fixation occurs when carbon dioxide becomes attached to a five-carbon molecule, ribulose bisphosphate, by the enzyme rubisco (ribulose bisphosphate carboxylase). Rubisco works slowly and is present in great amounts.

C. Energy, electrons, and hydrogens generated during the light-dependent reactions are added to the newly fixed carbon.

---

tpq 4. Which came first in evolution, noncyclic photosynthesis or aerobic respiration?

---

1. Six turns of the cycle are required to produce one glucose molecule.
2. The formation of each glucose molecule requires 18 ATPs and 12 NADPHs.

D. The five-carbon starting compound is regenerated at the cost of some ATP molecules.

E. Glucose is a better storage molecule than ATP.
1. One glucose molecule stores enough energy to regenerate 36 ATP molecules but is smaller than one ATP molecule.
2. Glucose is less reactive than ATP.
3. Carbohydrates can be converted to other biological molecules.

(p 133) (msg 8) VI. C_4 plants such as corn can continue photosynthesis even in hot, dry weather.

A. When non-C_4 plants (called C_3 plants) dehydrate, their stomata close, carbon dioxide is used up, and oxygen builds up.
1. Rubisco is inhibited and carbon is lost instead of fixed, a process called photorespiration.
2. This reduces the rate of photosynthesis.

B. Corn and certain other plants have special carbon dioxide pumps that maintain a high ratio of carbon dioxide to oxygen.

1. **Carbon dioxide is fixed to a three-carbon molecule and shunted to the photosynthetic cells.**
2. **<u>CAM (crassulacean acid metabolism) plants</u>, including succulents and cacti, have a slightly different carbon dioxide pump.**

(p 134) (msg 9) **VII. The global <u>carbon cycle</u> is the flow of carbon atoms from autotrophs to heterotrophs to the environment and back to autotrophs again.**

(tpq 5) **A. Carbon dioxide in the atmosphere traps heat, preventing it from escaping from the earth, a phenomenon called the <u>greenhouse effect</u>.**

B. <u>Nuclear winter</u> might result from nuclear war, leading to a global reduction in photosynthesis.

---

tpq 5. What consequences might result from a buildup of greenhouse gases?

INSTRUCTIONAL AIDS

VIDEOS, FILMS, SLIDES

Photosynthesis. Educational Images Ltd., Elmira, NY. Detailed explanation of light-dependent and light-independent photosynthesis for students with limited backgrounds in chemistry. Transparencies, filmstrip.

Photosynthesis: A Demonstration Series. International Film Bureau, Chicago, IL. Four films that discuss the effects of light and chlorophyll on the production of oxygen and starch. Film, video.

Photosynthesis: Life Energy. National Geographic Educational Services, Washington, DC. The basic process of photosynthesis, with discussion of the food pyramid and industrial uses. Film, video.

REFERENCES

Govindjee and W. J. Coleman. 1990. How plants make oxygen. *Scientific American* 262 (2):50–58. An illustrated discussion of the "water-oxidizing clock" that releases molecular oxygen.

Kite, Geoffrey. 1986. Evolution by endosymbiosis: the inside story. *New Scientist* July 3:50–52. An essay presenting molecular biology's evidence that chloroplasts were once free-living bacteria.

Martin, Frederick L. 1987. Radioactive CO_2 fixation in *Geranium* leaves. *American Biology Teacher* 49:433–435. A lab experiment for advanced biology classes investigating factors affecting photosynthetic rates.

Walsby, A. E. 1986. Prochlorophytes: origins of chloroplasts. *Nature* 320:212. A short, readable review of the discovery of free-living prochlorophytes, a group thought to include the ancestors of chloroplasts.

KEY TERMS

absorption spectrum	page 127	greenhouse effect	135
antenna complex	128	light-dependent reaction	124
C_3 plant	134	light-independent (dark) reaction	124
C_4 plant	134	noncyclic photophosphorylation	131
Calvin-Benson cycle	131	nuclear winter	135
CAM plant	134	photon	126
carbon cycle	134	photorespiration	134
carbon fixation	131	photosynthesis	123
carotenoid pigment	127	photosystem	128
chlorophyll *a*	127	reaction center	128
chlorophyll *b*	127	rubisco	132
chloroplast	124	thylakoid	125
cyclic photophosphorylation	131		
electromagnetic spectrum	126		

FUN FACTS

COUSINS OF CHLOROPLASTS

One of evolution's "missing links" has been found alive and well – and in great abundance. The discovery of photosynthetic microorganisms that may be close relatives of chloroplasts, the microscopic organelles that carry on photosynthesis in plants, demonstrates that there are still many surprises in biology.

Chloroplasts are thought to be the descendants of free-living photosynthetic bacteria that became incorporated into larger, more complicated eukaryotic cells in the process known as endosymbiosis. Chloroplasts capture the energy of sunlight and use it to build sugars and other energy-rich molecules that feed the host cell. The host in return supplies water, minerals, and protection to the chloroplast. When scientists examine the biochemistry of chloroplasts, they find that chloroplasts have their own DNA (which is more like DNA of bacteria than the DNA in the nucleus of the host cell) and their own protein-synthesizing machinery, indicating that they were once free-living and autonomous.

A problem has been that the biochemistry of chloroplasts of the plants we're familiar with, tulips and trees, is different from that of cyanobacteria, the most common photosynthetic bacteria. While chloroplasts have chlorophylls *a* and *b*, cyanobacteria have only chlorophyll *a*. They rely on other pigments in place of chlorophyll *b*. Until the mid-1970s, no photosynthetic bacterium with both chlorophylls had been found, and the ancestor of chloroplasts remained unknown.

In 1975, Ralph Lewin of the Scripps Institution of Oceanography discovered a photosynthetic bacterium living on the surface of sea squirts, marine animals encrusting the roots of mangrove trees in a lagoon in Baja California. When he analyzed their pigments, he found that they had both chlorophylls but lacked the pigments that distinguish cyanobacteria. He had discovered a new phylum of life!

Scientists found other "prochlorophytes" (from Greek, "before green plants") associated with sea squirts, but these chloroplast-cousins remained in the category of odd and unusual organisms for another decade. In the 1980's scientists found free-living prochlorophytes in lakes in the Netherlands. Other researchers soon discovered that prochlorophytes are abundant in the ocean – as many as 100,000,000 can be found in a liter of sea water. They had gone undetected for so long because they are very small and do not show up well on the instruments oceanographers use to count cyanobacteria.

Chloroplasts have changed during the billion years that they have been endosymbionts, becoming more and more dependent on their hosts, but they have relatives that have followed the path of independence.

See: Chisholm, Sallie W., et. al. 1988. A novel free-living chlorophyte abundant in the oceanic euphotic zone. *Nature* 334:340 – 343.

Kite, Geoffrey. 1986. Evolution by endosymbiosis: the inside story. *New Scientist* July 3:50 – 52.

Lewin, R. A., and N. W. Withers. 1975. Extraordinary pigment composition of a prokaryotic alga. *Nature* 256:735 – 737.

Walsby, A. E. 1986. Origins of chloroplasts. *Nature* 320:212.

CHAPTER 6 PHOTOSYNTHESIS: TRAPPING SUNLIGHT TO BUILD NUTRIENTS

MESSAGES

1. Photosynthesis is the metabolic process by which solar energy is trapped, converted to chemical energy, and stored in the bonds of organic molecules.

2. The reactants and products of photosynthesis are the products and reactants of aerobic respiration.

3. The general formula for photosynthesis is:
   $6\ CO_2 + 6\ H_2O + \text{light energy} \dashrightarrow C_6H_{12}O_6 + 6\ O_2$

4. The light-dependent reactions harvest light energy and convert it to ATP and NADPH energy. The light-independent reactions use this energy to make carbohydrates.

5. Pigment molecules operate in antenna complexes.

6. Noncyclic photophosphorylation involves two separate antenna complexes that operate in tandem to generate NADPH.

7. Water is split during photosynthesis, yielding electrons to replace those lost from chlorophyll and producing O_2 as a waste product.

8. C_4 plants can continue photosynthesis during hot, dry weather because of adaptations that reduce photorespiration.

9. Carbon circulates in a global cycle between the atmosphere, autotrophs, and heterotrophs.

7

CELL CYCLES AND LIFE CYCLES

PERSPECTIVE

This chapter presents the mechanisms by which cells reproduce. Chromosomes are the repositories of genetic information that direct cell growth and reproduction; their duplication and orderly division is the crux of cell reproduction. The regulation of the cell cycle is discussed. Mitosis is compared and contrasted with meiosis and the roles of these processes in an organism's life cycle and in evolution are presented. This chapter builds on cell theory, presented in Chapter 3, and provides the basis for Chapters 8 through 14, the genetics section.

The themes and concepts of this chapter include:

- Cell division is the basis of reproduction and growth.
- Cellular reproduction is carefully regulated. Cancer is the result of uncontrolled cell proliferation.
- The apportionment of genetic information is crucial to reproducing cells and organisms.
- Both cell cycles and life cycles include periods of growth.
- Genetic information, which lies in the chromosome, determines a cell's physical traits and helps guide cell division.
- The cell cycle of eukaryotes includes interphase, mitosis, and cytokinesis.
- Meiosis produces haploid daughter cells that become gametes. Meiosis includes mechanisms that increase genetic diversity.
- Reproduction may be asexual, through mitosis, or sexual, where gametes fuse to form diploid zygotes.
- Meiosis and mitosis differ in their roles in life cycles and in evolution.

CHAPTER OBJECTIVES

Students who master the material in this chapter will see that genetic information, which determines a cell's physical traits and helps guide cell division, lies in the chromosomes. They will understand the phases of the cycle of cell growth and reproduction and how the cycles of eukaryotes and prokaryotes differ. They will appreciate the regulation of the cell cycle. They will know the difference between mitosis and meiosis and the roles of these two types of cell division in the organism's life cycle and in evolution.

LECTURE OUTLINE

(p 138) **I. Skin cancer and surgery introduce cell cycles.**

- **A. Reproduction, a basic characteristic of life, requires cell division.**
- **B. Each organism has a life cycle.**
- **C. Cell division is usually precisely controlled.**
- **D. Once the genetic information has been copied, it must be apportioned correctly to reproducing cells and organisms.**
- (msg 1) **E. Both the cell cycle and the life cycle generally involve periods of growth.**

(p 140) (msg 2) **II. Chromosomes are the repositories of information for constructing each part of the cell.**

- **A. Experiments demonstrated that a eukaryotic cell's nucleus directs events in the cell and in the whole organisms and its life cycle.**
- (msg 3) **B. Chromosomes are molecules that contain the hereditary information.**

1. **Eukaryotic species have chromosomes of variable numbers, shapes, and sizes. They are wound around proteins.**
2. **Prokaryotes have a single long, circular DNA molecule.**

(p 142) **III. Cell cycles in prokaryotes and eukaryotes show similarities and differences.**

(tpq 1) **A. Within a given organism, each body cell has the same hereditary material and identical chromosomes.**

(msg 4) **B. Prokaryotes divide by <u>binary fission</u>.**

1. **The circular DNA chromosome <u>replicates</u> to form two side-by-side circles, each with its own attachment to the cell membrane.**
2. **The cell elongates and the attachment points separate, pulling the daughter chromosomes apart.**
3. **Cell contents are parceled out into the two daughters.**

(msg 5) **C. The cell cycle of eukaryotes includes four phases.**

1. **<u>Interphase</u> includes three growth phases. A cell usually doubles in size during interphase.**

---

tpq 1. How can cells be as different as liver cells and nerve cells but still have the same genetic material?

---

a. Gap 1 (G1) precedes the synthesis of new DNA. New cell components are built during this phase, which is variable in length.

b. During the S (synthesis) phase, DNA is replicated and histone proteins are produced.

c. During gap 2 (G2), the cell continues to make proteins and prepares to divide.

(msg 6) 2. M (mitosis) phase is the period of cell division.

(tr 34) a. During mitosis, the two sets of chromosomes are separated and two daughter nuclei form.

(tr 35) b. During cytokinesis, the daughter nuclei and cellular components are separated into daughter cells.

(p 145) IV. Mitosis and cytokinesis are presented in detail.

A. During M, chromosomes become tightly wound up and visible. Each consists of two chromatids held together at the centromere.

B. The dance of the chromosomes and their separation to poles of the cell includes five steps.

(tpq 2) 1. In prophase, chromosomes condense, the nucleolus disappears, and a mitotic spindle forms.

---

tpq 2. Some drugs, such as colchicine, block the formation of the spindle. What might be the effect on a cell?

---

2. In prometaphase, the nuclear envelope disappears and the chromosomes attach to the spindle.
3. In metaphase, the chromosomes become aligned in the middle of the cell.
4. In anaphase, the centromeres divide and the spindle pulls the chromatids toward the poles.
5. In telophase, the chromosomes arrive at opposite poles, nuclear envelopes reappear, and the spindle dissolves.

C. Cytokinesis differs between plants and animals.

1. A contractile ring pinches animal cells, forming a furrow which cleaves the cell in two.
2. Plant cells have rigid walls; a central cell plate forms and divides the cell in two.

(p 151) V. Cell division is controlled.

(tpq 3)
(msg)

A. Cells often exhibit contact inhibition and stop dividing when they touch other cells.

B. Proteins called growth factors can enhance the growth and division of particular cell types.

C. Researchers discovered that two types of regulatory proteins, pp34 and cyclin, appear to be regulators of cell cycles in diverse eukaryotes.

---

tpq 3. Normal human cells can divide only about 50 times then are unable to divide further. What might be the evolutionary advantage of this limit?

1. The amount of pp34 does not change during the cycle.
2. Different cyclins appear and disappear during different cycle phases.
3. During G_2, a cyclin binds to pp34, forming an active complex called M-phase-promoting factor or MPF, and the cell enters mitosis.

(tpq 4) D. Cancer is the result of uncontrolled cell multiplication.

1. Cancer cells do not stop growing or moving after contact with other cells.
2. Tumors are masses of cancer cells.

(p 152) VI. The life cycle is the process by which an immature individual matures and reproduces.

A. Asexual reproduction is the formation of an offspring genetically identical to its parent.

1. Asexual reproduction is common in plants.
2. Certain animals reproduce asexually, by budding or regeneration.

B. Sexual reproduction involves the fusion of gametes.

1. Fertilization results in a single cell, the zygote.

---

tpq 4. Cells don't become cancerous until they've had genetic mutations. What normal cellular function might these mutations affect?

(msg 8) 2. **Meiosis is cell division that reduces a diploid**
(tr 36) **cell's chromosome number by half to produce haploid gametes.**

3. **Gametes come from specialized cells called the germ line. The body cells of an organism are its somatic cells.**

(msg 9) 4. **Fertilization doubles the chromosome number, while meiosis divides it in half.**

(p 156) **VII. Meiosis includes mechanisms that create genetic variability.**

(msg 10) **A. Meiosis includes one chromosome replication with two cell divisions. The division stages are similar to the stages of mitosis.**

B. In meiosis I, a diploid parent cell divides to produce two haploid daughters. Homologous chromosomes are separated.

1. **No chromosome replication occurs between meiosis I and meiosis II.**
2. **During meiosis II, sister chromatids are separated.**

(tpq 5) **C. Genetic variation arises through genetic recombination during meiosis.**

---

tpq 5. What might be the evolutionary advantage of genetic variability?

---

1. **Crossing over, the exchange of parts between homologous chromosomes, occurs during prophase of meiosis I.**
2. **During meiosis I, homologous chromosomes are randomly distributed, a process known as independent assortment.**
3. **The resulting variations in offspring produce differences in fitness, the raw material on which natural selection acts.**

(p 161) **VIII. Asexual reproduction may be advantageous in constant environments. Genetic variability produced by sexual reproduction may be beneficial in changeable environments.**

INSTRUCTIONAL AIDS

VIDEOS, FILMS, SLIDES

Cell Division and the Life Cycle. Human Relations Media, Pleasantville, NY. In two parts, one covering mitosis and cytokinesis; the second, meiosis and sexual reproduction. Film, video.

Meiosis, Second Edition. Coronet/MTI, Deerfield, IL. Explains the roles of nucleus, DNA, chromosomes, crossover, and independent assortment in meiosis. Film.

Mitosis, Second Edition. Pennsylvania State University AV Services, University Park, PA. Microphotography and animation depict mitosis in plant and animal cells. Film.

REFERENCES

Balter, Michael. 1991. Cell cycle research: Down to the nitty gritty. *Science* 252:1253–1254. A report from a conference on the latest advances in cell cycle research.

Murray, Andrew W., and Marc W. Kirschner. 1991. What controls the cell cycle? *Scientific American* 264 (3):56–63. Describes the role of the cdc2 protein in cell cycle regulation.

Prescott, David M. and A. S. Flexer. 1986. *Cancer: The Misguided Cell*, Second Edition. Sunderland, MA: Sinauer Associates. Good general reading as well as a good undergraduate textbook.

Strickberger, M. 1985. *Genetics*, Third Edition. New York: Macmillan. A good source of information on chromosomes, meiosis, and recombination.

KEY TERMS

anaphase	page 145
asexual reproduction	152
binary fission	142
budding	153
cancer	152
cell cycle	139
cell division	139
cell plate	149
centromere	145
chromatid	145
contact inhibition	151
contractile ring	149
crossing over	157
cytokinesis	144
diploid	155
Down syndrome	157
fertilization	154
furrow	149
gamete	153
genetic recombination	157
germ line	154
growth factor	151
haploid	155
homologous chromosomes	157
independent assortment	160
interphase	143
life cycle	139
meiosis	154
metaphase	145
mitosis	144
mitotic spindle	145
nondisjunction	157
pole	145
prometaphase	145
prophase	145
regeneration	153
sexual reproduction	153
somatic cell line	154
telophase	149
tumor	152
zygote	154

FUN FACTS

HUMAN MELANOMA SUPPRESSED BY CHROMOSOME 6

Transformation of normal cells to malignant cells is a multistep process that often involves loss of specific chromosomes or parts of chromosomes, leading researchers to postulate the existence of tumor suppressor genes. Melanoma, a particularly deadly type of cancer, is often associated with the loss of the long arm of chromosome 6.

Researchers isolated chromosome 6 from normal human cells and produced many copies in human-mouse hybrid cells. Copies of chromosome 6 were transferred into human melanoma cell lines, resulting in dramatic changes in cell morphology and reductions in cell division rates and tumor-forming ability in laboratory experiments. Melanoma cells thus modified sometimes lost the transplanted chromosome 6 and resumed active multiplication typical of tumor cells. These results suggest that chromosome 6 carries one or more genes involved in suppressing the malignant expression of melanoma by controlling cell division.

See: Trent, Jeffrey M., et al. 1990. Tumorigenicity in human melanoma cell lines controlled by introduction of human chromosome 6. *Science* 247:568–571.

MULTIPLE MUTATIONS FOUND IN CANCER

Development of a fully malignant cancer requires many mutations within a cell. Oncogenes must be activated so that they drive cellular multiplication; tumor suppressor genes, which normally inhibit cell growth, must be lost or damaged. As many as ten distinct mutations may have to accumulate in a cell before it becomes cancerous.

Development of colon cancer, for example, requires at least five independent mutations, some associated with alterations of specific genes and others with loss of whole chromosomes or chromosome parts, especially of chromosomes 5, 17, and 18, which seem to carry tumor suppressor genes. Because humans are diploid, and one copy of a suppressor gene is enough to inhibit cell proliferation, both homologous chromosomes must be lost or mutated, thus raising the minimum number of mutations that must accumulate before producing colon cancer to eight or ten. Deletions of certain chromosomes are associated with specific cancers; e.g., lung cancers, but not colon cancers, are frequently associated with losses or deletions of parts of chromosomes 3 and 11.

The multiplicity of mutations may provide an important clinical tool in the detection and prevention of cancer. If precancerous cells could be detected early in the mutation sequence, individuals could be warned to avoid mutagens, such as cigarette smoke. Many kinds of cancers can be treated most effectively early in their progression. The techniques of molecular biology might be used against cells bearing specific mutations.

See: Marx, Jean. 1989. Many gene changes found in cancer. *Science* 246:1386–1388.

Sager, Ruth. 1989. Tumor suppressor genes: The puzzle and the promise. *Science* 246:1406–1412.

CHAPTER 7 CELL CYCLES AND LIFE CYCLES

MESSAGES

1. The cell cycle is an alternation of growth and division.
2. Chromosomes are repositories of genetic information for directing cell growth and reproduction.
3. The correct apportionment of genetic information is crucial to reproduction.
4. Prokaryotic cells divide by binary fission.
5. Eukaryotic cells have a four-phase cell cycle. Three phases are devoted to growth and DNA duplication, one to cell division.
6. During mitosis, duplicate chromosomes are pulled to opposite poles. During cytokinesis, the cell divides in two.
7. Cell contact and growth factors can control cell division. Cancer cells do not stop growing or migrating.
8. Mitosis produces diploid daughter cells. Meiosis yields haploid daughters which become gametes.
9. During fertilization, gametes fuse to produce a diploid zygote. Development and reproduction follow in the life cycle.
10. Meiosis involves two cell divisions with only one chromosome duplication. Crossing-over and independent assortment increase genetic variability.

8

MENDELIAN GENETICS

PERSPECTIVE

This chapter outlines the history of Mendelian genetics and the gene interactions that may obscure the classical rules of genetics. The story of Mendel's experiments illustrates how scientists can form concrete conclusions from indirect data and how incorrect theories may be overthrown. This chapter builds on the material presented in Chapter 7, which covers cell reproduction, and provides background for concepts presented in the remainder of the genetics section.

The themes and concepts of this chapter include:

- All inherited traits are controlled by one or more genes, which are discrete stretches of DNA.
- Alternate forms of genes may be more or less adaptive than others, providing the raw material on which natural selection acts.
- Diploid organisms have pairs of homologous chromosomes which are segregated into gametes. Pairs of alleles are separated as the chromosomes separate.
- Mendel's principle of independent assortment explains that the genes on different chromosomes are distributed into gametes independently of one another.
- Genes that are located close together on the same chromosome tend to be inherited together.
- Dominant alleles may mask the phenotypic expression of recessive alleles.
- Sex-linked genes show different patterns of expression in males and females.
- Interactions between genes or between genes and their environments may modify their phenotypic expression.

CHAPTER OBJECTIVES

Students who master the material in this chapter will gain an understanding of the course of scientific progress by studying the history of genetics. They should be able to explain the logical arguments that Mendel used to demonstrate particulate inheritance. They should understand the concepts of recessiveness and dominance, of segregation during meiosis, of phenotype and genotype, and of independent assortment of non-linked genes. They should be able to devise a testcross to determine genotypes, to construct a map showing the relative loci of linked genes, and to describe how the interaction of genes affects their phenotypic expression.

LECTURE OUTLINE

(p 164) **I. Albino tigers introduce the study of heredity.**

A. Genes, which are portions of chromosomes, control traits.

B. Observation of how genes are expressed in succeeding generations reveals the workings of evolution by natural selection.

C. Systematic study of inheritance permits the detection of phenomena that cannot be directly seen.

(p 166) **II. Gregor Mendel discovered and analyzed the rules of inheritance.**

(tpq 1) **A. The blending theory of inheritance held that the genetic information of parents blended to produce offspring of intermediate characteristics.**

tpq 1. How much variability would there be in natural populations if the blending theory of inheritance were true?

(tr 37) **B. Mendel proposed and tested the particulate theory of heredity.**

1. Mendel used garden peas for his experiments.

a. Pea strains showed clear alternatives for single traits.

b. Mendel could control the mating because peas normally <u>self-fertilize</u> but he could <u>cross-fertilize</u> them.

2. Mendel disproved the blending theory.

a. He began with strains of <u>pure breeding</u> peas.

b. Mendel carried out <u>monohybrid crosses</u>, matings between individuals that differ in only one trait.

c. The <u>first filial generation</u> (F_1) showed the <u>dominant</u> trait of the <u>parental generation</u> (P) but not the <u>recessive</u> trait.

(tpq 2) **d. Mendel allowed the F_1 hybrid plants to self-fertilize, producing the seed for the second filial generation (F_2), where the recessive traits reappeared, disproving the blending theory.**

C. Mendel found that most F_2 generations showed approximately a 3:1 ratio of dominant to recessive forms.

---

tpq 2. What happened to the recessive trait in the F_1 generation? Why couldn't Mendel perceive it?

(msg 1) 1. He concluded that the unit of inheritance, now called a gene, is particulate and that genes come in alternate forms, now called alleles.

a. We now know that genes are discrete sections of chromosomes.

b. Each chromosome will have just one allele for any individual gene.

2. Mendel decided that a hybrid has two different alleles, one dominant and one recessive.

a. Two organisms with different genotypes may have the same phenotype.

(msg 2) b. Hybrids with two different alleles for a trait are heterozygous for that trait; pure-breeding organisms are homozygous.

(tpq 3) (msg 3) D. Mendel suggested that each parent donates one allele to each offspring during the formation of gametes, leading to the segregation principle: Sexually reproducing diploid organisms have two copies of each gene, which segregate from each other during the formation of gametes that contain only one copy of each gene.

(tr 38) 1. A Punnett square can be used to visualize the results of hybrid crosses, demonstrating how a 3:1 ratio arises.

---

tpq 3. How does the behavior of chromosomes during meiosis resemble the inheritance pattern of dominant and recessive traits?

(msg 4) 2. Mendel developed the testcross to determine the genotype of an organism. The organism with unknown genotype is crossed with a homozygous recessive.

(p 175) III. Mendel made dihybrid crosses to determine how different genes are inherited.

(tr 39) A. The 9:3:3:1 ratio of phenotypes he observed in the F_2 generation convinced him that different genes are inherited independently.

1. Parental types have the same phenotypes as the individuals of the parental generation.
2. Recombinant types have phenotypes that combine traits of the individuals of the parental generation.

(tpq 4) (msg 5) B. He proposed the principle of independent assortment: Different genes segregate into gametes independently of each other.

(p 176) IV. Mendel's results were ignored until they were rediscovered in 1900.

A. In the study of cytology, scientists discovered parallels between the distribution of chromosomes during meiosis and the inheritance of genes, suggesting that genes were physically linked to chromosomes.

---

tpq 4. What might cause the principle of independent assortment to be violated?

(msg 6) **B. The study of sex chromosomes in fruit flies revealed**

(tr 40) **that a gene for eye color was X-linked, indicating that it was carried on the X chromosome.**

1. **Autosomes are chromosomes in identical pairs in both sexes.**

(tpq 5)

2. **Males in many animal species have one X chromosome and one Y chromosome.**
   a. **Whether recessive or dominant, a sex-linked gene in males will be expressed in the phenotype because it exists as a single copy.**
   b. **Females in these species have two X chromosomes and obey the normal rules of recessiveness and dominance.**

C. Other genetic studies confirmed that genes ride on chromosomes and that each chromosome carries many genes.

(msg 7) **D. Genes on the same chromosome tend to be inherited together.**

1. **Parental combinations of genes appeared more often than recombinants in some experiments.**
2. **A linkage group is a set of genes usually inherited together.**
3. **Crossing over during meiosis accounts for recombinants.**

---

tpq 5. How is it true that boys inherit more of their phenotypic traits from their mothers than from their fathers?

E. Crossing over reveals that a chromosome is a linear array of genes.

1. The farther apart two genes are on a chromosome, the more likely they are to recombine during meiosis.
2. Geneticists use the frequency of recombination in genetic mapping.
3. Each gene has its place, called a locus, on a chromosome.
4. A chromosome is a linear array of genes, like beads on a string.

(p 180)

(msg 8)

V. Gene interactions modify and obscure Mendelian principles. Alleles may modify the expression of other alleles.

A. In incomplete dominance, the phenotype of the heterozygote may be intermediate between those of the homozygotes.

(tr 41)

B. In hybrids that exhibit codominance, two alternative alleles are fully apparent in the hybrid. Human ABO blood types are an example.

C. Some genes, including the major histocompatibility complex, have many alleles.

D. Epistasis occurs when one gene masks the effect of another gene.

E. Polygenic traits are controlled by several interacting genes. Quantitative traits are often polygenic.

(tpq 6) **1. Skin color in humans is controlled by the interaction of four genes.**

2. Agricultural geneticists manipulate quantitative traits to increase productivity.

F. A single gene may determine several different phenotypic traits, a phenomenon known as pleiotropy.

G. The expression of a gene may be altered by its environment.

---

tpq 6. Could two dark-skinned people have a light-skinned child? Why or why not?

INSTRUCTIONAL AIDS

VIDEOS, FILMS, SLIDES

Chromosomal Basis of Heredity. Britannica Films and Video, Chicago. An animated film that illustrates mitosis and meiosis and the history of the chromosomal theory.

Inheritance in a Fungus. Films for the Humanities and Sciences, Princeton, NJ. Illustrates experimental techniques involved in analysis of inheritance in a fungus. Video.

Mitosis and *Meiosis*. International Film Bureau, Chicago. Two short animated films that describe the stages of cell division. Film, video.

REFERENCES

Hoffman, Michelle. 1991. How parents make their mark on genes. *Science* 252:1250–1251. Descibes gene imprinting, which makes genes from the mother behave differently from those coming from the father.

McKean, Heather R., and L. S. Gibson. 1989. Hands-on activities that relate Mendelian genetics to cell division. *American Biology Teacher* 51 (5):294–300. Lab exercises and card games that give students insights into mitosis and meiosis.

Stephens, J. Claiborne, et al. 1990. *The Human Genome Map 1990*. Available from *Science*, Washington, DC. This poster summarizes the results of gene mapping and sequencing, detailing the progress made for each of the chromosomes.

KEY TERMS

FUN FACTS

GENES THAT CHEAT

When a male *Drosophila melanogaster* has one chromosome that carries a segregation disorder (SD) gene and an ordinary allele on the homologous chromosome, the SD-carrying chromosome is inherited by almost all of the offspring. The chromosome carrying the ordinary allele virtually disappears. SD has this effect only in the production of sperm; offspring of female flies heterozygous for SD show the 1:1 ratio predicted by Mendel.

Experiments with heterozygous males showed that the sperm that inherited the normal allele were incompletely developed. Their nuclei did not condense properly, their tails did not develop, and the sperm remained stuck together. Investigators concluded that SD somehow causes its homolog to destroy itself.

See: Crow, James F. 1979. Genes that violate Mendel's rules. *Scientific American* 240 (2):134–146.

PARENTAL GENE IMPRINTING

In an apparent violation of Mendel's principles, identical genes contributed to an offspring by the father and the mother may have different effects. Experiments with mouse zygotes formed with two sets of maternally derived or paternally derived genes produce deformed embryos that seldom reach term, even though the male-derived and female-derived genes have identical sequences and would produce normal embryos if combined heterosexually.

The differences seem to arise from sex-specific genome imprinting that is reversible. A gene that produces a specific effect when passed through a female might have a different effect after passing through her son. The difference may be due to the temporary attachment of methyl groups to certain base pairs in a gene. Scientists suspect that methylation affects interactions between DNA and the proteins that transcribe it into RNA.

See: Sapienza, Carmen. 1990. Parental imprinting of genes. *Scientific American* 263 (4):52–60.

YEAST ARTIFICIAL CHROMOSOMES

Gene mapping often depends on the frequency of meiotic recombination between linked genes. Human gene mapping is difficult because of our long generation times, small families, and mating habits driven by motives other than experimental design. To counter these difficulties, scientists construct yeast artificial chromosomes (YACs) incorporating segments of human chromosomes. After being implanted in yeasts, the YACs are replicated as the yeasts reproduce. When yeasts containing YACs produce spores through meiosis, recombination frequencies can be used to determine the locations of the genes on the YAC. This approach allows gene mapping on larger segments of chromosomes than other techniques, and might be extended to whole-chromosome mapping.

See: Green, Eric D., and M. V. Olson. 1990. Chromosomal region of the cystic fibrosis gene in yeast artificial chromosomes: A model for human genome mapping. *Science* 250:94–98.

CHAPTER 8 MENDELIAN GENETICS

MESSAGES

1. **Inherited traits are controlled by genes, which are discrete stretches of DNA. Homologous chromosomes carry the same genes.**

2. **The phenotypic expression of dominant alleles masks the phenotypic effects of recessive alleles. True-breeding strains are homozygous; hybrids are heterozygous.**

3. **Diploids have two alleles of each gene. These alleles segregate at random to form gametes that contain only one allele of each gene.**

4. **A testcross involves mating an individual with unknown genotype to a homozygous recessive individual.**

5. **Different genes, on different chromosomes or separated by a distance on the same chromosome, segregate into gametes independently of each other.**

6. **In many species of animals, the X chromosome exists as a single copy in males.**

7. **Genes closely situated on the same chromosome tend to be inherited together. Crossing-over separates genes distant from each other on the same chromosome.**

8. **Alleles may show incomplete dominance, codominance, epistasis, pleiotropy, or environmental influence. Traits may be polygenic.**

9

DNA: THE THREAD OF LIFE

PERSPECTIVE

This chapter presents the molecular structure of DNA and the basis of gene function woven into a history of the discovery that genetic information is carried on chromosomes. The process of DNA replication is outlined. This chapter builds on the concepts introduced in Chapters 7 and 8 and provides the fundamentals of DNA structure and function that will be the basis for understanding the material in Chapters 10 through 12, which present how genes work, how they can be manipulated, and how human characteristics are inherited.

The themes and concepts of this chapter include:

- Genes have two major functions, replication and carrying information.
- Genes look and act fundamentally the same in all living things.
- Each gene specifies the amino acid sequence of one polypeptide.
- The key to gene function lies in the double-helix structure of DNA.
- DNA is a linear molecule composed of two DNA strands oriented in opposite directions.
- Complementary base pairing and error-correcting enzymes assure faithful DNA replication.
- DNA in eukaryotes is wound around histones and coiled repeatedly.
- DNA stores information in the order of its bases.

CHAPTER OBJECTIVES

Students who master the material in this chapter will understand what DNA is, what genes are, how genes replicate, and how they function. They will see how scientific knowledge is built bit-by-bit, through the analysis of natural phenomena and clever experiments. They will appreciate the structure of DNA and understand how that structure makes replication possible. They will understand how chromosomes are packaged in eukaryotic cells.

Students should realize that genes act by specifying amino acid sequences. They will see that mutations change the sequence of nucleotides and can change the sequence of amino acids in polypeptides.

LECTURE OUTLINE

(p 188) **I. The spread by plasmids of multiple drug resistance in bacteria introduces what DNA is, what genes are, how genes replicate, and how they function.**

(tpq 1) **A. Both the structure of genes and the way genetic information is encoded and replicated are fundamentally the same in all living organisms.**

B. The key to how DNA works lies in the molecule's double helix structure.

(msg 1) **1. DNA's structure enables it to serve as a template for its own replication.**

(msg 2) **2. Genes work by specifying the amino acid sequences of polypeptides.**

C. Knowledge of how genes are built and operate has led to new ways to manipulate DNA.

---

tpq 1. Is it possible that life on earth comes from two or more independent origins; that is, could the same hereditary mechanism have arisen twice?

(p 190) **II. Researchers discovered that DNA, not protein, is the hereditary material.**

A. Researchers found that they could effect <u>transformation</u>, the transfer of an inherited trait from one bacterium to another, by transferring DNA, thus proving that genes are made of DNA.

(tpq 2) **B. Experiments with radioactively labeled <u>bacteriophages</u> demonstrated that phage DNA enters bacterial cells but phage protein does not; thus DNA carries the information to make new phages.**

(msg 3) **C. Subsequent research has confirmed that DNA is the genetic material in all cells and many viruses.**

(p 192) **III. DNA is the nearly universal hereditary material.**

A. The variability inherent in DNA is the basis of natural selection and the explanation of life's diversity.

(msg 4) (tr 42) **B. DNA is a linear molecule whose structure provides almost infinite capacity for carrying information.**

(msg 5) **1. The four kinds of nucleotides are identical except for the bases they contain.**

2. Every species has equal amounts of A and T and equal amounts of C and G.

---

tpq 2. Given the ability to grow bacteriophages so that they incorporate radioactive atoms (P only in DNA, S only in proteins), how would you perform an experiment to see which was the hereditary material?

---

3. James Watson and Francis Crick determined the structure of DNA.

a. From previous experiments, they knew that DNA was a linear molecule with a sugar-phosphate backbone and the four bases dangling off to the side.

b. Rosalind Franklin and Maurice Wilkins analyzed the structure of DNA by X-ray diffraction, leading Watson to conclude that a DNA molecule consists of two paired strands.

(tr 43) c. The final model has the DNA molecule composed of two nucleotide chains oriented in opposite directions with the bases pointing inward.

i. Base pairs are held together by complementary base pairing: A with T, G with C.

ii. The two strands are twisted together to form a double helix.

(tr 44) C. In living creatures, DNA is packaged.

1. In prokaryotes, viruses, and plasmids, DNA forms a circle.

(tpq 3) 2. In eukaryotes, DNA molecules are long and are

(msg 6) wound around proteins called histones.

---

tpq 3. What evolutionary disadvantage might there be to eukaryotes with loose, unpackaged DNA molecules in the nucleus?

---

a. A single histone with two loops of DNA is called a nucleosome.

b. Adjacent nucleosomes pack together to form a longer coil, which is in turn looped and packaged with other proteins into chromatin.

(p 196) IV. DNA's structure makes accurate replication and information storage possible.

(tr 45) A. DNA acts as a template for its own replication.

(msg 7) B. DNA replication includes three stages.

1. The strands unwind and separate, a process catalyzed by enzymes. The Y-shaped junction where the two strands unwind is the replication fork, where DNA is copied.
2. Unpaired bases on the strands pair with complementary bases on unattached nucleotides. Thus the order of bases on the old strand specifies the order in the new strand.
3. The unattached nucleotides are joined by covalent bonds into a new strand. DNA polymerase catalyzes this process but can work in only one direction, leading to differences in the way the two newly made DNA strands are synthesized.
4. One strand of each new molecule is inherited intact and the other is newly synthesized in semiconservative replication.

(msg 8) C. **DNA synthesis is highly accurate due to the actions of proofreading enzymes.**

(tpq 4) 1. **A typical error rate in the replication of an animal genome is 1 error per 10^9 base pairs.**

2. **Mutations are changes in genetic information.**

D. **The sequence of bases in DNA is the hereditary information.**

1. **A study of alkaptonuria convinced Archibald Garrod that the inherited disease was due to an absence of enzyme function; thus, a gene functions by allowing a specific enzyme reaction to occur.**

2. **George Beadle and Edward Tatum induced mutations in Neurospora crassa and found that different genes control the inheritance of different nutritional requirements. They reached certain conclusions.**
   a. **Each step in a biochemical pathway is controlled by a specific gene.**
   b. **Each step in a biochemical pathway is catalyzed by a specific enzyme.**
   c. **Therefore, a gene causes a specific enzyme to be formed.**
   d. **They summarized their work with the phrase one gene-one enzyme.**

---

tpq 4. Would there be any evolutionary advantage or disadvantage to having higher or lower rates of mutation?

---

3. **Sickle-cell anemia is caused by a substitution of a single amino acid in beta hemoglobin, causing a change in the shape of the molecule. This led to the conclusion that a gene specifies the order of amino acids in a single polypeptide chain.**
4. **Other experiments led to the conclusion that the sequence of DNA bases contains the genetic message.**

INSTRUCTIONAL AIDS

VIDEOS, FILMS, SLIDES

Cracking the Code of Life. American Cancer Society. Animation and examples explain the structure and function of DNA and the nature of cancer. Film.

Enzyme Defects and DNA. Milner-Fenwick, Inc., Timonium, MD. A survey of DNA structure and replication emphasizing the symptoms of genetic disease and the one gene-one enzyme concept.

Mendel's Laws. Coronet/MTI, Deerfield, IL. Discusses the beginnings of scientific genetics; includes footage of Mendel's garden. Video, film.

The Sickle Cell Story. Milner-Fenwick, Inc., Timonium, MD. Describes the history and physiology of sickle-cell disease. Film.

REFERENCES

Bardell, David. 1987. The function of penicillin in nature and its use in medicine. *American Biology Teacher* 49:175 – 176. A lab exercise demonstrating that molds produce antibiotics and inhibit bacterial growth.

Cohen, Stanley N., and James A. Shapiro. 1980. Transposable genetic elements. *Scientific American* 242 (2):40 – 49. "Jumping genes" transfer genetic information among plasmids, viruses, and chromosomes in living cells. Clearly written, well illustrated.

Fedoroff, Nina V. 1984. Transposable genetic elements in maize. *Scientific American* 250 (6):85–99. A well-illustrated treatment of the history, actions, and molecular nature of jumping genes.

Gallo, Robert C. 1987. The AIDS virus. *Scientific American* 256 (1):46–56. A discussion of the life cycle of this RNA virus, the history of its discovery, and its epidemiology. An authoritative essay by a leading scientist.

Mendel, Gregor. 1965. *Experiments in Plant Hybridisation*. Cambridge, MA: Harvard University Press. The seminal paper in genetics and biographical notes about the author.

Moyzis, Robert K. 1991. The human telomere. *Scientific American* 265 (2):48–55. Discusses the specialized DNA cap at each end of the chromosome.

Thomson, Robert G. 1988. Recombinant DNA made easy: 1. "Jumping genes." *American Biology Teacher* 50:101–106. A laboratory exercise involving the intracellular transfer of bacterial DNA.

KEY TERMS

bacteriophage	page 190	nucleosome	196
chromatin	196	plasmid	188
complementary base pairing	194	semiconservative replication	199
double helix	189	template	197
genome	200	transformation	190
histone	196		

FUN FACTS

HUMAN SICKLE HEMOGLOBIN GENES TRANSPLANTED INTO MICE

Sickle cell anemia was the first genetic disease understood at the molecular level but has been difficult to study experimentally because it occurs only in humans. Gene transfer has been used to create an animal model that can be studied in the laboratory.

Sickle hemoglobin genes were transplanted into fertilized mouse eggs, which were then implanted into the uteruses of foster mother mice. Transgenic mice thus formed were bred to create a strain of mice with red blood cells that became sickle-shaped when exposed to low oxygen levels. Except for their smaller size, mouse sickle cells were indistinguishable from those of a human with sickle cell disease. Sickle-cell mice offer the opportunity to test drug and gene therapies that could not be tested in humans.

See: Ryan, Thomas M., et al. 1990. Human sickle hemoglobin in transgenic mice. *Science* 247:566–568.

ARTIFICIAL CHROMOSOMES

Chromosomes include many nucleotide sequences that are never translated into proteins. Some of these elements control chromosome replication and segregation during mitosis and meiosis. In an effort to understand these functional elements, scientists have devised artificial chromosomes. Three chromosomal elements seem necessary. Replication sites are where the synthesis of new DNA begins; there may be many replication sites on each chromosome. Each chromosome includes a centromere, to which the spindle fiber attaches. The ends of chromosomes are protected by repetitive DNA sequences called telomeres. Experiments involving the deletion of selected portions of these elements contribute to the understanding of their function.

Artificial chromosomes are being used to study meiosis with the hopes of acquiring greater understanding of genetic disorders such as Down syndrome. Artificial chromosomes also allow the cloning in yeast of large human genes such as the cystic fibrosis gene, affording new opportunities for studying the genetics of humans and other animals.

See: Murray, Andrew W., and Jack W. Szostak. 1987. Artificial chromosomes. *Scientific American* 257 (5):62–68.

Green, Eric D., and Maynard V. Olson. 1990. Chromosomal region of the cystic fibrosis gene in yeast artificial chromosomes: A model for human genome mapping. *Science* 250:94–98.

CHAPTER 9 DNA: THE THREAD OF LIFE

MESSAGES

1. Genes have two major functions: they replicate and they carry information.

2. Each gene acts by specifying the amino acid sequence of one polypeptide.

3. Genes are made of DNA, not protein. They look and act fundamentally the same in all living things.

4. DNA is a linear molecule composed of two nucleotide chains oriented in opposite directions. The chains are held together by bonds between complementary bases.

5. The four bases of DNA can occur in any order, allowing great diversity.

6. DNA in eukaryotes is packaged in orderly coils.

7. During DNA replication, the two strands separate. Free nucleotides form complementary base pairs with the bases in each strand and are joined by DNA polymerase.

8. Base complementarity and proof reading enzymes assure accurate replication.

10

HOW GENES WORK: FROM DNA TO RNA TO PROTEIN

PERSPECTIVE

This chapter continues the discussion, begun in Chapter 7, of how molecules transmit hereditary traits. It details how genes are translated into polypeptides, including the molecular nature of the genetic code, the roles of the various types of RNA, and the action of ribosomes. Gene mutations and their effects are discussed. This chapter also presents the ways in which organisms control gene expression.

The material presented in this chapter will reappear in Chapter 11, Recombinant DNA, and will provide a foundation for the topics covered in Chapters 12 through 14, which cover human genetics and reproduction.

The themes and concepts of this chapter include:

- The flow of information in a cell is DNA ---> RNA ---> protein.
- Base pairing is essential to protein synthesis.
- Protein synthesis consumes energy. Cells have evolved means of saving energy by controlling gene expression.
- Prokaryotes and eukaryotes regulate gene expression in similar ways but at different timescales.
- The basic genetic mechanisms, including the genetic code, are universal.
- Ribosomes assemble polypeptides according to mRNA messages.
- Transfer RNA molecules translate a nucleic acid message into an amino acid sequence.
- Mutations can result in genetic defects or cancer.
- All cells in a eukaryotic organism have the same genes but become differentiated through selective gene expression.

CHAPTER OBJECTIVES

Students who master the material in this chapter will understand how the sequence of bases in DNA specifies the sequence of amino acids in a polypeptide and how mutations can affect an organism. They will understand the differences between DNA replication and RNA transcription and will appreciate the importance of complementary base pairing. These students will understand how the genetic code is used to translate the nucleotide message into amino acids. They will see how gene activity is regulated and will appreciate the evolutionary and developmental advantages of gene regulation.

LECTURE OUTLINE

(p 208) **I. Cystic fibrosis, a fatal disease inherited as an autosomal recessive, introduces the topic of how genes function.**

(tpq 1) **A. Symptoms include a buildup of sticky mucus in lung passages and in ducts of other organs. No cure is now possible.**

B. The protein CFTR (cystic fibrosis transmembrane regulator), a plasma membrane protein that probably transports chloride ions out of the cell, is defective.

(msg 1) **C. Genetic information flows in one direction.**

D. Protein synthesis requires a large expenditure of energy, and cells have evolved ways that minimize that energy cost by regulating gene expression.

---

tpq 1. If one in every 20 people of northern European ancestry carries a copy of the mutant allele, what is the probability that a child of this racial stock will be born homozygous for the allele?

E. **Basic genetic mechanisms are essentially universal.**

(p 210) II. **Genetic information flows from DNA to RNA to protein.**

A. **In the process of <u>transcription</u>, information in a portion of the DNA molecule is transcribed into a molecule of <u>messenger RNA</u> (<u>mRNA</u>), called a message.**

B. **In the second step, <u>translation</u>, the mRNA is translated into a sequence of amino acids by <u>ribosomal RNA</u> (<u>rRNA</u>) and <u>transfer RNA</u> (<u>tRNA</u>).**

(tr 46) C. **RNA is a string of nucleotides that differs from DNA in four respects.**

1. **The base uracil replaces thymine.**
2. **It is called a <u>ribonucleotide</u> because the sugar is ribose instead of deoxyribose.**
3. (tpq 2) **RNA is usually a single strand.**
4. **RNA molecules are much shorter than DNA molecules. Each short mRNA molecule that carries information to make a polypeptide is called a <u>message</u>.**

(tr 47) D. **During transcription, a section of the DNA double helix is copied into a new strand of RNA.**

---

tpq 2. How might the double-stranded structure of DNA make it more stable and long-lasting than RNA?

(msg 2)

1. **The double helix of DNA unwinds and unattached ribonucleotides pair with the exposed DNA bases of one strand.**
2. **RNA polymerase joins the ribonucleotides into a chain.**
3. **Transcription is similar to DNA replication with some important differences.**

(tpq 3)

   a. **Only one strand is usually transcribed into RNA for any given region of DNA.**
   b. **Either strand of DNA might be transcribed, but the transcribed stretch is always the same for any given gene.**
   c. **During transcription, a limited number of DNA bases are transcribed at a time.**
   d. **A single gene may be transcribed thousands of times.**

E. The translation of an RNA message into a polypeptide chain requires a translator, a code, and a site.

1. **The genetic code determines which amino acids the cell will translate from each base sequence.**
2. **Three adjacent bases, a codon, specify one amino acid.**
   a. **The code is redundant; more than one codon might specify one amino acid.**

---

tpq 3. What problems might arise if both strands of a section of DNA were transcribed?

---

b. Codons do not overlap and no bases are skipped.

c. The place where translation begins, the start codon, determines the meaning of the message by defining reading frames.

d. Three codons act as stop codons.

(msg 3) e. Nearly all organisms use the same genetic code, strong evidence that all living things derive from ancestral cells with a singular origin in our planet's history.

(msg 4) (tr 48) 3. Transfer RNA (tRNA) molecules translate mRNA messages into amino acids and transfer those amino acids to the developing polypeptide chain.

a. The tRNA anticodon pairs with a codon of mRNA.

b. At the other end of the tRNA, an amino acid is carried.

4. Ribosomes support mRNA and tRNA molecules and contain enzymes that link amino acids together.

(tr 49) F. Protein synthesis has three stages: initiation, elongation, and termination.

(216) III. A mutation is a change in the base sequence of DNA.

(msg 5) A. Chromosomal mutations affect large regions of chromosomes or entire chromosomes, including changes in chromosome number and changes in chromosome structure.

(tpq 4) (tr 50) B. Gene mutations alter individual genes. Point mutations are changes in one or a few bases.

1. A base substitution is the replacement of one base by another.
   a. Amino acid substitution may result.
   b. An amino-acid-specifying codon may change to a stop codon, a nonsense mutation.
2. Base insertion and base deletion change the number of bases. A frameshift mutation is the insertion or deletion of one or more base pairs, changing the reading frame and all codons past the mutation.

C. A mutation can profoundly affect an individual.

D. Mutations arise through spontaneous errors during DNA replication or through the effects of mutagens on DNA.

(tr 51)
1. DNA repair enzymes detect errors and repair DNA.
2. Carcinogens cause cancer by generating mutations.

---

tpq 4. Many point mutations are neutral in their effect on an animal but almost all chromosome mutations are lethal. Why are the effects different?

(p 219) IV. Gene regulation controls each cell's gene activity and protein production.

(msg 6) A. Genes can be regulated at any of five levels.

(tr 52) 1. Transcriptional regulation: Cells control the amount of mRNA transcribed from a gene.

2. Posttranscriptional regulation: Some mRNAs must be chemically modified before they can be transported from the nucleus. If not modified, mRNA may fail to reach the ribosomes and will not be translated.

3. Translational regulation: Access between ribosomes and mRNA may be blocked or growth of the polypeptide may be stopped.

4. Posttranslational regulation: Cells may not process polypeptides that have to be modified before they become functional.

5. Regulation of protein activity: Environmental factors may affect the efficiency of enzymes.

(tpq 5) B. Prokaryotes regulate their gene activity to maximize their efficiency.

(tr 53) 1. An operon includes a protein signaller and a group of genes that work together.

---

tpq 5. What might be the evolutionary disadvantage to a human gut bacterium that always produced lactose-digesting enzymes?

---

2. The lac operon of E. coli includes a protein, the repressor, that binds to DNA (at the operator) and blocks transcription when lactose is absent, preventing the production of lactose-digesting enzymes.

3. Important features that allow gene regulation in bacteria to respond rapidly include transcription-level control, simultaneous transcription and translation, and short-lived mRNAs.

(msg 7) C. Gene regulation in eukaryotes differs from that in prokaryotes.

1. Operons do not occur in eukaryotes.

(tr 54) 2. Eukaryotes cannot transcribe and translate simultaneously.

3. Eukaryotic gene expression is slow and unresponsive to environmental factors.

(msg 8) 4. During cell differentiation, genes in eukaryotic cells may be permanently turned off as the organism matures. As a result, different cell types produce different sets of proteins and perform different activities even though all cells have the same genotype.

5. Eukaryotic transcription is controlled by regulatory proteins that bind to special regions (called enhancers) in the DNA.

(tr 55)

6. **Introns are stretches of DNA that intrude into the gene but do not appear in the final mRNA. Exons are expressed.**
   a. **The primary transcript is edited by cleaving out the introns and splicing the exons together.**
   b. **Some introns (ribozymes) splice themselves out.**

INSTRUCTIONAL AIDS

VIDEOS, FILMS, SLIDES

Functions of DNA and RNA. Coronet/MTI, Deerfield, IL. How DNA directs the various various types of RNA to produce proteins. Video, film.

Origins of change: Heredity and mutation and *Origins of change: The function of DNA.* Films for the Humanities, Princeton, NJ. Step-by-step demonstrations of classic experiments in genetics, with a focus on evolutionary mechanisms. Video.

REFERENCES

Hoffman, Michelle. 1991. RNA editing: What's in a mechanism? *Science* 253:136 – 138. A discussion of a new model of RNA editing.

Hollstein, Monica. 1991. p53 mutations in human cancers. *Science* 253:49 – 53. A review of cancers and the associated mutations in the p53 tumor suppressor gene.

Judson, Horace F. 1979. *The Eighth Day of Creation: Makers of the Revolution in Biology*. New York: Simon and Schuster. A history of the origins and development of molecular biology, written for a general audience.

McKnight, Steven Lanier. 1991. Molecular zippers in gene regulation. *Scientific American* 264 (4):54 – 64. Two protein molecules may "zip" together, allowing them to bind to DNA and regulate gene expression.

Miller, Julie Ann. 1990. Genes that protect against cancer. *BioScience* 40 (8):563–566. The study of tumor suppressor genes may yield new therapies and greater insight into normal development.

Nomura, Masayasu. 1984. The control of ribosome synthesis. *Scientific American* 250 (1):102–114. The assembly of ribosomes from rRNA and protein molecules is controlled by the cell's needs.

Ptashne, Mark. 1989. How gene activators work. *Scientific American* 260 (1):40–47. A discussion of gene regulation in prokaryotes and eukaryotes.

Ryan, Thomas M., et. al. 1990. Human sickle hemoglobin in transgenic mice. *Science* 247:566–568. Biological engineering is used to transplant human hemoglobin genes into mice, creating an experimental animal for the study of sickle-cell anemia.

KEY TERMS

anticodon	page 214	operon	222
chromosomal mutation	217	protein synthesis	215
codon	212	reading frame	213
elongation	216	ribonucleotide	210
exon	225	ribosomal RNA (rRNA)	210
gene mutation	217	RNA polymerase	211
gene regulation	219	start codon	213
genetic code	212	stop codon	213
initiation	216	termination	216
intron	225	transcription	210
message	211	transfer RNA (tRNA)	210
messenger RNA (mRNA)	210	translation	210
mutation	216		

FUN FACTS

MALARIA, ANEMIA, AND NATURAL SELECTION

One out of 500 American blacks suffers from sickle-cell anemia, a disease characterized by a point mutation that changes one amino acid in the beta hemoglobin molecule. As a result of this substitution, hemoglobin molecules form long, narrow crystals, distorting the red blood cell into a sickle shape. Sickle-shaped red blood cells may not be able to pass through capillaries, resulting in a constellation of symptoms that are likely to be fatal.

Natural selection has preserved the sickle-cell allele despite its adverse consequences. Heterozygotes with the sickle-cell allele (Hb^S) and the normal allele (Hb^a) can resist infection by the parasite that causes malaria while people who are homozygous for the normal allele (Hb^a/Hb^a) often die from malaria. The distribution of malaria in Africa and the distribution of the sickle-cell allele are almost the same. American blacks carry a legacy of natural selection's effect on their African ancestors.

See: Mange, Arthur P., and E. J. Mange. 1980. *Genetics: Human Aspects*. Philadelphia: Saunders College.
Novitski, Edward. 1982. *Human Genetics*. New York: Macmillan.

ANTISENSE RNA AND DNA

Molecular biologists have developed antisense nucleotide sequences, powerful new tools to investigate individual genes. Antisense sequences are complementary to, and bind with, RNA or DNA sequences found in cells or viruses. When an antisense sequence binds to a mRNA molecule, it forms a duplex that cannot be translated.

Regulation or inhibition of gene expression with antisense RNA seems to be ubiquitous among viruses and bacteria, controlling many aspects of cell metabolism. Genetic engineering techniques have been used to generate antisense RNA and DNA molecules and to introduce them into cells, inhibiting the expression of specific genes This strategy may be useful in controlling viral infections and cancers.

See: Weintraub, Harold M. 1990. Antisense RNA and DNA. *Scientific American* 262 (1):40–46.

CHAPTER 10 HOW GENES WORK: FROM DNA TO RNA TO PROTEIN

MESSAGES

1. The flow of information in a cell is DNA ---> RNA ---> protein.

2. Base pairing allows the transmission of information from DNA to mRNA and the specification of amino acids by mRNA codons.

3. The genetic code is universal in almost all organisms.

4. Transfer RNA molecules act as the translator between mRNA codons and amino acids.

5. A mutation is a change in the base sequence in DNA and may result in an altered, missing, or nonfunctional protein.

6. Cells regulate the expression of their genes to conserve energy and to perform specialized tasks.

7. Prokaryotes and eukaryotes regulate gene expression in similar ways but at different timescales.

8. All cells in a eukaryotic organism have the same genes but become differentiated through selective gene expression.

11

GENETIC RECOMBINATION AND RECOMBINANT DNA RESEARCH

PERSPECTIVE

This chapter describes the molecular basis of genetic recombination in nature and in laboratories. Building on Chapters 7 through 10, it describes how recombination during meiosis increases variability and discusses the techniques, promises, and problems of genetic engineering. Students will gain insight into the public debate over genetic engineering.

The material presented in this chapter will reappear in Chapter 12, Human Genetics, and Chapters 33 through 37, which cover ecology.

The themes and concepts of this chapter include:

- Genetic recombination is a natural process that occurs in all sexually reproducing organisms.
- New combinations of alleles can create novel phenotypes.
- Variations produced by mutation and recombination are the raw material of evolution by natural selection.
- During recombination, two DNA molecules are cut and a piece of one is joined to a piece of the other.
- Using enzymes, researchers cut, recombine, and splice genes and move DNA from one organism to another.
- Genetic engineering can treat human disease and improve crop species.
- Society must confront the questions of safety and ethics that recombinant DNA technology poses.

CHAPTER OBJECTIVES

Students who master the material in this chapter will understand how scientists cut and splice DNA, move genes from one organism to another, and produce quantities of a gene or protein. They will comprehend how variety can arise in nature through recombination and how geneticists speed up this process. They will appreciate the promises and problems posed by genetic engineering.

LECTURE OUTLINE

(p 228) **I. Recombinant DNA technology is introduced with mice that grew large because the gene for human growth hormone had been introduced into their genomes.**

(msg 1) **A. Genetic recombination occurs naturally, during the crossover phase of meiosis, and in laboratories, where scientists use genetic engineering techniques.**

B. Three themes emerge.

1. Genetic recombination depends on complementary base pairing.

2. Genetic recombination is a key feature in the evolution of life.

(tpq 1) **3. Genetic engineering offers tremendous potential for advancing science and technology.**

tpq 1. How has genetic engineering been in the news recently?

(p 230) **II. Genetic recombination is a central feature in the**
(msg 2) **reproduction of all organisms.**

(msg 3) **A. Novel phenotypes may result from recombinations**
(tr 56) **of existing alleles.**

(msg 4) **B. By creating new, potentially successful variants, recombination plays an important role in evolution by natural selection.**

(tpq 2) **C. Species in which genes recombine are more adaptable and more likely to persist and evolve than species that cannot undergo recombination.**

(p 232) **III. Genetic engineers can produce new genes faster**
(tr 57) **than nature can.**

(msg 5) **A. A <u>restriction enzyme</u> recognizes a specific base**
(tr 58) **sequence and cleaves the DNA at or near that sequence. Some restriction enzymes leave staggered ends.**

(msg 6) **B. <u>DNA ligase</u> can be used to bind the staggered ends of two DNA fragments together, allowing scientists to join two different DNA molecules.**

C. DNA fragments spliced into a plasmid and introduced into a bacterium will be replicated when the bacterium reproduces and becomes a <u>clone</u> of cells.

---

tpq 2. Bacteria don't reproduce sexually. Do they have any ways to recombine genetically?

---

1. A probe of RNA or DNA that forms base pairs with the desired gene identifies the bacterial colony containing the gene.
2. Plasmids carrying the desired gene are isolated from the bacteria, supplying many copies of the gene.

(p 234) **IV. Genetic engineering holds promises and problems.**

(tpq 3) (msg 7) **A. Engineered organisms can be used to manufacture virtually any protein, such as human growth hormone, insulin, tissue plasminogen activator, and gamma interferon.**

(msg 8) **B. Recombinant DNA can accelerate the improvement of crops and livestock. Transgenic plants and animals may be more resistant to diseases and more productive than their ancestors.**

C. Genetic engineering has been used experimentally in human therapy.

1. Somatic cell gene therapy treats an individual's disease. The modified gene would not be passed to the individual's offspring.

---

tpq 3. Are there ethical problems with using engineered bacteria to produce human insulin? With using engineered goats to produce pharmaceuticals?

(tpq 4) **2. Germ line gene therapy would affect the individual's offspring. The manipulation of hereditary material raises complex technical and ethical issues.**

(msg 9) **D. Questions about the safety and morality of recombinant DNA research and application persist.**

---

tpq 4. Are there ethical problems with using genetic engineering to replace a mutant CFTR gene in a person's germ cells so no descendants would risk getting cystic fibrosis? With replacing a gene for sickle cell hemoglobin? With adding genes for athletic prowess?

INSTRUCTIONAL AIDS

VIDEOS, FILMS, SLIDES

Chromosomes and Genes (Meiosis). Coronet Film and Video, Deerfield, IL. Models show that genetic differences in individuals reflect events in meiosis. Film, video.

Genetic Engineering. Educational Images Ltd., Elmira, NY. Presents genetic engineering techniques and the benefits and risks of gene manipulation. Transparencies, video.

Lights Breaking: Ethical Questions about Genetic Engineering. Bullfrog Films, Oley, PA. A discussion of the ethical and practical concerns raised by genetic engineering. Video.

REFERENCES

Flanagan, Dennis, ed. 1981. Industrial microbiology. *Scientific American* 245 (3):66–215. A special issue covering the methods used in industrial microbiology.

King, F. A., et. al. 1988. Transgenic animals. *Science* 240:1468–1474. Discusses the techniques used to introduce foreign genes into the germ lines of mammals.

Koshland, D. E., ed. 1987. Frontiers in recombinant DNA. *Science* 236:1223–1286. A special issue devoted to leading-edge research in genetic engineering.

Moffat, Anne Simon. 1991. Making sense of antisense. *Science* 253:510–511. Discusses novel methods of blocking gene expression with antisense RNA.

Moss, Bernard. 1991. Vaccinia virus: A tool for research and vaccine development. *Science* 252:1662–1667. Reviews the use of cowpox virus as a vector for expressing genes in eukaryotic cells.

Verma, Inder M. 1991. Gene therapy. *Scientific American* 263 (5):68–84. Discusses the problems and potential of gene therapy to treat human illnesses.

KEY TERMS

clone	page 233	recombinant DNA technology	228
DNA ligase	233	restriction enzyme	233
electrophoresis	236	transposon	231

FUN FACTS

LIGHTING UP TOBACCO

Scientists are lighting up tobacco, but not for nicotine. They transplant firefly genes into tobacco plants, making them glow in the dark. No parlor trick, the technique enables scientists to see how genes are controlled.

Fireflies have an enzyme, luciferase (Latin, "light-bearer"), that catalyzes the reaction of ATP, oxygen, and luciferin (a small organic molecule), producing a cold, green light. Scientists isolated the gene for luciferase and inserted it into tobacco plants, which are as widely used in plant research as lab rats and fruit flies are used in animal experiments.

The researchers spliced the luciferase gene into a piece of DNA from a plant virus and introduced the recombinant gene into a bacterium, *Agrobacterium tumefaciens*. The bacterium in turn infected plants and transferred some of its own DNA, including the recombinant gene, into its host's cells. The scientists then grew whole plants from infected cells, watered them with a solution containing luciferin, and watched the plants glow.

Luciferase genes make ideal "reporters" on gene activities. Scientists can attach the luciferase gene to the promoter of a gene they are interested in. When the promoter is activated, the gene of interest and the accompanying luciferase gene will be transcribed, and luciferase will be produced. Gene activity can be assessed by feeding the plant luciferin and seeing where and when cells light up. In a developing embryo, for example, a gene might be turned on at a specific time. By hooking the luciferase gene to the gene being studied, a worker could watch a developing embryo and see just when the gene is activated by observing when light production begins.

Scientists are optimistic that the luciferase gene can be transplanted into other organisms--fruit flies, fungi, bacteria, and mammals--to provide sensitive, easily-observed markers of gene activity. Since the luciferase gene would be inherited and can be observed in live organisms, it could be a useful tool to determine patterns of inheritance and development. The technique is likely to remain just a tool for research--don't expect self-lighting cigarettes or plants that can serve as night lights.

See: Anonymous, 1987. Lighting up. *Scientific American* 256 (1):60 – 62.
Ow, David W., et. al. 1986. Transient and stable expression of the firefly luciferase gene in plant cells and trangenic plants. *Science* 234:856 – 859.

CHAPTER 11 GENETIC RECOMBINATION AND RECOMBINANT DNA RESEARCH

MESSAGES

1. During genetic recombination, two DNA molecules are broken and rejoined.

2. Genetic recombination is a natural process that occurs in almost all organisms.

3. Recombination of existing alleles can produce novel phenotypes.

4. Mutation is the ultimate source of all new DNA sequences, but genetic recombination reshuffles those mutations, maximizing the expression of variation.

5. Restriction enzymes cut DNA strands. Some leave sticky ends that will base pair with complementary sticky ends.

6. DNA strands that are held to each other by sticky ends can be joined by covalent bonds made by DNA ligase.

7. Genes can be transferred from one organism to another and made to express their proteins in their new environments.

8. Genetic engineering can treat human disease and improve crop species.

9. Society must deal with problems of safety and ethics that genetic engineering poses.

12

HUMAN GENETICS

PERSPECTIVE

This chapter extends the discussion of inheritance and gene action to human matters. The material in this chapter bridges the transition from the basics of cell reproduction and genetics, presented in Chapters 7 through 11, to reproduction and development, in Chapter 13, and the human life cycle, Chapter 14.

The themes and concepts of this chapter include:

- People are unique genetic subjects.
- Gene mapping and other modern techniques grew out of research on more classical genetic subjects.
- Small genetic changes that alter a person's physiology usually have a negative impact on physical functioning.
- Human pedigrees can reveal whether a trait is dominant, recessive, autosomal, or sex-linked.
- Recessive autosomal traits, recessive sex-linked traits, and autosomal dominant traits have different inheritance patterns.
- Changes in chromosome number and shape can lead to improper body functioning.
- Somatic cell hybrids and RFLPs can reveal gene loci.
- Genetic diseases can be treated in various ways.
- Detection of defective genes and counseling can reduce the impact of genetic defects.

CHAPTER OBJECTIVES

Students who master the material in this chapter will understand the inheritance patterns of recessive, dominant, autosomal, and X-linked traits. They will comprehend how pedigrees can reveal these patterns and how karyotypes can indicate abnormalities in chromosome number and shape. They will see that only a few abnormalities in chromosome number are survivable and that chromosome translocations can result in cancer.

This chapter surveys the New Genetics, presenting how somatic cell hybrids and restriction fragment length polymorphisms can be used to localize genes and identify their products. Students should understand how these techniques can be used to help people with genetic diseases and to help prospective parents reduce their chances of having babies with genetic defects.

LECTURE OUTLINE

(p 244) **I. Phenylketonuria, or PKU, introduces the topic of human genetics.**

- **A. PKU is characterized by an inability to break down phenylalanine.**
- **B. Children with PKU may suffer mental retardation unless their dietary intake of phenylalanine is restricted.**
- (msg 1) **C. Human genes are difficult to study.**
  1. **People are unique genetic subjects.**
  2. (msg 2) **Gene mapping and other modern techniques applicable to human genetics grew out of research on more classic genetic subjects.**
- (tpq 1) (msg 3) **D. Small genetic changes that alter human physiology usually have negative impacts.**

---

tpq 1. Are humans still affected by natural selection?

(p 246) **II. Traditional techniques have revealed thousands of human genetic conditions.**

(tr 59) **A. <u>Pedigrees</u> allow analysis of family genetic histories.**

(msg 4) **1. Pedigrees allow geneticists to determine whether a given trait is dominant, recessive, autosomal, on the Y chromosome, or <u>X-linked</u>.**

2. Autosomal recessive traits, such as PKU and <u>albinism</u>, are not expressed in heterozygous <u>carriers</u>. They appear equally in males and females.

(tpq 2) **3. X-linked recessive traits, such as Duchenne muscular dystrophy and hemophilia, are expressed in males who carry the allele but not in heterozygous females, who are carriers and can pass the allele to their sons.**

4. Autosomal dominant traits, such as Huntington's disease, can be expressed when the individual is heterozygous. The allele passes from an afflicted parent to an afflicted child.

---

tpq 2. A man whose blood clots normally wants to have children but is concerned because his father had hemophilia. What are his children's chances of having hemophilia?

B. Karyotyping enables geneticists to study conditions caused by changes in chromosome number, rearrangement of chromosome parts, or easily broken chromosomes.

(tpq 3) (msg 6)

1. Changes in the number of sex chromosomes or chromosome 21 occur in humans.
   a. Down syndrome results from three copies of chromosome 21.
   b. Turner syndrome results when a person has one X chromosome and no Y chromosome (XO).
   c. Klinefelter syndrome is a product of a genome that includes two X chromosomes and one Y chromosome (XXY).
   d. XYY results in extra height and below-average intelligence.
   e. Embryos that have too many or too few of the autosomal chromosomes (except chromosome 21) die in the first few months of development.
   f. A special mechanism equalizes the dosage of X-linked genes in males and females.
      i. One or the other X chromosome is inactivated in each cell during embryonic development in females.

---

tpq 3. Why are people with an extra copy of a chromosome usually sterile?

---

ii. Women are mosaics of cells containing either an active paternal or maternal X chromosome.

2. Chromosome translocations occur when part of one chromosome moves to a new location on another chromosome.
   a. Translocations can cause cancer when oncogenes are moved to the wrong locus.
   b. In a condition known as the fragile X syndrome, X chromosomes break, causing mental retardation.
   c. A gene's location is crucial to its function because adjacent DNA sequences control the transcription and expression of genes.

(p 254) III. New techniques have revolutionized the mapping of human genes.

(msg 2) A. Somatic cell hybrids can be created by fusing a human cell with a mouse tumor cell.

1. When the hybrid cell divides, human chromosomes are gradually lost, leaving clones of cells with different subsets of human chromosomes.
2. Assays of these clones for specific gene products enable researchers to determine on which human chromosome the gene resides.

(tr 60) B. <u>Restriction fragment length polymorphisms</u>, or RFLPs, result when DNA from different individuals is cut by restriction enzymes.

1. Restriction sites can be used as <u>genetic markers</u>.

(tpq 4) 2. Because of base substitution, a restriction enzyme might cut one individual's DNA at a site where it will not cut another's.

3. RFLPs may be used in family pedigrees to determine the locus of specific genes.

4. Once a gene locus is determined, geneticists can clone the gene for further study.

(p 257) IV. The tools and techniques of the New Genetics allow sophisticated screening and treatment of genetic diseases.

(msg 5) A. Early identification of genetic diseases is important for prevention and treatment.

(msg 7) B. Treatment may be physiological, such as the removal of phenylalanine from an infant PKU sufferer's diet.

C. Some genetic disease can be treated by giving patients the protein they lack.

D. Someday scientists may be able to replace defective genes with normal DNA sequences.

---

tpq 4. RFLP analysis is sometimes called "genetic fingerprinting." How might it be used to identify people?

(p 261) **V. New cases of some genetic diseases can be prevented.**

(msg 8)

A. <u>Amniocentesis</u> allows fetal cells floating in the amniotic fluid to be analyzed for chromosome or gene defects.

B. <u>Chorionic villus sampling</u> involves the removal and examination of fetal cells from the placenta.

C. Options are limited mainly to detecting carriers and providing genetic counseling.

D. Decisions about reproduction and abortion entail emotional costs.

INSTRUCTIONAL AIDS

VIDEOS, FILMS, SLIDES

The Ascent of Man: 12 – Generation Upon Generation. Pennsylvania State University AV Services, University Park, PA. Examines human inheritance from Mendel to modern experiments. Video.

Heredity, Health and Genetic Disorders. Human Relations Media, Pleasantville, NY. Discusses the causes and effects of various genetic disorders, genetic screening, and medical technology. Filmstrip, video.

The New Genetics: Rights and Responsibilities. Carolina Biological Supply Company, Burlington, NC. Presents basic genetic principles, discusses ethical implications of genetic counseling and manipulation.

REFERENCES

Bittles, Alan H., et al. 1991. Reproductive behavior and health in consanguineous marriages. *Science* 252:789 – 794. Reviews morbidity and mortality resulting from mating of relatives.

Fraser, F. C., and J. J. Nora. 1986. *Genetics of Man,* Second Edition. Philadelphia: Lea and Febiger. A discussion of traditional and modern research intended for high school students and college undergraduates.

Lundberg, Doug. 1990. Human chromosome preparation. *American Biology Teacher* 52 (2):109 – 112. A step-by-step laboratory exercise during which students prepare, develop, and stain their own chromosomes.

McKusick, Victor A. 1988. *Mendelian Inheritance in Man: Catalogs of Autosomal Dominant, Autosomal Recessive, and X-Linked Phenotypes,* Eighth Edition. Baltimore, MD: The Johns Hopkins University Press. A massive volume intended for medical professionals and genetics researchers.

Sutton, H. Eldon, and R. P. Wagner. 1985. *Genetics, a Human Concern*. New York: Macmillan. A textbook for nonspecialists who want to learn of the genetics of humans and other organisms we depend on.

Watson, James D. 1990. The human genome project: Past, present, future. *Science* 248:44 – 49. A short history and discussion of the prospects of the most ambitious project ever undertaken by molecular biologists.

KEY TERMS

albinism	page 248	Klinefelter syndrome	251
amniocentesis	261	oncogene	252
carrier	248	pedigree	247
chorionic villus sampling	261	RFLP	255
chromosome translocation	252	somatic cell hybrid	255
fragile *X* syndrome	253	Turner syndrome	249

FUN FACTS

MARKERS ON THE CHROMOSOME MAP

The 46 human chromosomes contain some 100,000 genes and about 3 billion base pairs. A single defective gene can cause crippling disease or death, but locating a gene can be a formidable task. Markers, identifiable DNA segments that can be inherited with specific genes, provide tools to localize genes.

The markers most often used are RFLPs (restriction fragment length polymorphisms); linkage analysis establishes the connection between RFLPs and specific genes. Linkage studies work only if a person carries one copy each of the mutant and normal alleles and two different versions of a marker located near the gene. The frequency of recombination in gametes indicates the proximity of the gene and marker. Radioactive DNA probes may be used to locate and label specific DNA sequences with base sequences complementary to the probe, thus marking individual restriction fragments or loci on whole chromosomes. The loci of genes causing Duchenne muscular dystrophy, Huntington's disease, cystic fibrosis, and other inherited diseases have been isolated with this technique.

If a gene locus can be determined within one or two million base pairs, or less than 0.001 of the genome, the tools of molecular biology can be used to clone and test the gene. Labeled RNA complementary to the mRNA made from the gene can be used as a probe to determine in which tissues the protein is made. The composition of the protein encoded by the gene can be deduced. Labeled antibodies that react to the protein can indicate where the protein is found in tissues. The RFLP linked with a disease-causing allele may be used to screen carriers and affected individuals.

See: White, Ray, and Jean-Marc Lalouel. 1988. Chromosome mapping with DNA markers. *Scientific American* 258 (2):40–48.

INHERITING BEHAVIOR

Human inheritance of behavior is particularly difficult to study. Most behavioral traits are not "either/or" dichotomies like Mendel's round and wrinkled seeds. Most behaviors result from the additive effects of many genes and environmental influences. But studies of twins and adoptees reveal that a significant portion of human behavior is genetically determined.

In twin studies, the behavioral similarity of identical twins is compared with that of fraternal twins. If identical twins adopted into different families show more behavioral similarity than do fraternal twins similarly adopted, the behavior is thought to be inherited. Some identical twins show remarkable similarity despite being raised in different cultures.

In more than 30 studies, involving some 10,000 pairs of twins, IQ has been more strongly correlated between identical twins (0.85, where 1.0 is the maximum possible correlation) than between fraternal twins (0.60). For adopted children and their adoptive parents, the correlation is about 0.20, indicating that family environment plays a relatively small role in IQ. Verbal ability and spatial ability show as much genetic influence as IQ. Studies of personality traits, such as activity level, neuroticism, and shyness, indicate that half of the variation between people seems to be inherited genetically. Psychopathologies such as schizophrenia and depression also show significant heritability.

See: Plomin, Robert. 1990. The role of inheritance in behavior. *Science* 248:183–188.

CHAPTER 12 HUMAN GENETICS

MESSAGES

1. The study of human genetics is difficult but important because of genetic diseases. Pedigrees and karyotypes are traditional techniques of investigation.

2. New techniques, including somatic cell hybridization and RFLPs, enable researchers to map human genes and detect and treat genetic diseases.

3. Human mutations are usually deleterious.

4. Autosomal recessive, sex-linked recessive, and autosomal dominant traits show characteristic inheritance patterns.

5. Some genetic diseases appear by early childhood; others don't show up until adulthood.

6. Changes in chromosome number and structure can lead to improper body functioning.

7. Genetic diseases may be treated by physiological means, protein therapy, or the repair or replacement of defective genes.

8. Detection of genetic anomalies in adults or fetuses and genetic counseling help minimize the risk of birth of children with genetic problems.

13

REPRODUCTION AND DEVELOPMENT: THE START OF A NEW GENERATION

PERSPECTIVE

This chapter presents the life cycle of multicelled animals, from fertilized egg through adult to gametes. It emphasizes embryonic development, starting with the union of sperm and egg, through cleavage, gastrulation, and organogenesis. Growth and development continue in adulthood and gametes are formed, completing the cycle.

The material in this chapter refers to cell specialization, which is a product of selective gene expression, introduced in earlier chapters in Part II, Perpetuation of Life. The themes introduced in this chapter will reappear in Chapter 14, The Human Life Cycle, and in Part IV, How Animals Survive.

The themes and concepts of this chapter include:

- The egg is omnipotent; it gives rise to all other cell types.
- Mate recognition and hormonal and behavioral signals help organize and synchronize the release of gametes.
- Development is an ordered sequence of irreversible steps remarkably similar in almost all animals.
- Embryonic cells are committed to their tasks because they express specific genes at certain times.
- Eggs are storehouses of nutrients and developmental instructions, while sperm are specialized to deliver their haploid nuclei.
- Cleavage is a series of rapid cell divisions that transform the zygote into a ball of cells and that begin cell specialization.
- Gastrulation creates the three primary tissue layers.
- During organogenesis, organs assume their form and function.
- Cancer is development run amok.
- The formation of gametes and subsequent fertilization completes the cycle.

CHAPTER OBJECTIVES

Students who master the material in this chapter will appreciate the processes that lead from egg and sperm to an adult organism. The complexities of mating and the processes of fertilization should impress them. They will see how cytoplasmic determinants and cell location cause specialization and differentiation.

These students will understand cleavage, gastrulation, neurulation, and organogenesis. They will see how development continues through life and how cancer can result from errors in the control of development. They will see the cycle return to the beginning with the formation of new gametes.

LECTURE OUTLINE

(p 266) **I. The development of zebra fish embryos introduces the chapter.**

(tpq 1) **A. Similar molecules and mechanisms direct the development of all animals.**

B. Development is an ordered sequence of irreversible steps.

(msg 1) **C. Each cell expresses a distinctive set of genes at specific times in development.**

(p 268) **II. Mating and fertilization get egg and sperm together.**

(msg 2) **A. Cooperation between organisms must occur for them to mate.**

1. Mate recognition is the first step. Cues may be visual, olfactory, auditory.

2. Gamete production and release must be synchronized.

---

tpq 1. How would you explain the fact that virtually all animals develop by the same mechanisms?

a. External fertilization occurs in many water-dwelling species.

(tpq 2) b. Land-dwelling animals use internal fertilization.

i. During copulation, the male deposits sperm directly into the female's body cavity.

ii. Internal fertilization allows the retention of the young inside the mother's body.

(tr 61) B. Fertilization is the union of the egg and sperm.

(msg 1) 1. An ovum, or egg, is usually the largest cell in an animal's body.

a. The ovum nourishes the embryo with yolk, which contains many macromolecules.

b. Substances in the cytoplasm control the expression of genes.

2. The sperm is one of the smallest cells in the body, streamlined to deliver its haploid nucleus.

a. When a sperm contacts an egg, surface molecules bind to sperm receptor molecules on the egg's surface.

b. Then the acrosomal reaction occurs. The sperm releases digestive enzymes that enable it to penetrate the egg's plasma membrane.

---

tpq 2. Egg-laying frogs have external fertilization but egg-laying lizards have internal fertilization. How did natural selection lead to this difference?

c. When the sperm nucleus enters, a flurry of activity begins in the egg.

d. The egg's surface hardens, preventing the entry of more sperm.

e. The metabolic activity increases.

f. The haploid nuclei fuse and cytoplasmic differentiation may occur.

(tpq 3) 3. A few animal species reproduce by parthenogenesis.

(p 271) III. Vertebrate development follows regular patterns.

(msg 3) A. In a zebra fish, the sequence is:

1. The fertilized egg divides into many cells (cleavage).
2. These cells become rearranged into three layers (gastrulation).
3. The three layers interact and develop into tissues and organs (neurulation and organogenesis).
4. The fully developed embryo grows larger in size and matures sexually (growth and gametogenesis).

(msg 4) (tr 62) B. Cleavage consists of a series of cell divisions that transform the zygote into a morula and then a blastula.

---

tpq 3. Virtually all lizard species that reproduce parthenogenetically seem to have recently diverged from sexually-reproducing species. Why are these species evolutionarily young?

---

1. **No cell growth or transcription occurs between divisions.**
2. **The source of the genetic information that encodes new proteins is the mother's genome, not the embryo's.**
3. **The blastula includes several hundred <u>blastomeres</u>.**

C. **The amount of yolk determines the pattern of cleavage divisions.**

1. **The eggs of mammals are small.**
   a. **The developing embryo depends on the <u>placenta</u> instead of yolk for nutrition.**
   b. **Each cleavage division passes completely through the egg.** (tr 63)
2. **In eggs with large yolks, a group of cells forms the <u>blastodisc</u>. Cleavage furrows cannot pass all the way through the egg.** (tpq 4)

D. **Cleavage partitions developmental information.**

1. **Egg cytoplasm contains <u>developmental determinants</u>, localized in the yolk, that will act as instructions for the developing embryo.**
2. **When partitioned into embryonic cells, developmental determinants activate the expression of specific genes, causing different cell groups to develop differently.**

---

tpq 4. What evolutionary advantages or disadvantages might there be to egg-laying instead of carrying young internally?

(tpq 5)

a. **In frog eggs, the gray crescent appears opposite the point of sperm entry and is partitioned only into certain cells, eventually directing the formation of the dorsal region.**

b. **In mammals, the position of the cells relative to one another directs early development.**

c. **Toward the end of the cleavage stage, controlled gene expression becomes important.**

(msg 5)

E. Cleavage assigns a cell's position and its fate.

(tr 64)

1. **All cells of a mammalian embryo are equivalent at the four-cell and eight-cell stages.**
2. **At the morula stage, some cells form the trophoblast, which gives rise to the placenta (the afterbirth), and the inner cell mass, which becomes the embryo.**

(msg 6)

(tr 65)

F. Gastrulation transforms the ball of cells created by cleavage into the rudiments of an animal and establishes the three germ layers.

1. **Endoderm gives rise to the gut and associated organs.**
2. **Mesoderm produces muscles, bones, connective tissue, blood, and reproductive and excretory organs.**

---

tpq 5. What might happen to a frog embryo's development if you homogenized the contents of the egg before fertilization?

---

3. Ectoderm becomes the skin and the nervous system.

(tr 66) G. Neurulation is the formation of the neural tube.

1. The mesoderm forms the notochord, which stimulates the overlying ectoderm to roll up into the neural tube, a primary embryonic induction.
2. In some human births, the neural tube does not close, a birth defect known as spina bifida.

H. Determination begins in the blastula as cells become irreversibly committed, or determined, to a particular developmental pathway.

(p 278) IV. Organogenesis is the development of body organs.

(tpq 6) (msg 7) A. Organogenesis involves morphogenesis, the development of organ shape, and differentiation, the development of organ function.

B. Cells migrate in the developing embryo.

1. Neural crest cells migrate to form pigment cells, teeth, and structures of the lower face.
2. Cleft lip and cleft palate result if these cells fail to proliferate and grow together.

C. Differentiated cells are specialized to perform certain functions. Regulatory proteins select genes for use that are appropriate to the cell type.

---

tpq 6. Organogenesis in humans occurs primarily in the first trimester. Why would it be especially important for a pregnant woman to avoid drugs during this period?

(p 281) **V. Development continues throughout life.**

A. Growth is the most apparent aspect of postembryonic development.

1. Different parts of the body grow at different rates.

2. Stem cells are generative cells that retain the ability to divide throughout life.

3. Growth factors stimulate growth.

a. Growth hormone stimulates increased height.

b. Platelet-derived growth factor stimulates cells to divide and fill in wounds.

c. Activin is an embryonic organizer.

d. Nerve growth factor helps nerve cells survive and grow.

(tpq 7) (msg 8) **B. Cancer is development run amok. Cancer cells mimic normal embryonic cells.**

(tr 67) **1. A benign tumor is a clump of cells that grows unchecked but stays localized.**

2. A malignant tumor metastasizes, spreading throughout the body.

a. Cancer cells, or transformed cells, are not differentiated completely and thus preserve their capacity to divide. They are round in shape like immature cells.

---

tpq 7. Many cancers seem to be derived from mutant stem cells. Why would stem cells be especially prone to giving rise to cancer?

b. **Normal genes that regulate cell division can be mutated into oncogenes that stimulate cell division.**

c. **Tumor suppressor genes produce proteins that block tumor growth.**

(p 286) **VI. The formation of gametes begins the cycle anew.**

(msg 9) **A. Gametogenesis is the construction of germ cells.**

B. In many species, germ plasm is a cytoplasmic region in the egg that becomes the future germ cells.

INSTRUCTIONAL AIDS

VIDEOS, FILMS, SLIDES

Amphibian Embryo. Pennsylvania State University AV Services, University Park, PA. Time-lapse photography and animation illustrate embryogenesis in an amphibian. Film.

Patterns in Development. The Media Guild, San Diego, CA. Presents three systems of signals that organisms use to specify cell and tissue development. Video, film.

Reproduction: The Continuity of Life. Carolina Biological Supply Company, Burlington, NC. Covers a wide range of reproductive strategies and explains the effects on each species. Transparencies.

REFERENCES

Aral, Sevgi O., and King K. Holmes. 1991. Sexually transmitted diseases in the AIDS era. *Scientific American* 264 (2):62–69. A survey of STDs and how the new epidemics might be defeated.

Beardsley, Tim. 1991. Smart genes. *Scientific American* 265 (2):86–95. A discussion of gene regulation during embryonic development.

Igelsrud, Don. 1987. Sea urchins. *American Biology Teacher* 49:446 – 450. A discussion of the use of sea urchins to demonstrate fertilization, cleavage, and gastrulation.

Kenyon, Cynthia, and Bruce Wang. 1991. A cluster of *Antennapedia*-class homeobox genes in an unsegmented animal. *Science* 253:516 – 517. A perspective on the similarities of homeobox genes in insects and an unsegmented roundworm.

Rosner, Mitchell H., et al. 1991. Oct-3 and the beginning of mammalian development. *Science* 253:144 – 145. A review of the first regulatory gene to be identified by focusing on transcription factors expressed in embryonic stem lines.

Trinkaus, John P. 1984. *Cells into Organs: The Forces that Shape the Embryo,* Second Edition. Englewood Cliffs, NJ: Prentice-Hall. A review of the history and current status of embryology, clearly written and comprehensive.

KEY TERMS

benign	page 284
blastodisc	273
blastomere	272
blastula	272
cleavage	272
committed	278
copulation	269
development	267
developmental determinant	273
differentiation	280
ectoderm	275
embryo	267
endoderm	275
external fertilization	269
gametogenesis	286
gastrulation	274
germ layer	275
germ plasm	286
gray crescent	273
growth	281
internal fertilization	269
malignant	284
mate	268
mesoderm	275
metastasize	284
morula	272
neural tube	276
neurulation	276
notochord	276
organogenesis	279
ovum	269
parthenogenesis	270
placenta	273
primary embryonic induction	278
stem cell	283
yolk	269

FUN FACTS

SEX AND THE SINGLE LIZARD

Several species of whiptail lizard, in the genus *Cnemidophorus,* reproduce without fertilization. All individuals are female and all offspring are clones of their mother. In *C. uniparens,* although fertilization isn't necessary (or possible), reproduction requires courtship and mating behavior.

C. uniparens females act like "males," alternating sexual roles with other females. The male-impersonators' behavior closely resembles that of males of a closely related bisexual species, *C. inornatus*. Courtship among these lizards follows a well-defined ritual, imitated by *C. uniparens* females short of the final act, insertion of the penis.

Scientists raised *C. uniparens* females in isolation and with other females in various hormonal states. Isolated females laid only a third of the eggs laid by females that engaged in courtship and "mating" with other females. The researchers took blood samples from the lizards and found that their behavior was determined by their levels of hormones. Lizards showed typical female behavior when estrogen levels were high, just before they ovulated. Male-like behavior appeared in females just after ovulation, when estrogen levels dropped and progesterone levels reached a maximum. Apparently the different hormones activated two distinct neural circuits controlling sexual behavior, one active in males of bisexual species and the other in females.

See: Crews, David. 1987. Courtship in unisexual lizards: A model for brain evolution. *Scientific American* 257 (6):116–121.

Edwards, Diane D. 1987. Leaping lizards and male impersonators: Are there hidden messages? *Science News* 131:348–349.

HOMEOBOX GENES

Homeotic mutations in the fruit fly cause body parts to be replaced by structures normally found elsewhere in the body; for example, the antennae of mutants may be replaced by legs. Transformations may be caused by mutations in master homeotic genes that control the effects of many other genes. Homeotic genes in many different, distantly related organisms, such as frogs, flies, and mice, have a highly conserved DNA sequence, the homeobox.

The homeobox encodes a 60–amino acid sequence that recognizes and binds to specific DNA sequences in the genes regulated by the homeobox genes, thus controlling gene expression. During embryonic development, specific homeobox genes are active during the formation of specific tissues and organs. The order of homeobox gene loci on a chromosome is related to their site of expression in the embryo. Homeobox genes located to the left are expressed nearer the tail end of the body and genes to the right are active closer to the head. Gradients of the resulting proteins establish the head-tail axis and the positions of cells during limb development.

See: De Robertis, Eddy M., et al. 1990. Homeobox genes and the vertebrate body plan. *Scientific American* 263 (1):46–52.

CHAPTER 13 REPRODUCTION AND DEVELOPMENT: THE START OF A NEW GENERATION

MESSAGES

1. The egg cell is omnipotent; it gives rise to all other cell types. It is a storehouse of nutrients and developmental instructions. Sperm are minute carriers of haploid nuclei.

2. Mate recognition and hormonal and behavioral signals help organize and synchronize the release of gametes.

3. Development is an ordered sequence of irreversible steps remarkably similar in almost all animals.

4. Cleavage is a series of rapid cell divisions, without transcription or cell growth, that create the blastomeres.

5. Embryonic cells are committed to their tasks because they express specific genes at certain times. Cytoplasmic determinants and cell position trigger cell determination.

6. Gastrulation forms the three primary tissue layers. Neurulation forms the central nervous system.

7. During organogenesis, organs assume their form and function.

8. Cancer is development run amok.

9. The formation of gametes and subsequent fertilization complete the development cycle.

14

THE HUMAN LIFE CYCLE

PERSPECTIVE

This chapter presents an overview of the whole human life cycle from the formation of the zygote through embryonic development, birth, maturation, and aging. This chapter extends the material presented in Chapter 13, Reproduction and Development, and provides an underpinning for discussions in Part IV, How Animals Survive, that address organ systems and their functions.

The themes and concepts of this chapter include:

- Male and female reproductive systems are structurally and functionally similar. Differentiation occurs in response to hormones.
- Intricate networks of communication between body organs coordinate an individual's development, maturation, and gamete production.
- Male reproductive structures make sperm continuously and deliver them to the female reproductive tract.
- Female reproductive structures release eggs periodically and house and nourish the developing fetus.
- Pregnancy is a biological partnership between mother and embryo.
- Techniques have been devised to control pregnancy and combat infertility.
- Development of the human embryo is similar to the development of other mammalian embryos.
- The fetus programs itself and its mother for birth.
- Birth is the start of independent life but not the end of development. Infancy and childhood are periods of mental growth and physical development.
- Aging may be caused by genetic programming or wear and tear.

CHAPTER OBJECTIVES

Students who master the material in this chapter will gain an overview of the human life cycle from fertilization to senescence. They will understand the similarities and differences between male and female reproductive systems and how the maturation and maintenance of those systems are regulated by hormones.

The processes of fertilization, embryogenesis, fetal development, and sexual differentiation should be understood. Of special importance are the topics of birth control and the mother's role in maintaining the developing embryo. An understanding of sexual maturation and aging will help students in accepting the course of their lives.

LECTURE OUTLINE

(p 290) **I. In vitro fertilization introduces the human life cycle.**

(tpq 1) **A. Human embryos are similar to those of other vertebrate species in form and development.**

(msg 1) **B. Male and female reproductive systems are generally parallel in structure and function.**

C. Networks of communication between body organs coordinate the production and release of eggs and sperm.

D. Pregnancy is a biological partnership between mother and fetus.

E. Birth is the start of independent life but not the end of development.

(p 292) **II. Male and female reproductive systems include primary and accessory reproductive organs.**

---

tpq 1. Why do human embryos have tails and gill slits at certain stages?

A. Primary sexual characteristics are reproductive organs. Secondary sexual characteristics are external features not directly involved in sexual intercourse.

(tr 68) B. The male system makes sperm and transfers them to the female reproductive tract.

1. Some organs produce sperm.
   a. Testes are held in the scrotum.
   b. Each testis is subdivided into compartments that contain seminiferous tubules.
   c. Spermatogenic cells and supporting (Sertoli) cells line the walls of the tubules.
   d. Interstitial (Leydig) cells are embedded in connective tissue that surrounds the tubules. These cells produce testosterone.

(tpq 2) 2. High temperature suppresses sperm development.

3. As they mature, sperm pass down the tubules and collect in the epididymis.

C. Some organs transport sperm.

1. Sperm are propelled from the epididymis into the vas deferens (ductus deferens) which leads to the ejaculatory duct.

---

tpq 2. Why is the fertility of male bicycle racers low?

2. The <u>seminal vesicle</u> secretes sperm-supporting fluid into the ejaculatory duct.

3. The ejaculatory duct then passes through the <u>prostate gland</u> which secretes fluid that helps neutralize the acidity of the female tract.

4. The sperm passes into the <u>urethra</u> and out through the <u>penis</u>.

   a. The <u>bulbourethral gland</u> secretes lubricating mucus.

   b. <u>Erectile tissue</u> fills with blood and stiffens the penis.

(tpq 3) 5. <u>Ejaculation</u> ejects <u>semen</u> (including <u>seminal fluid</u>).

(msg 2) (tr 69) D. Hormones from the brain and testes work in a feedback loop that controls sperm production.

1. <u>Testosterone</u> is secreted by the testes.

2. Luteinizing hormone (LH) and follicle-stimulating hormone (FSH) are made in the pituitary.

3. A releasing hormone is made in the hypothalamus.

4. Releasing hormone stimulates the pituitary to produce LH and FSH.

---

tpq 3. What is the value in natural selection of the pleasure that accompanies orgasm?

---

a. LH and FSH activate cells in the testes to produce testosterone and sperm.

b. Testosterone causes supporting cells in the testes to release inhibin, which helps inhibit FSH production.

c. The production of these hormones is a self-regulating negative feedback system.

d. Testosterone induces the development of secondary sexual characteristics at puberty.

(tr 70) E. The female reproductive system produces and transports gametes, nourishes embryos, and delivers babies.

1. Ovaries contain follicles, each containing an oocyte, which develops into an egg, and follicular cells, which support the oocyte.
2. Ovulation involves the maturation and release of an ovum.
3. The ovum is swept into an oviduct, where a sperm may fuse with it.
4. The oviduct leads to the uterus, where the embryo will be nourished.
5. The cervix separates the uterus from the vagina, which receives the penis and acts as the birth canal.
6. External organs include the clitoris and Bartholin's glands, which produce lubricant.

(msg 3) **F. The endometrium, which lines the uterus, breaks down if no embryo arrives.**

(tpq 4) **1. Most women have a 28-day menstrual cycle.**

(tr 71) **2. Ovaries produce estrogen and progesterone and the pituitary makes LH and FSH in an interacting cycle.**

a. Menstrual flow begins when progesterone and estrogen levels are lowest.

b. The drop in hormones prompts the hypothalamus to secrete releasing hormone.

c. Releasing hormone stimulates the pituitary to produce FSH and LH.

d. FSH stimulates follicles to grow.

e. The follicle secretes estrogen, which causes the endometrium to build up.

f. About day 14, the pituitary secretes a pulse of LH and FSH, which triggers oocyte maturation and release of the ovum.

g. The ruptured follicle becomes the corpus luteum, which secretes estrogen and progesterone.

h. If the ovum is not fertilized, the corpus luteum degenerates on day 24.

---

tpq 4. The lunar cycle and the menstrual cycle are both 28 days long. Is this coincidence?

---

i. With the decline in estrogen and progesterone, the endometrium begins to slough off.

(p 298) III. Fertilization and pregnancy may or may not occur.

A. <u>Coitus</u> and <u>orgasm</u> result in the passage of sperm from male to female.

1. Only one sperm penetrates the egg.

a. The egg erects barriers to further sperm penetration.

b. The egg nucleus completes meiosis II.

2. The haploid sperm nucleus unites with the haploid egg nucleus.

3. <u>Pregnancy</u> begins with implantation of the embryo in the uterine wall.

(tpq 5) B. The exploding human population is the greatest threat to the planet's life-support systems.

(msg 4) C. <u>Contraception</u> can be achieved by preventing the release of an egg, blocking fertilization, inhibiting implantation, or preventing embryonic growth after implantation.

1. Behavioral and chemical techniques and physical barriers may prevent fertilization.

2. Oral contaceptives block FSH and LH production and thus suppress egg maturation.

---

tpq 5. How does overpopulation hurt the planet's life support systems?

---

3. Intrauterine devices interfere with implantation.
4. Abortion and induced menstruation terminate pregnancies.
5. Sterilization is achieved by tubal ligation or vasectomy.

D. Infertility occurs equally in women and men. Surgery, in vitro fertilization, and hormones are used to treat infertility.

(p 302) IV. Development of the fetus takes about nine months.

(tpq 6) (msg 5) A. The blastocyst implants into the uterine wall about six days after fertilization.

(tr 72)
1. Trophoblast cells develop into the <u>chorion</u>, which surrounds the embryo, absorbs nutrients from the mother's blood, and develops into the placenta.
2. The chorion produces <u>human chorionic gonadotropin</u> (hCG), which prevents a new menstrual cycle by preventing the corpus luteum from degenerating.
3. The mother's blood does not mingle with the blood of the fetus, but nutrients are exchanged at the placenta, which makes estrogen and progesterone.

---

tpq 6. Some contraceptive methods prevent the release of eggs while others prevent the implantation of blastocysts. Is one kind of method ethically different from the other?

(msg 6) **B. The human embryo goes through developmental stages.**

1. **Cell growth in the embryo creates the <u>amniotic cavity</u> and the <u>umbilical cord</u>.**
2. **Gastrulation begins in the third week of pregnancy.**
3. **Neural tube formation occurs in the third week.**
4. **Organogenesis begins as <u>somites</u> begin to pinch off. They will differentiate into muscles, bones, and other tissues.**
5. **By the eighth week, the fetus has the rudiments of all its organs.**

C. At eight weeks, the primary sex organs are <u>indifferent</u>: They haven't become specialized as male or female.

1. **In the absence of specific biochemical signals, these indifferent structures become female.**
2. (tpq 7) **The Y chromosome bears a gene, the testis determining factor (TDF), that causes male differentiation by initiating a cascade of events leading to the development of male organs.**
3. **The presence or absence of testosterone brings about sexual differentiation by the twelfth week of development.**

---

tpq 7. Some people who have a Y chromosome are phenotypically female. What sort of mutation might cause this to happen?

---

4. Mullerian inhibiting hormone kills the cells that would develop into oviducts and uterus.
5. Some testosterone is converted to 5DHT, which causes the external genitals to become penis and scrotum.

D. During the second and third trimesters, growth is the main activity. Physical and behavioral development continues.

(msg 5) E. The mother's body cooperates with the fetus during pregnancy.

1. Hormones from the fetus maintain the pregnancy; the mother's body changes to accomodate the fetus.

(tpq 8) 2. The mother's diet, health, and life-style habits influence fetal development. Drugs and diseases can affect the fetus.

(msg 7) F. Birth is a coordinated effort for which both fetus and mother are biologically prepared.

1. The mother's pituitary secretes oxytocin and her uterus releases prostaglandins, hormones that induce uterine contractions in a positive feedback loop.
2. During labor, the cervix widens, the amniotic sac bursts, and the contractions force the baby and placenta out of the birth canal.

---

tpq 8. During which period of pregnancy are drugs likely to have the most significant impact on organ development?

---

3. The baby must make the transition to breathing air and countering the loss of body heat.

(p 309) V. Development continues with growth, maturation, and aging.

(msg 8) A. Infancy and childhood are times of profound mental and physical changes. Growth is fastest in the head and body center and slowest in the periphery.

B. Puberty is the maturation of the reproductive system and the development of secondary sexual characteristics, which is dependent on sex-specific hormones.

C. Adulthood is the longest stage in the life cycle.

(msg 9) D. Aging is a progressive decline in the maximum functional level of cells and organs.

1. Females experience menopause around the age of 50 while males may lose some potency.

2. Senescence is the more rapid decline in cell and organ function.

(tpq 9) 3. Some scientists support the idea that there is a genetic clock that programs aging.

a. Evidence includes the 50-division limit on cell reproduction.

b. Organs seem to age in a programmed way.

---

tpq 9. What might be an evolutionary advantage to a species in which aging was genetically programmed to occur?

---

c. **Certain genetic conditions, such as progeria, can accelerate aging.**

d. **Some people inherit a dominant gene that causes Alzheimer's disease.**

4. **Other scientists believe aging is due to wear and tear.**

5. **Humans have very long life spans and life expectancy, which can be increased by good health habits.**

INSTRUCTIONAL AIDS

VIDEOS, FILMS, SLIDES

Embryology of Human Behavior. International Film Bureau, Chicago, IL. Reveals the physical and mental development of human fetuses. Film, video.

Human Sexuality and the Life Cycle. Human Relations Media, Pleasantville, NY. A three-part discussion of sexuality at all stages of human development. Filmstrip, video.

Well Conceived. MTI Film and Video, Deerfield, IL. Examines recent research on prenatal care and suggests diet and habits that promote healthy mothers and babies. Video.

REFERENCES

Lagercrantz, Hugo, and T. A. Slotkin. 1986. The "stress" of being born. *Scientific American* 254 (4):100 – 107. During birth, a baby's endocrine system produces stress hormones that help it make the transition to independent life.

Ulmann, André, et al. 1990. RU-486. *Scientific American* 262 (6):42 – 48. The investigators' own story of the discovery and development of this controversial pregnancy-terminating drug.

Weiss, Rick. 1988. Forbidding fruits of fetal-cell research. *Science News* 134:296–298. A discussion of the controversies regarding the use of fetal-cell transplants to treat diabetes, Parkinson's disease, and other maladies.

KEY TERMS

aging	page 311
amniotic cavity	304
Bartholin's gland	295
bulbourethral gland	294
cervix	295
childhood	310
chorion	302
clitoris	295
coitus	298
contraception	299
corpus luteum	298
ejaculation	294
ejaculatory duct	293
endometrium	296
epididymis	293
erectile tissue	294
estrogen	296
follicle	295
follicular cell	295
human chorionic gonadotropin (hCG)	302
infancy	310
interstitial cell	292
in vitro fertilization	290
life expectancy	312
life span	312
menopause	311
menstrual cycle	296
oocyte	295
orgasm	298
ovary	295
ovulation	295
oxytocin	309
penis	294
pregnancy	299
progesterone	296
prostaglandin	309
prostate gland	294
puberty	311
scrotum	292
semen	294
seminal fluid	294
seminal vesicle	293
seminiferous tubule	292
senescence	311
somite	304
spermatogenic cell	292
supporting cell	292
testis	292
testosterone	294
trimester	306
umbilical cord	304
urethra	294
uterus	295
vagina	295
vas deferens	293

FUN FACTS

ATHLETES AND STEROIDS

In the 1988 Olympics, sprinter Ben Johnson was stripped of his gold medal because a urine test revealed that he had taken anabolic steroids, banned muscle-enhancing drugs. Widely used, these hormones can lead to trouble.

Steroids have a characteristic structure of four interlocking carbon rings. By modifying the side groups attached to the rings, natural selection has created many different molecules, including the sex hormones estradiol, progesterone, testosterone, and androsterone. Among the effects of these hormones is the development of secondary sexual characteristics: in men, muscle bulk, body and facial hair, sexual appetite, and aggressive behavior.

Many athletes believe that testosterone and its synthetic analogs (anabolic steroids) can give a winning edge. The side effects of steroid use may be unpleasant. Athletes have been known to use up to 2000 milligrams a day, while the body naturally produces less than 10 mg. Overuse can lead to liver problems, suppression of the body's production of hormones, enlarged breasts, atrophied testes, and suppressed sperm count. Increases in sex drive, aggressiveness, and mood swings can lead to social problems: men charged with rape and attempted murder have claimed that steroid use instigated their actions. Women who use steroids may end up with facial hair, baldness, a deeper voice, and menstrual irregularities. Constellations of symptoms are involved because so many different hormones have similar structures and interact with receptors on many different cell types; cell activity must be carefully choreographed if development and homeostasis are to be maintained.

See: Haupt, Herbert A., and George D. Rovere. 1984. Anabolic steroids: A review of the literature. *American Journal of Sports Medicine* 12 (6): 469–483.

Lamb, David R. 1984. Anabolic steroids in athletics: How well do they work and how dangerous are they? *American Journal of Sports Medicine* 12 (1): 31–37.

MALENESS GENE DISCOVERED

A gene on the Y chromosome, called SRY, apparently makes a DNA-binding protein that controls the expression of secondary genes determining gender. Researchers found that in mice embryonic gonads, the gene produced mRNA for only two days preceding testis development, then shut down. The scientists spliced the gene into mouse zygotes; some genetically female mice, with two X chromosomes, developed as sterile males. This result demonstrated that the gene is sufficient for sex-reversal.

See: Ezzell, C. 1991. Maleness gene may be a master gene switch. *Science News* 140:70.

CHAPTER 14 THE HUMAN LIFE CYCLE

MESSAGES

1. Male and female reproductive systems are structurally and functionally similar. They arise from the same embryonic organs and differentiate in response to hormones.
2. Sex hormones function together in interlocking negative feedback loops to regulate gamete production and sexual maturation.
3. The menstrual cycle includes the buildup and loss of endometrium.
4. Birth control techniques block the release of gametes, or block fertilization, implantation, or embryonic growth.
5. Pregnancy is a partnership between embryo and mother.
6. The human embryo develops much as other mammalian embryos do.
7. The fetus programs itself and its mother for birth.
8. Infancy and childhood are periods of rapid physical growth and mental development.
9. Aging may be caused by genetic programming or by wear and tear.

15

LIFE'S ORIGINS AND DIVERSITY ON OUR PLANET

PERSPECTIVE

This chapter outlines the ultimate connections between life and its environment by presenting a concise history of the formation of the universe, the formation of earth, and the origin and diversification of life. The topics refer to the basics of chemistry and physiology, presented in Chapters 2 and 3, and provide an overview of the material that will be presented in Chapters 16 through 19, which deal with life's variety.

The themes and contents of this chapter include:

- Methane-producing prokaryotes may be similar the earliest cells on the planet.
- Life arose as a direct result of the physical conditions of the early earth.
- We might never know how life started.
- Life and the earth coevolve.
- The universe started with the Big Bang 18 billion years ago. The earth was formed about 4.6 billion years ago.
- The early earth had a combination of features, unique in our solar system, that allowed life to arise.
- It is thought that life arose from the pre-existing monomers created by nonbiological processes.
- Milestones of life's evolution are recorded in sedimentary rocks.
- Tectonic activity has affected the evolution of life.
- Taxonomy reflects evolutionary relationships.

CHAPTER OBJECTIVES

Students who master the material in this chapter will have an overview of the great evolutionary progression, from the poorly-understood origin of life through the elaboration of complexity and diversity that characterizes life today. They should understand that life and earth have coevolved and that each step in evolution depends on the preceding steps. They will appreciate the immensity of time that preceded the evolution of multicellular animals. These students will understand that taxonomic classifications are not arbitrary but reflect our understanding of evolutionary relationships.

LECTURE OUTLINE

(p 316) (msg 1) **I. Methanogens, anaerobic chemoautotrophs, may be the closest living relatives to the first life on earth.**

(tpq 1) **A. Early in life's history, when the planet was devoid of free oxygen, methane prokaryotes and similar species may have been the only life forms.**

B. The chemicals they now metabolize were common in the earth's first atmosphere.

C. Methanogens cannot tolerate free oxygen.

D. Structurally and biochemically, they are different from true bacteria.

E. Three themes will emerge in this chapter:

(msg 2) **1. Life arose as a direct result of the physical conditions on our planet, the molecules that were present, and the ways those molecules act and interact.**

2. We may never know exactly how life originated.

tpq 1. What were conditions like on earth during the first billion years after its formation?

(msg 3) 3. Life and earth coevolve.

F. Taxonomy helps reveal evolutionary relationships.

(p 318) II. The exploration of space has answered many questions about earth.

(msg 4) A. The universe began with the Big Bang, 18 billion years ago.

B. Stars form as gravity pulls clouds of hydrogen and helium together, compressing them and initiating nuclear fusion.

(tpq 2) 1. The heavier elements are formed by these reactions.

2. A cloud of heavier elements orbiting the young sun condensed to form the planets and moons about 4.6 billion years ago.

C. After its formation, the earth heated into a molten mass, stratified, and cooled.

D. The first atmosphere formed as gases emerged from the earth's interior: CO, N_2, H_2S, HCl, NH_3, and CH_4.

E. Only earth, of all the sun's planets, had the right combination of conditions for life to arise.

(msg 5) F. Organic compounds occur in space and are believed to have existed on the primordial earth.

---

tpq 2. Could life likely arise around a star formed only of the primordial matter of the universe?

---

1. These molecules can combine to form sugars, amino acids, and other molecules of life.
2. Carbonaceous chondrites may have brought prebiotic organic compounds to earth.
3. Natural forces such as sunlight, lightning, and volcanic heat could have driven endergonic reactions that formed organic monomers.

(p 320) (tr 73) **III. Life must have arisen before the oldest extant rocks were formed 3.8 billion years ago.**

A. The earth was heavily bombarded by giant meteors for its first 800 million years of existence.

(msg 6) B. Polymers formed.

1. Evaporation could have concentrated the primordial soup.
2. Clay particles may have acted as substrates and catalysts.

C. Polymers copied themselves.

1. RNA may have been the first self-replicating polynucleotide.
2. Ribozymes can act as primitive enzymes.

(tpq 3) D. Molecular interactions took place. There is no widely-accepted hypothesis that would explain how polynucleotides came to code for polypeptides.

---

tpq 3. How might the genetic code have arisen?

E. Cell-like compartments took shape.

1. Polypeptides and phospholipids spontaneously form tiny spheres, which might have been "protocells."

2. One group of these protocells, now called the <u>progenotes</u>, were the earliest common ancestors of all living things that survive today.

F. Coordinated cell-like activities emerged. The earliest cells were probably anaerobic heterotrophs.

(p 322) **IV. Earth and life coevolve.**

A. The earliest signs of life can be found in 3.8 billion-year-old rocks.

B. Stromatolites, massive colonies of several species, existed 3.5 billion years ago. They may have included autotrophs.

(msg 7) **C. Oxygen-producing photosynthesis arose by 2.8 billion years ago.**

(tpq 4) **D. Fossils dated at 2 billion years ago suggest that oxygen was accumulating in the atmosphere.**

1. Iron deposits were formed as iron oxidized and fell to the ocean floor.

---

tpq 4. How is it correct to say that the most significant atmospheric pollution was not caused by humans?

---

2. Cells evolved mechanisms for avoiding oxygen poisoning.

3. Ozone accumulated, blocking much of the ultraviolet light that would damage the DNA of exposed organisms.

E. By 1.5 billion years ago, eukaryotic life had arisen.

(tpq 5) (tr 74)

1. The <u>endosymbiont hypothesis</u> suggests that eukaryotic cells acquired their mitochondria, chloroplasts, and flagella by engulfing free-living bacteria.

2. The nucleus may have originated as invaginations of the plasma membrane that came to surround and protect the host cell's DNA.

F. <u>Metazoans</u> appeared about 670 million years ago. The oxygen level may have reached seven percent of the atmosphere, about one-third of today's oxygen level.

G. Hard-shelled animals appeared about 550 million years ago.

H. By 400 million years ago, the oxygen in the atmosphere had reached 20 percent.

1. The ozone screen was fully formed.

2. Large multicelled organisms dominated the seas and shorelines.

---

tpq 5. What evidence indicates that mitochondria are the descendants of free-living bacteria?

I. By 200 million years ago, multicelled plant and animal life had spread over the land.

(tpq 6) (tr 75) J. The surface of the earth is divided into plates that drift apart, forming rifts, or that collide, forming deep trenches and mountain chains.

(msg 8) 1. Plate tectonics began more than 4 billion years ago.

2. Geologists divide the earth's history into four geological eras, each of which is divided into geological periods.

3. During the Paleozoic era (570 to 225 million years ago), the continental masses aggregated into one supercontinent, called Pangaea.

4. At the beginning of the Mesozoic era (225 to 65 million years ago), Pangaea broke apart, leading to massive extinctions.

   a. Marsupials thrived all across Pangaea. After the breakup, placental mammals evolved on most continents but not on Australia.

   b. Giant reptiles and great swampy forests arose.

5. During the Cenozoic era (65 million years ago to the present), birds, mammals, and flowering plants diversified.

---

tpq 6. How would the evolutionary history of life have been different if the continents had been fixed in place?

(p 328) **V. Biologists classify life forms by their structural similarities and relatedness.**

A. Carolus Linnaeus devised the binomial system of nomenclature.

1. **Each organism is assigned a two-word name: genus and species.**
   a. **A genus is a group of very similar organisms related by common descent to a recent ancestor and sharing similar physical traits.**
   b. **A species includes members that share the same structural traits and can interbreed.**
2. **Each organism is placed in a series of taxonomic groups: species, genus, family, order, class, phylum, and kingdom.**

(msg 9) **B. Biologists generally recognize five kingdoms.**

(tr 76)
1. **Monera includes prokaryotes.**
2. **Protista encompasses single-celled eukaryotes.**
3. **Fungi are multicellular heterotrophs, including molds that decompose biological tissues.**
4. **Plantae are multicellular autotrophs.**
5. **Animalia includes multicellular heterotrophs that usually exhibit movement.**

C. Genetic comparisons suggest that life falls into three domains, the Eucarya, Bacteria, and Archaea. The Eucarya are more closely related to Archaea than to Bacteria.

INSTRUCTIONAL AIDS

VIDEOS, FILMS, SLIDES

The Beginnings. Pennsylvania State University AV Services, University Park, PA. Examines the common elements, chemical substances, and conditions under which life evolves. Film.

Organic Evolution. TVO Video, TVONTARIO, Chapel Hill, NC. This video includes 10-minute segments that discuss six different aspects of evolution, from the historical background to mutations and natural selection.

The Origin of Life: Chemical Evolution. Britannica Films and Video, Chicago, IL. Presentation of the evidence supporting evolution of life in the primordial sea, including footage of scientists who pioneered the study of the origins of life. Film, video.

The Search for Life. Time-Life Video, Paramus, NJ. Discusses Stanley Miller's experiment and other research on the origin of life. Film, video.

REFERENCES

Cloud, Preston. 1988. *Oasis in Space: Earth History from the Beginning.* New York: W. W. Norton. Extensive coverage of earth history from the beginning, entertainingly written and profusely illustrated.

Day, William. 1984. *Genesis on Planet Earth: The Search for Life's Beginnings,* Second Edition. New Haven, CT: Yale University Press. The history of the search for life's beginning and a readable presentation of the fundamentals of life.

Horgan, John. 1991. In the beginning... *Scientific American* 264 (2):116 – 125. A review of the theories explaining life's origin.

Margulis, Lynn, and Dorion Sagan. 1986. *Microcosmos: Four Billion Years of Microbial Evolution*. New York: Summit Books. A clearly written, readable discussion of the first four billion years of earth life.

KEY TERMS

Animalia	page 331	methanogen	316
binomial system of nomenclature	328	Monera	329
class	329	order	329
domain	329	phylum	329
endosymbiont hypothesis	324	Plantae	331
family	329	plate tectonics	326
Fungi	331	Protista	331
genus	328	ribozyme	321
geological era	326	rift	326
geological period	328	species	328
kingdom	345	taxonomic group	329
metazoan	325	taxonomy	317

FUN FACTS

CRYSTALS, CLAY, AND THE ORIGIN OF LIFE

One of the biggest unknowns in the origin of life is the beginning of self replication. A. G. Cairns-Smith has proposed that clay crystals were the first self-replicating organic system.

The first "organisms" must have had two contradictory qualities, faithful self-replication and mutability. Crystals may have provided the original template that gave rise to life. Although pure crystals are boringly regular, they are capable of directing their own duplication. In most natural crystals, there are defects; some units may be missing, others may be replaced by contaminants, or the units may be misaligned. These defects allow natural crystals to have more varied structures, to contain much more information, than their pure counterparts.

Clay minerals are formed by the crystallization of dissolved molecules released from weathering rocks. Commonly, clay particles are formed of stacks of layers. Within a layer, the atoms might be arranged in patches of different crystalline patterns. As more dissolved molecules settle onto a crystal, they pick up information, the pattern of crystallization, from the original layers. In a process much like natural selection, crystal selection would result in the proliferation of the fastest-growing, most stable patterns. Random defects, like mutations in organisms, could create new, more successful forms.

Many carbon-based macromolecules stick to clay crystals, often altering the physical properties of the clay. Crystals that were partially mineral-based and partially carbon-based might have evolved in the primordial soup. Perhaps RNA, with its self-replicating ability, formed a partnership with clay minerals to form a crystal that was particularly successful. Gradually, says Cairns-Smith, RNA became more elaborate and the clay minerals lost their importance; eventually, the self-replicating system became entirely carbon-based, losing its mineral origin entirely: our earliest living ancestor.

See: Cairns-Smith, A. G. 1985. The first organisms. *Scientific American* 252 (6):90–101.

IRON AND OXYGEN

The accumulation of oxygen produced by photosynthesis had widespread effects on earth's deposits of iron-rich minerals. Some, such as pyrite (FeS_2), are formed only in anaerobic conditions and disintegrate when exposed to oxygen. Thus they are most abundant in buried deposits formed before oxygen gas appeared in the atmosphere. Red beds display the opposite pattern, and are not found in deposits older than 2.2 or 2.3 billion years. The red color comes from hematite (Fe_2O_3), a highly oxidized iron mineral that forms only in the presence of oxygen gas. Thus, mineral deposits may be used to determine when photosynthesis-produced oxygen first accumulated in the atmosphere.

See: Stanley, S. M. 1986. *Earth and Life Through Time*. New York: W. H. Freeman.

CHAPTER 15 LIFE'S ORIGIN AND DIVERSITY ON OUR PLANET

MESSAGES

1. Methanogens may represent the oldest living group on earth.
2. Life arose as a direct result of the physical and chemical conditions on earth, conditions that are unique in the solar system.
3. Earth and life coevolve.
4. The universe began with the Big Bang 18 billion years ago. The solar system arose 4.6 billion years ago.
5. The organic building blocks of life preceded life.
6. Polymerization, self-replication, compartmentalization, molecular interactions, and the origin of the genetic code were steps in the origin of life.
7. Free oxygen was absent from the early atmosphere. Photosynthesis produced oxygen gas and the ozone layer.
8. Tectonic activity modifies the earth's surface, causing extinction and diversification.
9. Scientists classify organisms into five kingdoms or three domains.

16

LIFE AS A SINGLE CELL

PERSPECTIVE

This chapter elaborates and expands on the introduction to single-celled life presented in Chapter 15. The diversity, importance, and abundance of prokaryotes, protists, and allied organisms are explained. The material builds on the concepts in Chapters 3 through 6, which introduced cells and cellular physiology. Organisms and activities that are presented in this chapter will reappear in Chapters 35 through 37, which deal with ecology.

The themes and concepts of this chapter include:

- Single-celled organisms are ubiquitous because of their success in obtaining nutrients and in reproducing.
- Prokaryotes and protists have tremendous importance in ecology, medicine, economics, and science.
- Prokaryotes may be heterotrophs or autotrophs. Many are pathogenic.
- Prokaryotes can be motile or nonmotile, may have cell walls, and do not have membrane-bound nuclei or cellular organelles.
- Genetic evidence indicates that prokaryotes should be divided into two domains, Archaea (or Archaebacteria) and Bacteria (or Eubacteria).
- Viruses, viroids, and prions are nonliving agents of disease.
- Protists are single-celled eukaryotes with specialized organelles.
- Protozoa are animal-like protists. Many protists are autotrophs.
- Slime molds are funguslike protists that carry out extracellular digestion of organic matter.

CHAPTER OBJECTIVES

Students who master the material in this chapter will appreciate the diversity, abundance, and importance of single-celled organisms. They will gain an understanding of the causes of diseases and of the importance of prokaryotes and protists in environmental cycles. These students will understand the ways in which single-celled organisms make their livings and the characteristics that unite and separate them.

LECTURE OUTLINE

(p 336) **I. *Didinium nasutum* introduces single-celled life.**

A. Prokaryotes and protists have the ability to carry on life processes as independent, single cells.

(msg 1) **B. Single-celled organisms are ubiquitous on our planet because of their success in obtaining nutrients and in reproducing.**

(tpq 1) **C. Single-celled organisms have ecological, medical, economic, and scientific importance.**

(p 338) **II. Prokaryotes share certain traits.**

A. Prokaryotic cells share certain structural characteristics.

1. They have an outer cell wall made of peptidoglycans, a plasma membrane, and a noncompartmentalized cytoplasm dotted with ribosomes.

tpq 1. How have single-celled organisms affected your life today?

a. **Gram-positive** cells have a single broad layer of cell wall. They stain purple in the Gram test.

b. **Gram-negative** cells have a protein-lipopolysaccharide outer layer covering the peptidoglycans. They stain red in the Gram test.

c. The Gram test suggests the types of antibiotics that might be effective in fighting a **pathogenic bacterium**.

2. A single circular strand of DNA is coiled in one region.

3. Prokaryotes occur in various shapes, including **cocci** (spheres), **bacilli** (rods), **spirilla** (spirals), and **vibrios** (curved rods).

(msg 2) B. Prokaryotes can be heterotrophs or autotrophs. They can survive on an enormous range of energy sources.

1. **Saprobes** are decomposers.

2. **Parasites** and **symbionts** live in or on other organisms.

(tpq 2) 3. Autotrophs include green and purple photosynthetic bacteria, cyanobacteria, and chemoautotrophs.

---

tpq 2. How is it true that prokaryotes are the only organisms absolutely necessary to ecological systems?

(tpq 3)

C. **Prokaryotes usually reproduce by binary fission.**

1. **Rapid reproduction and high mutation rates allow bacteria to adapt quickly to changes in the environment.**
2. **Conjugation is the direct exchange of DNA between prokaryotes.**
3. **DNA may be exchanged indirectly.**
   a. **Transformation is the uptake of DNA directly from the surrounding medium.**
   b. **Transduction is the transfer of genes via a virus.**
4. **Some species form <u>endospores</u> or <u>spores</u> to withstand adverse conditions.**

(msg 3)

D. **Some prokaryotes are motile and are very sensitive to gradients in their environments. Some contain iron oxide crystals as compasses.**

(msg 4) (tr 77)

E. **Bacteria may be classed in three divisions. Modern classification techniques rely on the similarity of ribosomal RNA.**

1. **Archaebacteria include two lineages.**
   a. **Methanogens and <u>halophiles</u> (salt lovers) make up one group.**
   b. **Sulfur-dependent thermophiles compose the second.**

---

tpq 3. Why has natural selection led to a higher mutation rate in bacteria than in humans?

---

2. RNA analysis suggests that there are eleven main lineages among the true bacteria.

   a. The largest and most diverse group are the purple bacteria.

(tpq 4)

      i. Many are autotrophs with chlorophylls different from those in green plants.

      ii. Others are heterotrophs, including E. coli.

      iii. Mitochondria in eukaryotes probably arose from purple phototrophic bacteria.

   b. Cyanobacteria have chlorophyll a and often occur in chains or colonies.

      i. Some fix nitrogen in heterocysts, one of the few examples of division of labor among prokaryotes. The thick heterocyst walls protect nitrogen-fixing enzymes from oxygen generated in adjacent cells.

      ii. Cyanobacteria are widespread and ecologically important.

      iii. Cyanobacteria may have been the ancestors of eukaryotic chloroplasts.

   c. Gram-positive bacteria cause tooth decay, strep throat, and staph infections such as toxic shock syndrome.

---

tpq 4. What does the existence of different chlorophylls in different photosynthesizers indicate about the evolutionary history of photosynthesis?

---

d. Gram-positive actinomycetes produce substances that fend off competing microorganisms. These substances may be used as antibiotics.

e. Mycoplasmas are simplified, Gram-positive parasitic bacteria that lack cell walls.

f. The most frequent sexually transmitted disease in North America is caused by species of the genus Chlamydia.

g. Spirochetes include the agents that cause Lyme disease and syphilis.

F. Bacteria are important in environmental cycles, medicine, food preparation, industry, and science.

(p 344) III. Viruses, viroids, and prions are noncellular agents of disease.
(msg 5)
(tpq 5)

A. Viruses are minute packages of DNA or RNA wrapped in a protein capsid. They rely on their hosts' cellular machinery for reproduction.

B. Viroids consist of small RNA molecules without protein coats. They seem to infect mainly plants.

C. Prions lack genetic material and consist of nothing but proteins. Biologists are not sure how they reproduce or cause disease.

---

tpq 5. Because viruses are so dependent on other organisms, some people argue that viruses are not alive. Is this a valid argument?

(p 345) (tr 78) **IV. Kingdom Protista (Protoctista) includes single-celled eukaryotes.**

(msg 6) **A. Protists are much larger than prokaryotes and are compartmentalized. The expanded membrane surface is the site of many physiological processes.**

(msg 7) **B. Protozoa are animal-like protists that usually stalk and consume other cells or food particles.**

1. **Members of phylum Mastigophora bear flagella and do not have cell walls. Most are parasites or symbionts. African sleeping sickness is caused by a mastigophoran.**
2. **Members of phylum Sarcodina have pseudopodia for locomotion.**
   a. **Foraminiferans secrete calcium-based shells.**
   b. **Radiolarians produce silicon-based shells.**
3. **Sporozoans are internal parasites with a sporelike stage that lacks locomotion.**
   a. **Malaria is caused by a sporozoan, Plasmodium vivax.**
   b. **Microspora are parasites that live inside cells of almost every group of animals. They have no mitochondria and were one of the earliest branches off the main line of eukaryotic evolution.**

4. Ciliates have rows of cilia and organelles analogous to animal organs of excretion, support, musculature, and digestion.
   a. Some have tiny toxic darts (trichocysts) used in hunting prey.
   b. A polyploid macronucleus directs cell activities. Diploid micronuclei are exchanged during conjugation.

C. Photosynthetic protists occur as part of the phytoplankton, the base of the aquatic food chain.
   1. Euglenoids are motile. Each has a stigma, or eyespot.
      a. They diverged from the main eukaryotic line eons ago.
      b. Their chloroplasts may have had an origin independent from the chloroplasts of green plants.
   2. Dinoflagellates have two flagella that cause them to spin as they swim. They can cause red tides.
   3. Golden brown algae and diatoms have silicon instead of cellulose in their cell walls. They are abundant and diverse. Many have beautiful shells.

(msg 8) D. Slime molds are bizarre funguslike heterotrophs.

1. True slime molds exist as plasmodia, large masses of continuous cytoplasm containing many diploid nuclei and surrounded by one plasma membrane. When food is scarce they send up fruiting bodies.
2. Cellular slime molds are single-celled, amoeba-like organisms that aggregate to produce spores.
3. Water molds may be single-celled or may form hyphae.
   a. They have several nuclei in a common cytoplasm and form large, immobile eggs.
   b. After fertilization, they form spores that disperse by swimming.

E. Protists are a polyphyletic group with many different ancestors.
   1. One of the oldest lines was Microsporidia.
      a. They are parasites that contain a membrane-bound nucleus but no organelles.
      b. They probably diverged from the main line while the atmosphere was still anaerobic.
      c. They probably resemble the nucleated cell that became host to the first mitochondria.
   2. Flagellates diverged, then the slime molds, and over a more recent and shorter period, the ciliates, some amoebas, fungi, plants, and animals.

INSTRUCTIONAL AIDS

VIDEOS, FILMS, SLIDES

The Microscopic Pond. Educational Images Ltd., Elmira, NY. An introduction to microscopic plant and animal life in ponds. Filmstrip, video, transparencies.

Professor Bonner and the Slime Molds. Films Incorporated, Chicago, IL. A presentation of the life history of a slime mold. Video.

Protozoa: Structures and Life Functions. Coronet Film and Video, Deerfield, IL. A microscopic examination of four classes of protozoa. Film, video.

REFERENCES

Gest, Howard, 1987. *The World of Microbes*. Madison, WI: Science Tech Publications. An easy-to-read guide designed for those with little background in microbiology.

Gillen, Alan L., and R. P. Williams. 1988. Pasteurized milk as an ecological system for bacteria. *American Biology Teacher* 50 (5):279 – 282. Laboratory exercises demonstrating bacterial growth and ecology.

Lee, John J., et. al. 1985. *An Illustrated Guide to the Protozoa*. Lawrence, KS: Society of Protozoologists. A guide for all readers, copiously illustrated with photographs and diagrams and with an extensive bibliography.

Prusiner, Stanley B. 1991. Molecular biology of prion diseases. *Science* 252:1515 – 1522. A review of the pathology and etiology of prion-caused diseases.

KEY TERMS

bacillus	page 338	prion	345
capsid	344	protozoan	346
cellular slime mold	352	pseudopod	348
ciliate	349	red tide	352
coccus	338	saprobe	338
cyanobacterium	344	slime mold	352
endospore	340	spirillum	338
fruiting body	352	stigma	350
halophile	341	symbiont	339
macronucleus	350	true slime mold	352
micronucleus	350	vibrio	338
parasite	339	viroid	345
pathogenic	338	virus	344
phytoplankton	350	water mold	352

FUN FACTS

PLANTS, ANIMALS, AND EUGLENA

Protists such as *Euglena* don't fall into the neat categories of "plant" and "animal." *Euglena* includes both photosynthetic and heterotrophic forms and acts like a combination plant and animal. The common laboratory organism *Euglena gracilis* is cigar-shaped with a deep depression in its surface from which two flagella emerge. It has six to twelve chloroplasts, containing chlorophylls *a* and *b*, the same pigments found in higher plants, and a red eyespot at the anterior end of the cell near the base of the flagellum. The eyespot absorbs light from specific directions and conducts the light to a light-sensitive swelling at the base of the flagellum, stimulating the organism's response to light. It moves toward light unless the light intensity is excessively high, in which case it swims away.

Euglena gracilis is photosynthetic in sunlight but can grow in the dark if provided with organic nutrients. Unlike other euglenoids that are active predators, it does not ingest solid foods, but absorbs nutrients from the environment. Some biologists have classified euglenoids as plants, but others have pointed out that many species are never photosynthetic. This once led to the absurd situation that some *Euglena* species were grouped into the kingdom Plantae when they were grown in light and into the kingdom Animalia when they were cultured in the dark. Establishment of the kingdom Protista enabled scientists to lump all euglenoids together and to avoid worrying about the differences between plants and animals.

See: Sze, Philip. 1986. *A Biology of the Algae*. Dubuque, IA: Wm. C. Brown Publishers.

SULFUR BACTERIA

While bacteria are structurally simple, they are biochemically diverse. Their ability to exploit different forms of chemical energy is astonishing. In anaerobic environments, they can extract oxygen from ions such as nitrate, carbonate, and sulfate, and use the oxygen in cellular respiration.

Sulfates are present in soil, water, organic material, and many minerals. Sulfate-reducing bacteria release hydrogen sulfide, a smelly, toxic gas, as a waste product. Sulfide increases the alkalinity of the environment, precipitates metals such as iron (creating "black sands"), removes oxygen from the environment, and kills aerobic organisms. Sulfate-reducers cause economic problems for the petroleum industry by corroding machinery and pipelines, spoiling petroleum, and producing toxic gas. They can cause fish kills by producing sulfide in lakes with anaerobic lower layers and they make the canals of Venice stink like rotten eggs.

See: Postgate, John. 1988. Bacterial worlds built on sulphur. *New Scientist* July 14:58–62.

CHAPTER 16 LIFE AS A SINGLE CELL

MESSAGES

1. Single-celled organisms are diverse, ubiquitous, and important in ecology, medicine, economics, and science.

2. Prokaryotes may be saprobes, symbionts, parasites, or autotrophs. Many are pathogens.

3. Prokaryotes can be motile or nonmotile, may have cell walls, and do not have membrane-bound nuclei or cellular organelles.

4. Genetic evidence indicates that prokaryotes should be divided into two domains, Archaea and Bacteria.

5. Viruses, viroids, and prions are noncellular agents of disease.

6. Protists include one-celled eukaryotes. Many are complex cells with organelles analogous to the organs of animals.

7. Protists may be heterotrophic or autotrophic, parasitic or free-living.

8. Slime molds and water molds are funguslike protists that carry out extracellular digestion of organic matter.

17

PLANTS AND FUNGI: DECOMPOSERS AND PRODUCERS

PERSPECTIVE

In this chapter, we see another act in the evolutionary drama: the rise of land plants and fungi. Chapter 15 began the play with the origin of life. Chapter 16 revealed the origins and spread of single-celled life, setting the stage for the events presented in this chapter.

The material presented in this chapter will be relevant to the two chapters that follow, which describe the rise of invertebrates and vertebrates; to Part V, How Plants Survive; and to Part VI, which covers ecology.

The themes and concepts of this chapter include:

- Plants and fungi play reciprocal roles in environmental cycles as producers and decomposers.
- Fungi carry out extracellular digestion.
- Plants and fungi alternate between haploid and diploid phases.
- Plant evolution is characterized by increasing dominance of the diploid phase. The fungal life cycle is dominated by the haploid phase.
- Mycorrhizae and lichens are symbioses between fungi and green plants.
- Fungi are classified by their reproductive structures. Algae are classified by their pigments.
- Green algae are the probable ancestors of land plants.
- As plants evolved on dry land, a series of physical systems and physiological trends emerged that helped them meet the challenges of life on land.
- Flowering plants and animal pollinators coevolve.

CHAPTER OBJECTIVES

Students who master the material in this chapter will have an overview of the great chain of evolution that stretches from the simplest and smallest photosynthetic eukaryotes to the largest and most complex plants found today. They will see that the evolution of land plants has paralleled the evolution of terrestrial vertebrates.

These students will appreciate the diversity found in plants and fungi and will understand the importance of fungi to plants: decomposers recycle materials and mycorrhizal fungi feed roots.

LECTURE OUTLINE

(p 356) **I. The giant sequoia, Sequoiadendron giganteum, and its subterranean fungal partners introduce plants and fungi.**

(msg 1) **A. Fungi are the great decomposers, plants the great producers.**

(tr 79) **B. Fungi and plants are highly interdependent.**

C. Fungi and plants have life cycles based on alternation of haploid and diploid phases.

D. Evolution has led to physical and physiological systems that helped plants colonize land.

(p 358) **II. Fungi are the major decomposers, returning simple minerals and elements to the environment.**

A. Most fungi are saprobes that decompose nonliving organic matter. Many cause disease in plants and animals.

(tpq 1) B. Fungi are heterotrophs that secrete digestive

(msg 2) enzymes into the environment and absorb the released nutrients.

C. The body of a fungus is made of filamentous cells called hyphae packed into a mass called a mycelium.

1. The cross walls between cells are perforated or absent, allowing cytoplasm to flow freely.
2. The extensive meshwork of hyphae gives a fungus a large surface-to-volume ratio.

D. Fungi reproduce asexually by fragmentation of hyphae or by producing asexual spores.

(msg 3) E. Fungi also produce spores sexually.

(tr 80)

1. The haploid phase usually dominates the life cycle.
2. Haploid cells of opposite mating types fuse, forming a dikaryon with two haploid nuclei.
3. During sexual reproduction, dikaryons produce fruiting bodies.
   a. Fruiting bodies produce gametes.
   b. Gametes fuse, forming a diploid zygote.
   c. The zygote undergoes meiotic division to produce haploid cells, completing the cycle.
   d. The diploid phase is often limited to a single cell, the zygote.

---

tpq 1. Are saprobes beneficial or harmful to human activities?

F. Fungi are more beneficial than harmful to plants.

(tpq 2) 1. <u>Mycorrhizae</u> are symbiotic associations

(msg 4) between roots and fungi. Some 90 percent of land plants rely on mycorrhizal fungi.

2. The fungus may grow outside the root, sheathing it, or may penetrate and grow inside the root.

3. The fungus expands the effective surface area of the roots, takes up nutrients and water, and delivers them to the roots.

G. <u>Lichens</u> are associations between fungi and algae. The alga photosynthesizes, feeding itself and the fungus; the fungus protects the alga and supplies it with water.

H. Fossils show that fungi grew on land before plants left the ocean. Fossilized mycorrhizae are found on the roots of the oldest fossilized land plants.

(p 360) III. Fungi are classed by the shapes of their spore-

(msg 5) producing structures. Spores are their only means of dispersal.

A. Division <u>Zygomycota</u> includes filamentous fungi that form dark, thick-walled spore-producing structures. This group includes bread molds and many mycorrhizal species.

---

tpq 2. Why might crop scientists work on improving fungi as well as green plants?

B. Ascomycota is the largest division of fungi.

1. During sexual reproduction, they produce spores in a sac called an ascus which may be borne in an ascocarp.
2. This class includes yeasts, some mushrooms, and organisms that cause many plant diseases.

C. Division Basidiomycota includes most of the familiar mushrooms and some agents of plant disease.

1. They produce basidiospores on club-shaped basidia in fruiting bodies known as basidiocarps.
2. The haploid basidiospores grow into primary mycelia.
   a. Cells of two mycelia of opposite mating types may fuse to form a single cell with two nuclei.
   b. The binucleate cell forms a secondary mycelium, or dikaryon, which makes basidiocarps.
   c. Each basidium has two haploid nuclei, which fuse to form a diploid zygote that quickly undergoes meiosis to form haploid basidiospores.

D. Division Deuteromycota, or Fungi Imperfecti, are fungi that seem to lack sexual reproduction.

(tpq 3)

1. **Most appear to be ascomycotes that reproduce only with asexual spores (conidia) on hyphae called conidiophores.**
2. **These organisms produce antibiotics, are used in food production, and can produce carcinogens.**

(p 363)

IV. Plants are multicellular autotrophs.

A. Plants have a wide range of reproductive and somatic structures but all share alternation of generations.

(msg 3)

1. **Both diploid and haploid phases are multicellular and may be free-living.**
2. **The haploid phase is the <u>gametophyte</u>, the diploid, the <u>sporophyte</u>.**
3. **Meiosis occurs in the sporophyte and results in the production of female and male spores, not gametes.**
4. **These spores germinate and form male and female gametophytes, which produce gametes that fuse, forming a new sporophyte.**
5. **In many algae, the gametophyte dominates the life cycle.**

---

tpq 3. Zygomycota, Ascomycota, and Basidiomycota are thought to be monophyletic groups. Why isn't Deuteromycota?

---

6. **In some algae and many simple land plants, both haploid and diploid stages are conspicuous green plants.**
7. **In more complex land plants, the diploid phase dominates.**

(tpq 4) **B. Plant evolution is a product of a changing planet and adaptations to life on land.**

(msg 6) (tr 81)

1. **Green algae, photosynthetic protists, were the probable ancestors of land plants some 400 million years ago.**
2. **Repeated submergence and emergence of the land would have favored organisms that could survive on dry land.**
3. **Evolutionary advances in plants include the development of transport vessels, dominance of the sporophyte, and the production of seeds.**

(p 364) **V. Algae remained aquatic.**

A. The term "algae" refers to organisms as diverse as photosynthetic prokaryotes, protists, and plants.

1. **Algal plants probably arose about 600 million years ago from photosynthetic protists.**
2. **They diversified and inhabit many environments today.**

---

tpq 4. Why couldn't plants colonize the land until the amount of oxygen in the atmosphere reached a high level?

---

3. They may be the most important photosynthesizers on earth, capturing 90 percent of all solar energy trapped by plants.
4. Aquatic plants do not need rigidity, roots, circulatory systems, or waxy coverings.
5. A range of photosynthetic pigments absorbs light of different wavelengths that penetrate to different depths.

(msg 7) B. Algal plants are classified in three divisions on the basis of their photosynthetic pigments.

1. Red algae have phycoerythrins that absorb blue-green light, which penetrates deepest into water. The haploid gametophyte dominates the life cycle.
2. Brown algae have chlorophylls a and c and fucoxanthin that collects blue and violet light penetrating medium-deep water.
    a. Kelps are large, with the sporophyte the dominant stage.
        i. In most other brown algae, haploid and diploid stages are similar and small.
        ii. The sporophyte produces mobile zoospores, whose flagella disperse them.
    b. Fronds are like leaves; stipes are like stems; holdfasts are anchors, like roots.
    c. Tubelike conducting cells carry sugars from fronds to deeper parts of the plant.

3. <u>Green algae</u> have carotenoids and chlorophylls <u>a</u> and <u>b</u>, pigments that maximize photosynthetic efficiency in shallow waters.

   a. Land plants also have this combination of pigments, leading to the belief that they are descendants of green algae.

   b. Some green algae are coenocytic; they have many nuclei in a single continuous cytoplasm.

(tpq 5) c. In most species, e.g., the unicellular <u>Chlamydomonas</u>, the haploid phase dominates. In <u>Ulva</u>, the haploid <u>thallus</u> and diploid thallus are nearly identical.

(p 368) **VI. Simple land plants, including <u>bryophytes</u> and <u>seedless vascular plants</u>, are still tied to bodies of water.**

(msg 8) A. Challenges of life on land include:

1. Preventing dessication.
2. Supporting the plant body.
3. Absorbing water and minerals from the substrate.
4. Reproducing sexually without shedding gametes into water.

(tr 82) B. Bryophytes are diverse, including mosses, liverworts, and hornworts.

---

tpq 5. Why might some green algae be considered protists?

---

1. They have waterproof coatings, pores in their leaves, and rigid tissues, but lack internal vascular systems.
2. They have rhizoids that act as anchors but do not absorb water and minerals. Water reaches the cells by diffusion.
3. They have swimming sperm that must have liquid water in which to reach the egg.
4. Most mosses have conspicuous gametophytes and sporophytes.
    a. Gametophytes are haploid green threads that may develop into dense mats.
        i. The antheridium produces sperm; the archegonium produces eggs.
        ii. Flagellated sperm cells swim to the eggs.
    b. The diploid zygote divides and forms a sporangium, which remains attached to the gametophyte and represents the entire diploid generation.

C. Seedless vascular plants evolved tissues that lend vertical support and conduct water, minerals, and products of photosynthesis.
1. These plants formed vast forests during the Paleozoic, today preserved as coal beds.
2. They lack true roots but have rhizomes from which stems or fronds grow. The spread of the rhizomes is a type of vegetative reproduction.

(tr 83) 3. Sori, spore-bearing structures, dot the underside of fern leaves. Spores are produced by meiosis.

4. In horsetails, lycopods, and ferns, the sporophyte is the conspicuous adult and the gametophyte is a small free-living green plant.

   a. Spores develop into tiny gametophytes, which produce eggs and/or swimming sperm.

   b. Fertilization must take place in standing water.

   c. The zygote develops into a sporophyte.

(p 371) VII. Gymnosperms appeared around the time that Pangaea broke up.

(msg 9) (tr 84) A. Their reproductive innovations include pollen, seeds, and further shifting toward dominance by the diploid generation.

(tpq 6) 1. Pollen grains are immature male gametophytes composed of two nonmotile cells and a dry outer coat.

2. Both male and female gametophytes are reduced to small nonphotosynthetic structures housed by the sporophyte.

---

tpq 6. How did the development of pollen help plants colonize dry land?

---

3. Unlike ferns, which develop from homospores, gymnosperm gametophytes develop from heterospores: male microspores and female megaspores.
4. The ovule sits naked on the surface of the sporophyll.
5. The embryo develops into a seed, which includes food for the developing plant and a tough outer coat.

B. Cycads are the remnants of a group that flourished at the time of the dinosaurs. They resemble palm trees.

C. The deciduous ginkgo (division Ginkgophyta) and the gnetophytes (division Gnetophyta) are odd relics of two formerly widespread divisions.

D. Conifers are today's most abundant gymnosperms.
   1. They are characterized by their narrow, cuticle-covered leaves and woody cones.
   2. Sporangia inside male cones undergo meiosis and produce pollen grains, the male gametophytes.
   3. Each scale of a female cone bears an ovule, which is a structure housing a sporangium.
      a. Meiosis inside the ovule gives rise to megaspores, which produce the female gametophytes.

b. **After landing on an ovule, the pollen grows a <u>pollen tube</u> in which two sperm nuclei develop.**

c. **A sperm nucleus unites with an egg, leading to the formation of an embryo within the ovule's protective coat. The rest of the female gametophyte forms a food supply for the developing embryo.**

(p 373) **VIII. During the Cenozoic era, mammals and flowering plants, in the division Anthophyta (angiosperms or anthophytes), came to dominate the land.**

(msg 10) **A. Flowering plants developed the ovary, which houses and protects the ovules, as well as flowers and fruits, which allow efficient fertilization and seed dispersal.**

B. Flowering plants are classified as <u>monocots</u> and <u>dicots</u>.

(tr 85) **C. The diploid stage dominates the life cycle.**

D. The flower includes female carpel, made up of the <u>stigma</u>, <u>style</u>, and ovary; and the male stamen, made up of the <u>anther</u>, which produces pollen, and the filament.

1. **Each megaspore mother cell in the ovary gives rise, via meiosis, to four haploid megaspores. Three degenerate and one divides into the female gametophyte, which produces the egg.**

2. **Microspore mother cells in the anthers undergo meiosis, forming microspores, which develop into pollen grains.**

(tpq 7)

3. **Pollen may be carried to the stigma by wind or animal pollinators.**
4. **A pollen tube grows down inside the style, carrying two sperm nuclei.**
   a. **Double fertilization involves the fusion of one sperm nucleus with the egg while the other unites with two polar nuclei to form a triploid nucleus.**
   b. **The triploid nucleus divides to form nutritive endosperm.**
5. **The wall of the ovary enlarges into a fruit.**

E. **The coevolution of flowering plants with pollinators and fruit eaters has led to efficient pollination and seed dispersal and animal specializations to exploit these resources.**

---

tpq 7. What are the relative advantages and disadvantages of wind pollination? Are there habitats where one is more advantageous than the other?

INSTRUCTIONAL AIDS

VIDEOS, FILMS, SLIDES

The Algae. Carolina Biological Supply Company, Burlington, NC. A survey of algal divisions in the Kingdoms Monera, Protista, and Plantae. Transparencies, filmstrip, video.

The Kingdom of Plants. Human Relations Media, Pleasantville, NY. A three-part discussion of the life processes of plants. Filmstrip, video.

Mushrooms. Films for the Humanities and Sciences, Princeton, NJ. Illustrates the growth and reproductive cycles of an edible mushroom. Video.

REFERENCES

Collins, Don, and T. Weaver. 1988. Measuring vegetation biomass and production. *American Biology Teacher* 50 (3):164 – 166. A lab exercise to demonstrate the measurement of biomass in a three-layered forest.

Dobey, Daniel C., and Ellie Gilbert. 1987. Decomposition in nature. *American Biology Teacher* 49 (4):234 – 238. A discussion of lab procedures for determining rates of decomposition.

Kendrick, Bryce. 1985. *The Fifth Kingdom*. Owen Sound, Ontario: Stan Brown Printers. This comprehensive but readable text offers an overview of fungi, from fungal classification to fungi as food.

Raven, Peter H., et. al. 1981. *Biology of Plants,* Third Edition. New York: Worth Publishers. A botany text emphasizing basic biological concepts; readable and well illustrated.

KEY TERMS

anther	page 384	megaspore	371
antheridium	369	microspore	371
archegonium	369	monocot	373
ascocarp	360	mycelium	358
ascus	360	mycorrhiza	359
basidiocarp	362	ovule	372
basidiospore	362	pollen tube	372
basidium	362	pollinator	374
bryophyte	368	rhizoid	369
coevolution	376	rhizome	370
cuticle	372	seed	371
cycad	371	seedless vascular plant	368
deciduous	372	sorus	370
dicot	373	sporangium	369
double fertilization	374	spore	358
endosperm	374	sporophyte	363
frond	366	stem	370
fruit	374	stigma	374
gametophyte	363	stipe	366
gymnosperm	371	style	374
holdfast	366	thallus	367
hypha	358	vascular plant	369
kelp	366	vegetative reproduction	370
lichen	360		

FUN FACTS

TRUFFLES AND TREES

Truffles have been prized since the days of the ancient Greeks. Gourmets call them "the black diamonds of gastronomy" and are willing to pay hundreds of dollars a pound for the best French truffles. They are the fruiting body, or mushroom, of a mycorrhizal fungus. Truffles have a completely subterranean fruit that relies on animals for dispersal. To attract animals, they produce a musty, earthy odor that pigs, rodents, and other mammals (including many humans) find almost irresistible. Truffle spores survive their passage through the animal's gut and, like apple seeds, are dispersed as the animal defecates.

Since the 1970s, when scientists succeeded in cultivating the truffle mycelium and inoculating tree roots with the fungus, truffles have become a cultivated crop. Oak and hazelnut seedlings are inoculated, planted, and cultivated for five years before the first truffle appears. Truffle orchards have been planted in several areas in the United States. If and when they reach full production, truffles may be a taste sensation available to all.
See: *International Wildlife* 10 (May-June, 1980):12 – 15.

PLANTS WITHOUT CHLOROPHYLL

Although photosynthesis is one of the primary characteristics of plants, there are many plants that do not have chlorophyll. Many obtain their nutrients by parasitizing green plants, often through fungal intermediates.

Many orchids, including coral-roots and bird's-nest orchids, rely on soil fungi that draw carbohydrates and other substances from plant roots. Often their flowers emerging from the soil are the only evidence of their existence. The stalks of silkweed plants wind around their host plants, driving special sucking outgrowths called haustoria into the hosts and sucking nutrients out. Broomrape plants take their nutrition from the roots of their hosts, into which they drive haustoria. Some parasitic plants cause widespread damage to cultivated crops.
See: Hohn, Reinhardt. 1980. *Curiosities of the Plant Kingdom*. New York: Universe Books.

CHAPTER 17 PLANTS AND FUNGI: DECOMPOSERS AND PRODUCERS

MESSAGES

1. Plants and fungi play reciprocal roles in environmental cycles as producers and decomposers.

2. Fungi are multicellular heterotrophs that carry out extracellular digestion.

3. The life cycles of fungi and plants alternate between haploid and diploid phases. One phase may dominate.

4. Mycorrhizae and lichens are symbioses between plants and fungi.

5. Fungi are classified by their reproductive structures.

6. Green algae were probably the ancestors of land plants.

7. Algae are classified by their pigments.

8. The evolution of land plants has led to structures that reduce the plant's dependence on water.

9. In higher plants, the diploid phase is dominant. The haploid phase is reduced to a few cells.

10. Flowering plants are the newest and most successful land plants. They coevolve with pollinators and seed-dispersers.

18

INVERTEBRATE ANIMALS: THE QUIET MAJORITY

PERSPECTIVE

This chapter introduces the bizarre world of invertebrates, from sponges that barely cross the threshold of multicellularity to giant squids that are highly specialized and intelligent. Invertebrates have traits representing successive evolutionary modifications that increased their mobility, efficiency, and specialization.

This chapter continues the presentation of the drama of evolution that began with Chapter 15 and leads to Chapter 19, The Chordates. The material presented in this chapter will be relevant to Part VI, Interactions: Organisms and Environment.

The themes of this chapter include:

- Three lines of animals arose from single-celled progenitors. Radiation of the lineages has led to more than a million animal species.

- Important anatomical trends evolved in invertebrates, including bilateral symmetry, cephalization, a two-ended gut, a coelom, and segmentation.

- Sponges have cellular specialization but no organs. Cnidarians have radial symmetry, nerve nets and stinging tentacles.

- Flatworms are bilaterally symmetrical, cephalized, and have three tissue layers and true organs. Roundworms have a two-ended gut and a pseudocoelom.

- The main line of animals split into two lineages, protostomes and deuterostomes, that differ in embryonic development.

- Mollusks are diverse, soft-bodied animals. Annelids are segmented and display all five major anatomical advances.

- Arthropods have an exoskeleton and make up the largest animal phylum. Echinoderms are deuterostomes with endoskeletons and water vascular systems.

CHAPTER OBJECTIVES

Students who master the material in this chapter will see the spectrum of diversity among invertebrates and understand that living phyla display the evolutionary advances that led from the simplest to the most complex animals. These advances include bilateral symmetry, cephalization, a two-ended gut, motility, segmentation, and specialized appendages. Students should be able to describe the traits that characterize the major invertebrate phyla and to state the differences between protostomes and deuterostomes.

LECTURE OUTLINE

(p 378) **I. Termites introduce invertebrate animals, which lack backbones, the characteristic feature of vertebrates.**

(tpq 1) (msg 2) **A. Evolutionary trends in animals include adoption of bilateral symmetry, cephalization, one-way movement of food through the gut, development of a coelom, and segmentation.**

B. Although these features allow specialization, they are not prerequisites for survival.

(p 380) (msg 1) **II. Animals are multicellular heterotrophs that move about under their own power.**

A. Animals are diploid. Their primary mode of reproduction is sexual.

B. The earliest fossil traces of animals are dated at 700 million years in age. These creatures probably arose from protists.

---

tpq 1. Are vertebrates or invertebrates more important ecologically? More diverse? More numerous?

(msg 3) **C. The sponges and cnidarians probably diverged in two independent lines from the main line very early.**

D. The earliest fossils of animals with hard parts were found in 580-million-year-old rocks.

(p 382) **III. Sponges, the phylum Porifera, are the simplest animals.**

(tpq 2) **A. Sponges are asymmetrical and sessile.**

B. Each central cavity has an opening, through which water exits, and hundreds of incurrent pores.

1. **Cells lining the cavity remove and digest suspended food particles.**
2. **Canals permeate the body, allowing sponges to attain large sizes.**

C. Sponges lack specialized tissues.

1. **The sponge's body wall has three layers supported by a scaffolding of protein, and/or spicules of silica or calcium carbonate.**
2. **The sponge is covered by a layer of flattened cells for protection.**
3. **An inner layer of collar cells, or choanocytes, each with a flagellum, circulates water.**

---

tpq 2. How do sponges make their living?

---

4. A gelatinous filling contains several different kinds of amoeboid cells that are involved in digestion and reproduction.
5. Sponge cells can function independently.

D. Collar cells and amoebocytes take in food particles via phagocytosis and digest them intracellularly.
1. Amoebocytes distribute materials to all body cells.
2. Sponges have no gills, kidneys, or other organs.
3. Sponges can reproduce asexually by budding off gemmules.

E. Some zoologists believe that sponges evolved from protozoa called choanoflagellates, which have collars and flagella, live in colonies, and produce spicules.
1. The evolutionary advances of the sponges include cellular specialization, primitive cell layers, and the division of labor among cells.
2. Sponges are not directly related to other animal groups.

(p 383) IV. The phylum Cnidaria includes aquatic animals with a radial body plan and stinging capsules.

A. Cnidarians have two body plans: the upright polyp and the umbrella-shaped medusa.

B. They have three-layered body walls with an outer epidermis, an inner gastrodermis, and a jellylike mesoglea in between.

C. Cnidarians have three adaptations related to the capture and digestion of food.

   1. Nematocysts are stinging capsules that can poison prey or enemies.
   2. Food is digested in the coelenteron, or gastrovascular cavity. Food is broken down by extracellular digestion and absorbed.
   3. Nerve cells are arranged in a loose network, coordinating movement and directing the capture of prey.

D. Cnidarians are capable of regenerating lost body parts.

E. They have complex reproductive cycles.

   1. Sexual reproduction involves the release of eggs and sperm into the water.
   2. The gametes unite and develop into planulae, swimming larvae.
   3. The planula eventually settles down and, in polyp types, develops the adult body.
   4. In medusa types, the polyp develops into a strobila, a stack of incipient medusae, which develop and swim away then mature sexually.

(p 385) V. Flatworms, the phylum Platyhelminthes, display bilateral symmetry and cephalization.

(msg 4) A. The head contains a brain and sense organs and is the part of the body that leads when the animal moves.

B. Flatworms have three distinct tissue layers, epidermis, mesoderm, and endodermis.

1. The mesoderm gives rise to organs, structures made of two or more tissues that function together.

2. Flatworms have five organ systems.

a. The digestive system has a pharynx and intestine with only one opening.

b. The excretory system includes flame cells with cilia that drive water into the adjacent tubules and out of the body.

c. A muscular system lies below the epidermis.

d. The nervous system includes the light-detecting cells of the eye, longitudinal nerve cords, and lateral nerves.

e. The reproductive system includes both ovaries and testes in most flatworms.

f. Because the flatworm is so thin, gases and nutrients diffuse to all cells easily.

(tpq 3) **C. Flatworms may be free-living or parasitic.**

1. Free-living flatworms such as planarians have <u>eyespots</u> and well-developed nervous systems.

(tr 86) **2. Parasites like flukes and tapeworms have reduced sensory and digestive systems but greatly elaborated reproductive systems.**

a. They may have different hosts at different parts of their life cycle; blood flukes, which cause <u>schistosomiasis</u>, alternate between snails and humans.

b. A parasitic tapeworm has a head, or scolex, that serves for little more than attachment to the host tissues.

(p 387) **VI. <u>Roundworms</u>, in the phylum Nematoda, are the most numerous animals on earth. They are free-living or parasitic.**

(tr 87) **A. They have a <u>pseudocoelom</u> and a digestive system with mouth and anus.**

1. The fluid-filled pseudocoelom lies between the innermost tissue layer and the outer two.

2. All animals more complex than roundworms have a true <u>coelom</u> within the mesoderm. The pseudocoelom or coelom confers several advantages.

---

tpq 3. How do flatworms affect human health?

---

a. **Internal organs can grow larger and more complex.**

b. **Internal organs are cushioned and protected.**

c. **The cavity acts as a hydroskeleton.**

d. **The activities of the gut are undisturbed by events that occur at the body's outer wall.**

B. **Many roundworms are economically and medically important parasites.**

(p 389) **VII. Soon after the origin of bilateral symmetry, two great**
(msg 5) **lines diverged. They differ in embryonic development.**

(tpq 4) A. **Protostomes include mollusks, annelids, and arthropods. The blastula's first indentation becomes the mouth.**

B. **Deuterostomes include echinoderms and chordates. The blastula's first indentation becomes the anus.**

(p 389) **VIII. The phylum Mollusca arose before the early Cambrian.**

(msg 6) A. **The generalized body plan includes a foot, a head, a visceral mass, and a mantle that covers the visceral mass. The gills are in the mantle cavity.**

B. **Evolutionary advances include an open circulatory system with a heart and vessels.**

---

tpq 4. Why is a seemingly trivial detail, which embryonic opening becomes the mouth, thought to indicate a fundamental difference between lineages?

---

1. Some mollusks have a well-developed nervous system.

(tpq 5) 2. Most have a <u>radula</u> and a motile trochophore larva.

C. <u>Gastropods</u> (class Gastropoda) include snails, garden slugs, and nudibranchs.

1. Most snails exhibit torsion, a twisting of the body during embryonic development.

2. Land snails and slugs move on <u>slime trails</u>.

D. <u>Bivalves</u> include oysters, clams, and relatives.

1. The two shells, or valves, protect the body.

2. They are <u>filter feeders</u>.

E. <u>Cephalopods</u> include the squid, octopus, and nautilus.

1. The foot is modified into sucker-studded arms.

2. Water can be forced out of the mantle cavity through the <u>siphon</u>, providing jet propulsion.

3. They have large brains, acute senses, and complex nervous systems.

(p 391) IX. The <u>segmented worms</u> are in phylum Annelida.

A. There are three classes.

1. The class Polychaeta includes marine worms.

(tr 88) 2. The class Oligochaeta includes earthworms.

3. The class Hirudinea includes leeches.

---

tpq 5. Both mollusks and segmented worms have trochophore larvae. Is this indicative of convergent or divergent evolution?

B. Annelids generally reproduce sexually.

1. Marine worms release their gametes into the water, where the gametes unite and develop into trochophore larvae.
2. Earthworms and leeches are hermaphrodites; pairs reciprocally fertilize each other.

C. Annelids show all five evolutionary trends: segmentation, bilateral symmetry, cephalization, a tubular gut, and a coelom.

1. Annelid bodies are segmented.
   a. Segments are separated by septa.
   b. Each segment contains two excretory nephridia.
   c. Each segment contains a fluid-filled coelom chamber surrounded by circular and longitudinal muscles.
   d. Setae are bristles attached to each segment.
2. The gut tube runs the full length of the worm and includes a crop for storing food and a gizzard for grinding it.
3. The closed circulatory system contains the blood, which does not empty into sinuses. (tpq 6)

---

tpq 6. Humans and segmented worms have closed circulatory systems. Does that mean that the most recent common ancestor must have had a closed circulatory system?

(p 393) (msg 7)

X. Arthropods (phylum Arthropoda) include the great majority of living animal species.

A. Arthropods are characterized by an exoskeleton that protects the body and provides muscle attachment.

1. Joints, where the exoskeleton is thinner, allow movement.
2. The animal must molt in order to grow.

B. Body segments and parts are specialized.

1. The insect body is divided into three parts: head, thorax, and abdomen. Two parts may be fused in other groups.
2. Specific appendages for walking, swimming, and flying arose from the thorax.
3. Mouthparts and antennae grew from the head.

C. Arthropods have high metabolic rates, supported by efficient respiratory organs.

1. Insects have tracheae, branching networks of air passages.
2. Spiders and relatives have book lungs.
3. Aquatic arthropods have gills.

D. Arthropods have acute senses that detect chemicals, sounds, sights. Their compound eyes consist of many facets. They can detect pheromones for many miles.

E. There are many classes of arthropods.

1. Centipedes and millipedes, of the class Chilopoda, have two or four legs per segment.
   a. They may be predators or scavengers. Centipedes have poison claws.
   b. A related phylum, Onychophora, may be a transitional form between annelids and arthropods.
2. Crustaceans include crabs, barnacles, sowbugs, and many other types.
   a. Almost all have two pairs of antennae and a carapace hardened with calcium salts.
   b. The lobster has a cephalothorax and abdomen.
3. Class Arachnida includes spiders, ticks, mites, and scorpions.
   a. They lack antennae.
   b. The first pair of legs is modified into chelicerae, or poison fangs.
   c. The second pair of legs holds the prey.
   d. Spiders have silk-producing spinnerets at the rear of the abdomen.
4. Insects are the largest class of animals on earth.
   a. They are very diverse in shape, size, and habitat.
   b. Insects have one pair of antennae, three pairs of legs, and usually one or two pairs of wings.

c. **Mouthparts may be highly modified for specific diets.**

d. **Some insects, such as grasshoppers, have similar body plans as juveniles and adults.**

(tpq 7) e. **Some insects go through metamorphosis while in a transitional stage known as the pupa, sometimes in a cocoon.**

f. **Social insects, such as termites, may be specialized into castes that perform specific tasks.**

(p 398) (msg 8) **XI. Echinoderms (phylum Echinodermata) include starfish and relatives.**

A. **They have an endoskeleton and a water vascular system for locomotion.**

B. **They are deuterostomes that are related to phylum Chordata.**

C. **Adults are radial, headless, and brainless, and without excretory and respiratory systems. The larvae are bilaterally symmetrical.**

D. **They reproduce sexually by shedding gametes into the water, and they can regenerate missing parts.**

---

tpq 7. What evolutionary advantage might there be to having a larval form very different from the adult form?

INSTRUCTIONAL AIDS

VIDEOS, FILMS, SLIDES

Adaptive Radiation – the Mollusks. Pennsylvania State University AV Services, University Park, PA. Investigates adaptive radiation in the body plan of mollusks. Film.

Notes of a Biology Watcher: A Film with Lewis Thomas. Time-Life Video, Paramus, NJ. The biologist and award-winning author examines individuality and interconnectedness among invertebrates. Film, video.

The Structure of Animals, Part 1: Invertebrates. National Geographic Society, Washington, DC. Outstanding photography portraying the major groups of invertebrates, with a focus on insects. Filmstrip.

REFERENCES

Brusca, Richard C., and Gary J. Brusca. 1990. *The Invertebrates*. Sunderland, MA: Sinauer Associates. A comprehensive, well illustrated text.

Buchsbaum, Ralph, et al. 1987. *Animals Without Backbones*, Third Edition. Chicago: University of Chicago Press. A text that has been the standard for more than 50 years.

Hölldobler, Bert, and E. O. Wilson. 1990. *The Ants*. Cambridge, MA: Harvard University Press. A comprehensive review of the systematics, ecology, behavior, and evolution of ants.

Kaneshiro, Kenneth Y., and A. T. Ohta. 1982. The flies fan out. *Natural History* 91 (12):54 – 58. A discussion of adaptive radiation among Hawaiian fruit flies.

Kingsolver, Joel G. 1985. Butterfly engineering. *Scientific American* 253 (2): 106 – 113. The life-sustaining functions of the butterfly are analyzed according to engineering principles.

Ternes, Alan, ed. 1991. Mosquitoes unlimited. A special edition of *Natural History* devoted to mosquitoes. July, 1991.

KEY TERMS

abdomen	page 394	mantle cavity	390
animal	378	medusa	383
antenna	394	mesoglea	384
bilateral symmetry	386	metamorphosis	397
bivalve	390	mollusk	389
book lung	395	molt	394
carapace	396	mouthpart	394
caste	397	nematocyst	384
central cavity	382	nephridium	392
cephalization	386	nerve cell	384
cephalopod	391	nerve cord	387
cephalothorax	396	nudibranch	390
chelicera	396	organ	387
circulatory system	390	organ system	387
cnidarian	383	poison claw	395
coelenteron	384	polyp	393
coelom	388	pore	382
crop	392	protostome	389
deuterostome	389	pseudocoelom	388
echinoderm	398	pupa	397
endoskeleton	398	radula	390
exoskeleton	394	roundworm	387
eyespot	387	schistosomiasis	387
facet	395	segmented worm	391
filter feeder	391	septum	392
flame cell	387	sessile	382
flatworm	385	seta	392
foot	389	siphon	391
gastropod	390	slime trail	390
gastrovascular cavity	384	social insect	397
gill	390	spicule	382
gizzard	392	spider	396
head	390	spinneret	397
heart	390	sponge	382
hydroskeleton	388	thorax	394
insect	397	trachea	395
intestine	387	vertebrate	378
invertebrate	378	vessel	390
joint	394	water vascular system	398
mantle	390		

FUN FACTS

DEEP-SEA SYMBIOSIS

In 1977, geologists in the research submarine *Alvin* were startled to find an oasis of life on the floor of the Pacific Ocean 2,600 meters below the surface. Clustered around vents from which hot, sulfur-laden water flowed was a colorful and dense community of invertebrates.

Further investigations showed that there were many bizarre and exotic creatures around the hot springs: giant tube worms, ghostly crabs, shrimps, and fishes clustered around white clams. The food chain was supported by sulfur bacteria that gained their energy by oxidizing the hydrogen sulfide that emerged from the vents. Hydrogen sulfide is a highly toxic (and very smelly) compound that poisons hemoglobin and cellular respiration. But sulfur bacteria can oxidize it, capturing the energy released to power a process similar to photosynthesis. Like chloroplasts, sulfur bacteria have become endosymbionts of larger organisms.

The giant tube worm, *Riftia pachyptila,* is strikingly odd. Its body is a closed, thin-walled sac, without a mouth or digestive system or any other way to ingest food. Its plume-like gills are bright red because of the great amount of hemoglobin in its blood. The largest of its internal organs, the trophosome ("feeding body"), is full of sulfur bacteria, which feed their host. The tube worm supplies its endosymbiotic bacteria with the raw materials they need to fuel their metabolism: carbon dioxide, oxygen, and hydrogen sulfide.

Its hemoglogin helps the tube worm avoid being poisoned by hydrogen sulfide, which is as toxic as cyanide. Sulfide can block the oxygen binding sites on hemoglobin. It poisons cytochrome *c,* an enzyme that is crucial to cellular respiration. But *Riftia* can thrive in sulfide concentrations that would kill other animals. Its hemoglobin molecule is much larger than our hemoglobin. It binds sulfide tightly, preventing the poisoning of cytochrome *c*. It can carry sulfide and oxygen simultaneously without allowing them to react with each other.

See: Childress, James J., et al. 1987. Symbiosis in the deep sea. *Scientific American* 256 (5):115 – 120.

A "MISSING LINK"

As different as annelids and arthropods seem, their evolutionary links are revealed by a rare phylum, the Onychophora, known as "walking worms." These shy creatures are found under logs and leaf litter in moist tropical forests. At first glance they resemble soft-bodied, nonsegmented millipedes with stout, fleshy legs. Their anatomy demonstrates a mixture of annelid and arthropod characteristics, indicating that they are descendants from a line that branched off the primitive annelid-arthropod stock.

They are covered with a chitinous cuticle like arthropods but beneath lie layers of muscles, as in annelids, contrasting with arthropods' discrete muscle bundles. The open circulatory system and tubular respiratory system resemble those of arthropods. The excretory system consists of units that resemble the nephridia of annelids.

See: Buchsbaum, Ralph, et al. 1987. *Animals Without Backbones*, Third Edition. Chicago: University of Chicago Press.

CHAPTER 18 INVERTEBRATE ANIMALS: THE QUIET MAJORITY

MESSAGES

1. Animals are multicellular, motile heterotrophs. Three lines of animals arose from protists.

2. Important anatomical trends evolved in invertebrates, including bilateral symmetry, cephalization, a two-ended gut, a coelom, and segmentation.

3. Sponges have cellular specialization but no organs. Cnidarians have nerve nets and extracellular digestion.

4. Flatworms are bilaterally symmetrical, cephalized, and have three tissue layers and true organs. Roundworms have a two-ended gut and a pseudocoelom.

5. The main line of animals split into two lineages, protostomes and deuterostomes, that differ in embryonic development.

6. Mollusks are diverse, soft-bodied animals. Annelids are segmented and display all five major anatomical advances.

7. Arthropods have an exoskeleton and are the largest animal phylum.

8. Echinoderms are deuterostomes with an endoskeleton and a water vascular system. They are related to chordates.

19

THE CHORDATES: VERTEBRATES AND THEIR RELATIVES

PERSPECTIVE

This chapter presents the evolution of chordates, from the simplest tunicate to *Homo sapiens*. This chapter is the culmination of Part III, Life's Variety, and builds on the concepts presented in earlier chapters of this part. Material from this chapter will reappear in Part IV, How Animals Survive, in Part VI, Interactions: Organisms and Environment, and in Chapter 38, Animal Behavior.

The themes and concepts of this chapter include:

- Members of phylum Chordata have notochords, spinal cords, gill slits, a tail, and muscle blocks at some point in their lives.
- Tunicates and lancelets display the five chordate traits.
- Fish were the earliest class of vertebrates. Their innovations include a skull, bones, cartilage, vertebrae, jaws, and paired appendages.
- Amphibians were the first vertebrates to live on land.
- Reptiles were the first vertebrates fully adapted to dry land. They were dominant during the Mesozoic era.
- Birds, mammals, and modern reptiles arose from reptilian ancestors.
- Primate adaptations include opposable thumbs, cooperative social systems, complex brains, extended parental care, cultural transmission, and upright gait.
- Humans evolved from the Old World monkey lineage.
- Human evolution is marked by increasing brain size, decreasing bone strength, and the development of complex culture.

CHAPTER OBJECTIVES

Students who master the material in this chapter will be able to trace their evolution back to small, fish-like creatures. They will see how early chordates gave rise to fishes and subsequently to amphibians, reptiles, birds, and mammals. They will understand that the evolution of vertebrates includes increasing independence from standing water and from constant environmental conditions.

These students should be able to list the great adaptive steps in the evolution of vertebrates and, more specifically, in the evolution of humans from our prosimian ancestors.

LECTURE OUTLINE

(p 402) **I. Bats, the misunderstood vertebrates, introduce the phylum Chordata.**

(tr 89) **A. Chordates have a central nerve cord and a notochord or a backbone of vertebrae.**

B. Bats, humans, and others are in the class Mammalia of the subphylum Vertebrata.

C. Four unifying themes emerge:

1. Chordate evolution is a history of innovations that build upon the major invertebrate traits.

2. Chordate evolution is marked by physical and behavioral specializations.

3. Evolutionary innovations and specializations led to adaptive radiations.

(tpq 1) **4. The evolutionary process that led to humans was the same as for all other organisms.**

---

tpq 1. How is human evolutionary history different from that of other mammals?

(p 404) **II. Tunicates and lancelets represent the origins of chordates.**

A. Deuterostomes include echinoderms, acorn worms (phylum Hemichordata), and chordates.

(msg 1) **B. All chordates display bilateral symmetry, cephalization, a coelom, a two-ended gut, and body segmentation, features shared by invertebrate ancestors.**

(tpq 2) **C. New features include a notochord (absent in adult tunicates and replaced by the vertebral column in adult vertebrates), a spinal cord, gill slits, a tail, and myotomes.**

1. **The notochord and vertebral column allowed chordates to become large in size.**
2. **Centralized nerve coordination and control allowed a wide range of behaviors.**

(msg 2) **D. Tunicates are sessile as adults and have a larval form with all five chordate traits.**

1. **The larva is motile; it undergoes metamorphosis into the adult.**
2. **The adult has an outer tunic enclosing a large pharynx, which filters food particles from the water.**

E. Lancelets are filter feeders. They have a notochord. Larvae and adults can swim.

---

tpq 2. Do humans have gill slits, tails, and myotomes? If not, how can they be considered vertebrates?

---

(407) **III. Fishes were the earliest vertebrates.**

(msg 3) **A. Fishes emerged at the dawn of the Paleozoic era, about 550 million years ago.**

1. They had circular mouths with no jaws; they were <u>agnathans</u>.

2. Their advances over simpler chordates included a <u>skull</u>, muscled gill slits, bone supports for the gill openings, and bony armor plates.

B. Modern jawless fishes include the <u>lamprey</u>. They have lost the bony plates and skulls.

C. The placoderms appeared about 425 million years ago.

1. They had <u>hinged jaws</u>, derived from gill support bones. This adaptation allowed them to consume large chunks of food, which supported great body size.

2. They also had the first vertebrae and first <u>fins</u>. The sockets in which the fins fit gave rise to hip and shoulder joints of later land animals.

D. <u>Chondrichthyes</u>, which include sharks, skates, and rays, have skeletons made of cartilage.

E. <u>Osteichthyes</u> are bony fishes.

(tpq 3) **1. <u>Lobe-finned fishes</u> arose 400 million years ago. They have muscular fins and <u>lungs</u>.**

---

tpq 3. What would be the evolutionary advantage to a fish with lungs?

---

a. Survivors include lung fishes and coelacanths.

b. Rhipidistians, now extinct, were probably the first vertebrates to be adapted to land life.

2. Teleosts, or spiny-finned fishes, are the largest group of vertebrates today. They have delicate fins and swim bladders but have lost their lungs.

(p 410) IV. Amphibians were the first vertebrates to live on land.

(tpq 4) (msg 4) A. Fossilized tracks and bones show that early amphibians had legs with which they dragged themselves around on land some 375 million years ago.

B. They had lungs and skin that allowed gas exchange.

C. Amphibians are restricted to life near water because their skin must remain moist and their eggs cannot resist dessication.

(p 411) (msg 5) V. Reptiles were the first vertebrates fully adapted to life on land.

A. Cotylosaurs, which resembled crocodiles, had four innovations.

1. Dry, scaly skin provided a barrier to evaporation.

---

tpq 4. How might natural selection have led to the formation of legs?

---

2. Respiratory and circulatory advances increased gas exchange.

3. Internal fertilization and amniote eggs allowed reproduction outside of standing water.

4. Their legs extended beneath their bodies, allowing faster and easier locomotion.

(tr 90) B. Cotylosaurs were ancestors of modern reptiles, birds, and mammals.

(msg 6) 1. Thecodonts were ancestors of dinosaurs and birds.

(msg 7) 2. Therapsids led to mammals.

C. Most of the reptiles died out at the end of the Cretaceous period. A few smaller species survived and radiated.

(p 413) VI. Birds had arisen by 150 million years ago.

A. Most avian adaptations lead to efficient flight.

1. Feathers have tiny barbules that interlock.

2. Birds have hollow bones and a sternum that anchors the pectoral muscles.

(tpq 5) 3. Birds are warm-blooded, or homeothermic, maintaining a constant interior temperature. (Cold-blooded animals are poikilothermic.)

---

tpq 5. What advantage do birds gain from being warm-blooded instead of cold-blooded?

---

4. Birds have a series of connected air sacs as well as lungs.
5. A bird's heart is divided into four chambers, separating oxygenated and unoxygenated blood.
6. Birds have hard-shelled eggs.

B. Birds have radiated into many distinct modes of life.

(p 415) (tr 91) VII. Mammals arose at least 180 million years ago but didn't come to dominate until after the end of the Cretaceous.

A. In addition to homeothermy and a four-chambered heart, mammals have milk, produced in mammary glands, and body hair or fur for insulation. Blubber insulates some mammals.

B. Most mammals have a placenta that supports the growth of the embryo.
1. Monotremes lay eggs and suckle their young.
2. Marsupials give birth to tiny young that are nourished in the mother's pouch.

C. Mammals maintain a constant body temperature through their homeothermic metabolism, insulation, and behavior.

D. Mammals have specialized limbs and teeth.

(tpq 6) **E. Mammals, especially primates, care for their young for extended periods.**

F. Mammals have highly developed nervous systems and senses.

(p 418) **VIII. Humans are simply one branch of evolution, separated from the apes by no more than 6 million years.**

A. The order Primates is divided into two suborders.

1. Suborder Prosimii includes the <u>prosimians</u>.

a. Tree shrews are the most primitive primate. Other prosimians include lemurs, tarsiers, and lorises.

b. Like all primates, they have <u>opposable thumbs</u> and an acute sense of sight.

2. Suborder Anthropoidea includes <u>anthropoids</u>.

a. This group includes humans, monkeys, and apes.

b. New World monkeys usually have prehensile tails and flat noses.

c. Old World monkeys usually have downward-pointing nostrils and their tails are not prehensile. The apes (often called hominoids) are our closest living relatives.

(msg 8) **B. Primates evolved several traits for arboreal life.**

---

tpq 6. What advantages do primates gain from caring for their young for extended periods?

---

1. Stereoscopic vision is an adaptation that allows enhanced depth perception.
2. A large brain enhances locomotion and social behavior.
3. Primates give birth to small litters and care for their offspring long after birth.
4. Upright posture improves visibility and leaves the hands free for other activities.
5. Primates have teeth modified for an omnivorous diet.

(tr 92) C. Monkeys arose from prosimian ancestors about 50 million years ago.

1. By 20 million years ago, hominoids had arisen.
2. Proconsul africanus and Sivapithecus preceded the divergence of humans and apes.
3. By comparing DNA similarities, scientists have determined that chimpanzees are more closely related to humans than they are to gorillas.

(p 423) IX. Hominid evolution dates back at least 3.75 million years.

(msg 9) A. Australopithecus afarensis had an apelike skull and teeth and a brain about the size of a chimpanzee's. The posture was upright and adapted to bipedalism.

1. Several species of australopithecines lived simultaneously.

2. They were probably all vegetarians and may not have made tools.

(tpq 7) (msg 10) B. Homo habilis appeared by 2 million years ago, with larger brains than their predecessors and the ability to make tools.

C. Homo erectus spread throughout northern Africa, southern Asia, and southern Europe more than 1.5 million years ago.

1. Their brains had enlarged and their toolmaking techniques improved.
2. They used fire by 500,000 years ago.
3. Cooperative bands hunted and prepared food.
4. Their behavior was probably developed and maintained through cultural transmission.
5. Their anatomy indicates that they were born quite immature and had a long period of parental nurturing.

D. Homo sapiens gradually arose from Homo erectus by about 500,000 years ago.

1. Modern humans superseded archaic humans between 30,000 and 100,000 years ago.
2. Neandertals had larger brains than modern humans and a rich esthetic and social life.

---

tpq 7. Did humans use fire or make tools first? Are there other animals that make or use tools?

3. **Evidence suggests that modern humans are not descendants of Neandertals.**
4. **Cro-Magnons had a more modern appearance than Neandertals, more sophisticated tools, and more complex culture.**
5. **By 15,000 to 20,000 years ago, people had occupied virtually all the inhabitable regions of the earth.**
6. **Mitochondrial DNA analysis indicates that all modern humans can trace their ancestry to a single female who lived 200,000 years ago.**
7. **Cultural evolution has allowed humans to prosper.**

INSTRUCTIONAL AIDS

VIDEOS, FILMS, SLIDES

Comparative Vertebrate Anatomy. Educational Images Ltd., Elmira, NY. Dissection of an herbivore and a carnivore, demonstrating anatomical differences related to nutrition and mode of life; comparisons with dogfish, frog, and bird. Transparencies.

Fossils: Reptiles and Mammals. Films for the Humanities and Sciences, Princeton, NJ. Presents fossil evidence of evolution, shows field and lab techniques, and traces evolution of some modern mammals back through time. Video.

Putting Animals in Groups. International Film Bureau, Chicago. Explains methods of classification, points out characteristics of various animal taxa. Film, video.

REFERENCES

Gregory, Ed. 1991. Tuned-in, turned-on platypus. *Natural History* 5/91:31–37. Reveals how the platypus uses electrical fields to detect prey.

Kavanagh, Michael. 1984. *A Complete Guide to Monkeys, Apes, and Other Primates*. New York: Viking Press. A popular treatment of primates that emphasizes behavior. Well illustrated, readable.

Lewin, Roger. 1988. *In the Age of Mankind*. Washington, DC: Smithsonian Books. This well-illustrated book is written for a general audience and spans the field of inquiry into human evolution.

Rismiller, Peggy D., and R. S. Seymour. 1991. The echidna. *Scientific American* 264 (2):96–103. A discussion of the natural history and reproductive behavior of the spiny anteater.

Tattersall, Eric D., et al., eds. 1988. *Encyclopedia of Human Evolution and Prehistory.* New York: Garland Publishing. This comprehensive volume includes more than 12,000 entries written by experts in paleontology, geology, anthropology, anatomy, invertebrates, and primates.

KEY TERMS

adaptive radiation	page 403
agnathan	407
air sac	415
amniote	412
amphibian	410
anthropoid	418
body hair	416
blubber	416
Chondrichthyes	408
chordate	402
coelacanth	409
dinosaur	413
feather	413
fin	408
fur	416
gill slit	405
hinged jaw	408
homeothermic	415
Homo sapiens	403
leg	408
lobe-finned fish	408
lung	408
lungfish	409
mammal	416
marsupial	416
milk	416
monotreme	416
opposable thumb	419
Osteichthyes	408
pharynx	406
poikilothermic	415
prosimian	418
reptile	411
rhipidistian	409
skull	407
spinal cord	404
spiny-finned fish	409
stereoscopic vision	421
sternum	415
tail	405
teleost	409
tunic	406
vertebra	403

FUN FACTS

THE PLATYPUS

The platypus ("flat foot") is not a living fossil or a "missing link"; it is a highly evolved animal that shares features with mammals, reptiles, and sharks. It has the bill of a duck and the fur of a mammal. Its tail and webbed feet resemble a beaver's. Like a reptile or a bird, it lays eggs. Like a mammal, it suckles its young. Like a shark, it has sensors that detect electrical fields in water, enabling it to find prey.

Given the platypus's bizarre anatomy, it's no surprise that it was considered a missing link. Among its reptilian features is a cloaca ("sewer"), a single body opening for defecation and egg-laying. It has several bones known only from the fossils of therapsids. Its embryonic development is more like a reptile's than a mammal's. Its sperm is long and slender, with a threadlike head, like the sperm of reptiles. Its cell nuclei include tiny microchromosomes, which are found in reptiles but not in mammals, as well as the more familiar macrochromosomes. But its fur, mammary glands, and ability to regulate its body temperature brand it as a mammal.

Despite its seemingly archaic features, the platypus is highly specialized and well adapted to its life. It lives in freshwater streams and lagoons in Australia, dining on aquatic arthropods and clams at night and retiring to its burrow in the bank of the waterway in the day. During the breeding season, males establish territories and fight for females, using their poison spurs against other males. Courtship is an elaborate affair that may take days to consummate. The female lays one to three eggs and incubates them, holding them between her tail and body. The tiny newborns suck milk from the female's two nipples; in three months they are weaned and begin diving for food as the adults do.

The platypus's ability to find prey is remarkable. When it dives, its eyes, ears, and nostrils are tightly shut. As it searches along the bottom of a stream, it wags its bill from side to side two or three times a second. When it gets close to prey, it moves its bill over a small area and snaps up the hapless invertebrate. Scientists hypothesized that it detects electrical fields generated by its prey. When they put a tiny 1.5-volt battery in the mud, the platypus found it immediately, even from as far away as 30 centimeters. They examined the platypus's bill and discovered that it had many nerve endings sensitive to electrical fields. Such electroreceptors had previously been known only among fish.

The platypus's mixture of mammalian and reptilian traits suggests that it has a long evolutionary history, but few fossils that could show its lineage have been discovered. There can be little doubt that its ancestors diverged from the mammalian line more than 100 million years ago.

See: Griffiths, Mervyn. 1988. The platypus. *Scientific American* 258 (5):84–91.

CHAPTER 19 THE CHORDATES: VERTEBRATES AND THEIR RELATIVES

MESSAGES

1. Members of phylum Chordata have notochords, spinal cords, gill slits, and tails with muscle blocks at some point in their lives.
2. Tunicates and lancelets represent the ancestral chordates.
3. Fish were the earliest class of vertebrates. Their innovations include a skull, bones, vertebrae, and jaws.
4. Amphibians were the first vertebrates to live on land. They must return to water to breed.
5. Reptiles were the first vertebrates fully adapted to dry land. They were dominant during the Mesozoic era.
6. Birds arose from thecodont ancestors. Their adaptations enhance flight.
7. Therapsid reptiles gave rise to mammals, which were minor species until the Cenozoic era.
8. Primate adaptations include opposable thumbs, cooperative social systems, complex brains, extended parental care, cultural transmission, and upright gait.
9. Humans evolved from the Old World monkey lineage.
10. Human evolution is marked by increasing brain size, decreasing bone strength, and development of complex culture.

20

AN INTRODUCTION TO HOW ANIMALS FUNCTION

PERSPECTIVE

This chapter bridges the transition from cellular biology, presented in Parts I and II, to organismic biology, which is the subject of Part IV, How Animals Survive. The material in this chapter provides an overview of the integrated physiological activities and anatomical adaptations that allow an animal to survive in a changing and sometimes inhospitable environment, matters that are presented in Chapters 21 through 29.

The themes and concepts of this chapter include:

- Survival requires extracting energy from the environment, exchanging and distributing materials, and maintaining homeostasis.
- Multicellular organisms have physiological systems that bring about exchanges between cells and the environment.
- Homeostatic mechanisms, operating through feedback loops, keep the contents of body fluids constant.
- In multicellular organisms, cells are organized into tissues, organs, and organ systems.
- Large animals need structural support and systems that coordinate and integrate the body's activities.
- Anatomical and physiological adaptations, produced by natural selection, allow organisms to exploit their environments.
- Negative feedback systems help maintain homeostasis.
- Positive feedback loops bring about rapid change.

CHAPTER OBJECTIVES

Students who master the material in this chapter will gain an overview of the integration and control necessary for the maintenance of animal life. They will see that anatomy and physiology are intertwined and that many organ systems work together to maintain homeostasis. They will understand that integrated systems are necessary to support cellular life. This chapter provides an introduction to the exchange of materials via body tubes, to the organization of cells into tissues, organs, and organ systems, and to the four main types of tissues.

LECTURE OUTLINE

(p 430) **I. The blue whale introduces how animals function.**

A. Maintaining a high metabolic rate requires enormous food intake.

(tpq 1) **B. Each cell must be supplied with nutrients and oxygen and be rid of its wastes.**

C. Cells, tissues, and organs must be coordinated.

D. Anatomy is the science of biological structure; physiology is the study of how such structures work.

(msg 1) **E. Homeostasis is the maintenance of constant internal conditions despite fluctuations in the external environment.**

(p 432) (msg 4) **II. Maintenance of a steady internal state in a fluctuating external environment is a central problem for a living thing.**

tpq 1. In what ways do the cells of whales face the same tasks and challenges as protist cells? How are the tasks and challenges different?

A. **All cells and organisms must exchange materials with the environment.**

(msg 2) 1. **Diffusion is the means of transport at the cellular level.**

(msg 3) 2. **The digestive, respiratory, and excretory**
(tr 93) **systems have tubes that link the interior of an animal's body to the environment.**

a. **Each system has special regions where substances can be exchanged between the contents of the tube and nearby body fluids.**

b. **Bulk flow circulates fluids throughout the body.**

c. **Through diffusion, substances can move into or be released from each cell.**

d. **Most multicellular animals have two or three different types of fluid: extracellular fluid, intracellular fluid, and blood.**

(msg 5) B. **Cells are organized into tissues, organs, and**
(tr 94) **organ systems, all of which make up an organism.**

1. **Epithelial tissue covers body surfaces.**

a. **Epithelial tissue makes up exocrine and some endocrine glands.**

b. **Epithelial sheets typically have one surface facing a space.**

c. **The cells facing the space link tightly to one another by means of cell junctions, preventing leakage.**

d. **Junctions link the inner surface of the epithelium to the basement membrane or lamina.**

2. **<u>Connective tissue</u> binds tissues together and supports flexible body parts.**
   a. **Cartilage, bone, blood, and fat cells are connective tissues.**
   b. **Connective tissue generally lays down a network of protein fibers that holds other tissue together.**
   c. **Collagen is the most abundant protein in the body.**

3. **<u>Muscle tissue</u> enables the animal to move.**
   a. **Smooth muscle operates glands, blood vessels, and internal organs.**
   b. **Cardiac muscle makes the heart beat.**
   c. **Skeletal muscle moves the bones.**

4. **<u>Nervous tissue</u> transmits electrochemical signals.**

(tpq 2) (msg 6) C. **Large animals require structural support and avenues of coordination and communication to tie together body parts.**

---

tpq 2. One hypothesis to explain the extinction of the dinosaurs is that they became too big for coordination and communication to tie body parts together efficiently. Is this a reasonable hypothesis?

(p 437) **III. Specialized anatomical adaptations allow different**
(msg 7) **organisms to exploit their environment in different ways.**

A. Mouth parts may be adapted to prey of specific size.

(tpq 3) **B. Body shape may minimize or maximize heat exchange with the environment or contribute to efficiency in movement.**

C. Anatomical and physiological adaptations arise by natural selection.

(p 438) **IV. Physiological systems continuously adjust to aspects of the environment in and around cells.**

(msg 8) **A. Homeostasis requires three separate elements.**

1. A receptor senses environmental conditions.

2. An integrator evaluates the situation and makes decisions.

3. An effector executes the commands.

(tr 95) **B. Each set of elements constitutes a feedback loop.**

1. Negative feedback loops resist change by sensing a stimulus and activating mechanisms that oppose the trend away from a baseline condition.

a. A thermostat and heater control the temperature of a house.

---

tpq 3. Which is likely to have bigger ears, an arctic fox or a desert fox?

---

b. In humans, the hypothalamus acts as a thermostat and controls mechanisms for increasing or dissipating heat, preventing hypothermia and hyperthermia.

(tpq 4)

2. Positive feedback loops bring about rapid change. A change triggers more change.
3. Whales provide an example of feedback loops.
   a. Blubber and countercurrent heat exchange in the fins reduce a whale's heat loss.
   b. Its hypothalmus triggers activity that produces heat when the whale's internal temperature drops.
   c. When the whale is overheated, blood is shunted into the blubber layer, where it loses heat to the water.
4. Poikilotherms have behavioral adaptations that help regulate body temperature.

---

tpq 4. Are there feedback loops in the growth of the human population? Are the loops negative or positive?

INSTRUCTIONAL AIDS

VIDEOS, FILMS, SLIDES

Metabolism: Structure and Regulation. Carolina Biological Supply Company, Burlington, NC. Examines the organization of metabolism and explains the principles of metabolic control. Transparencies, filmstrip, video.

The Sunbaskers. The Media Guild, San Diego, CA. A survey of reptile behavior that maintains body temperature. Video, film.

Systems Working Together. Coronet Film and Video, Deerfield, IL. A portrayal of the complex interactions among the human body's various systems. Video.

REFERENCES

Bernd, Heinrich. 1987. Thermoregulation in winter moths. *Scientific American* 256 (3):104 – 111. Discusses endothermy in winter-active moths, which can fly, feed, and mate at near-freezing temperatures.

Guyton, Arthur C. 1991. Blood pressure control – special role of the kidneys and body fluids. *Science* 252:1813 – 1816. Reviews the feedback systems that control blood pressure and body fluid volume.

Smith, Anthony. 1985. *The Body*. New York: Viking. A popular account of human anatomy and physiology.

Vander, A., J. Sherman, and D. Luciano. 1985. *Human Physiology: The Mechanisms of Body Function*, Fourth Edition. New York: McGraw-Hill. A well-written introduction to human physiology and homeostasis.

KEY TERMS

anatomy	page 431	nervous tissue	435
connective tissue	435	organ	435
epithelial tissue	435	organism	435
feedback loop	438	organ system	435
homeostasis	438	physiology	431
muscle tissue	435	tissue	433

FUN FACTS

THE BEHAVIOR OF WHALES

Whales might seem quite different from deer and antelope, but they have many similarities. Deer, antelope, and related ungulates (as hoofed animals are called) are terrestrial grazers that are highly social, live in herds, care for young, and often migrate from one area to another. Whales evolved some 50 million years ago from ungulate ancestors. They took to the sea, lost their hind legs and fur, and adapted to life in water, but they demonstrate many of the behaviors of their terrestrial relatives. They travel in groups, which are often tightly knit, long-term associations. Males fight for dominant status to assure mating rights.

The behavior of mothers and calves is remarkably similar to the interactions of terrestrial ungulates that migrate long distances. The southern right whale is typical. During the first year of life, a calf nurses frequently and never strays far from its mother, who protects it against predators such as sharks and killer whales. When the whales return to the shallow-water calving grounds, parent-offspring conflict becomes apparent. The mother seeks to wean the calf so she can produce additional offspring. The calf maximizes its chances of success by exploiting the mother for as long as possible. Among terrestrial mammals, parent-offspring conflict may result in the mother's driving the yearling calf away; similar behavior may occur among whales. Family behavior among such widely divergent groups as ungulates, whales, and primates shows that social mammals have much in common.

See: Wursig, Bernd. 1988. The behavior of baleen whales. *Scientific American* 258 (4):102–107.

SNAKE BLOOD CIRCULATION

Snakes living in trees must counteract the effects of gravity on their circulation systems and lungs. A snake climbing a tree experiences blood pooling in its tail and blood loss from its head. The longer the snake, the more pronounced the pressure difference between its head and tail. Arboreal snakes have adaptations that maintain relatively constant blood pressures in the brain. Sea snakes lack these adaptations and would probably die of brain anoxia if held vertically.

Arboreal snakes maintain high blood pressure and have narrow, muscular bodies sheathed by tight-fitting skin, which counters blood pooling in the same way as a fighter pilot's G-suit. Vasoconstriction inhibits the blood flow to the lower half of the body when the snake is climbing. The hearts of arboreal snakes are located near their heads and their lungs are vascularized only near their hearts.

In contrast, sea snakes have fatter bodies with looser skin and musculature, lower blood pressure, reduced vasoconstriction, centrally located hearts, and lungs that are vascularized for nearly the entire length of the animal. Sea snakes do not have to accommodate gravity-induced blood pressure differences even when swimming vertically because the surrounding water has a similar, counteracting pressure gradient. Terrestrial snakes that do not climb trees, such as rattlesnakes, have traits intermediate between arboreal and aquatic snakes.

See: Lillywhite, Harvey B. 1988. Snakes, blood circulation, and gravity. *Scientific American* 259 (6):92–98.

CHAPTER 20 AN INTRODUCTION TO HOW ANIMALS FUNCTION

MESSAGES

1. Survival requires extracting energy from the environment, exchanging and distributing materials, and maintaining homeostasis.

2. Single cells and small multicellular organisms can rely on diffusion to exchange and distribute materials.

3. Multicellular organisms have physiological systems that bring about exchanges between cells and the environment.

4. Homeostatic mechanisms, operating through feedback loops, keep the contents of body fluids constant.

5. In multicellular organisms, cells are organized into tissues, organs, and organ systems.

6. Large animals need structural support and systems that coordinate and integrate the body's activities.

7. Anatomical and physiological adaptations, produced by natural selection, allow organisms to exploit their environments.

8. Negative feedback systems help maintain homeostasis. Positive feedback loops bring about rapid change.

21

CIRCULATION: TRANSPORTING GASES AND MATERIALS

PERSPECTIVE

This chapter presents an organ system that ties the functions of the animal body together: the circulatory system, which will reappear in subsequent chapters in Part IV, How Animals Survive. The material in this chapter is based on concepts presented in Part I, Life's Fundamentals.

The themes and concepts of this chapter include:

- The circulatory system moves body fluids and helps maintain homeostasis by delivering oxygen, nutrients, and other materials to each body cell and by removing metabolic wastes.
- Blood plasma contains dissolved proteins that function in immunity and clotting and carries red blood cells, white blood cells, and platelets.
- Open circulatory systems, which are relatively inefficient, occur in arthropods and mollusks. Closed circulatory systems make possible greater pressure and greater efficiency.
- The evolution of vertebrate hearts has led from a two-chambered heart to one having four chambers, reducing the mixing of oxygenated and deoxygenated blood.
- Capillaries are thin-walled sites of nutrient and gas exchange.
- Arteries carry blood away from the heart. They are thick-walled, elastic, and pressurized. Veins carry blood back to the heart. They are low-pressure blood reservoirs.
- Cardiac muscle fibers generate and transmit impulses that stimulate heart contraction.
- The clotting cascade is a multi-step process that stops the flow of blood from a wound.
- The lymphatic system drains the body of extracellular fluid and is a component of the immune system.

CHAPTER OBJECTIVES

Students who master the material in this chapter will understand how the circulatory system moves gases, nutrients, and other materials throughout the body. They will see that the evolution of animals has been accompanied by increasing complexity and efficiency in the circulatory system. They will gain an overview of the roles of the circulatory and lymphatic systems in immunity.

LECTURE OUTLINE

(p 446) **I. Giraffes, with numerous evolutionary adaptations that compensate for their height, introduce <u>circulatory systems</u>.**

(tpq 1) **A. High blood pressure helps distribute blood and its cargo throughout the body.**

B. Thick-walled blood vessels surrounded by high-pressure fluid prevent leakage and swelling.

C. The rete mirabile lowers blood pressure in the brain and eyes.

D. Three themes emerge:

1. The physical principles of diffusion and bulk flow underlie material transport.

2. The unique properties of blood allow many of an animal's homeostatic activities to take place.

3. Adaptations for material transport reflect the demands and constraints of an animal's environment and way of living.

---

tpq 1. Why is high blood pressure in humans called "the silent killer"?

(p 448) **II. Blood is a multipurpose liquid tissue for internal**
(msg 1) **transport.**

(msg 2) **A. Blood includes solids and a liquid.**

1. **Plasma makes up more than half the volume of blood.**
   a. **It is mostly water.**
   b. **It contains dissolved proteins, including globulins, antibodies, albumins, and fibrinogen.**
   c. **It transports salts, sugars, and fats from the food we eat.**
2. **Cells and cell fragments include white blood cells, platelets, and red blood cells.**

B. Red blood cells, or erythrocytes, transport oxygen.

1. **The disk shape allows the hemoglobin molecules to lie close to the outer membrane, enhancing diffusion.**
2. **Because the red blood cell does not have a nucleus, its life is limited to 120 days.**
3. **A negative feedback loop regulates red blood cell production.** (tpq 2)

---

tpq 2. Why might long-distance runners train by sprinting until they run out of oxygen?

---

a. **When there are fewer red blood cells, the liver and kidneys receive less oxygen; they produce the hormone erythropoetin.**

(tpq 3)

b. **Erythropoetin stimulates stem cells in bone marrow to divide more quickly and produce more red blood cells.**

c. **When the red blood cell count returns to normal, the kidneys and liver slow their production of erythropoetin, resulting in decreased red blood cell production.**

C. **White blood cells, or leukocytes, defend the body against microbes and other foreign material. There are five types of white blood cells.**

1. **Neutrophils and monocytes are phagocytes.**
2. **Eosinophils carry enzymes that break down foreign proteins and blood clots.**
3. **Basophils release an anticlotting agent and histamine.**
4. **Lymphocytes are active in immune system activity.**

D. **Platelets, or thrombocytes, are cell fragments that have broken off larger cells in the bone marrow. They plug small leaks in the circulatory system.**

---

tpq 3. In an attempt to gain a competitive advantage, some bicycle racers inject themselves with erythropoetin. How might this help them? How might it hurt them?

(p 450) **III. Circulatory systems include a transport fluid and a pump.**

(msg 3) **A. In open circulatory systems, hemolymph is**
(tr 96) **contained in vessels only part of the time.**

1. The fluid sloshes freely around and through the tissues.

(tpq 4) **2. Open circulation is not efficient.**

B. In closed circulatory systems, blood is completely contained in a system of vessels.

1. Blood can be shunted to specific areas where it is needed.

2. Blood pressure can be high.

3. Closed systems are more efficient than open systems.

(msg 4) **C. In vertebrates, oxygenated blood is separated from deoxygenated blood.**

(tr 97) **1. In fishes, the heart has two chambers and blood flows in one loop.**

a. The atrium receives blood from the veins.

b. The ventricle receives blood from the atrium and pumps it to the gills and the rest of the body.

---

tpq 4. In a science fiction movie, a mad scientist mutates a spider so that it grows as big as a truck. Would a spider's open circulatory system allow such great size?

---

c. After blood passes through the gills, pressure drops. This system could not maintain a high enough pressure to meet the needs of land vertebrates.

2. Land vertebrates have a two-loop system.
   a. Blood is first pumped to the lungs, then back to the heart, where it is pumped a second time.
   b. Amphibians and reptiles have two atria and a single ventricle.
      i. One atrium receives blood from the lungs and the other, from the body.
      ii. Mixing of oxygenated and oxygen-depleted blood is minimized by the position of the vessels leading out of the ventricle.
      iii. In some reptiles, a membranous septum divides the ventricle into two chambers.

(tr 98)

   c. In birds and mammals, this septum enlarged to divide the ventricle into two separate chambers.
      i. Pulmonary circulation transports deoxygenated blood through the right side of the heart to the lungs and then to the left side of the heart.

ii. Systemic circulation moves oxygenated blood through the left side of the heart and to the body, from whence it returns to the right side of the heart.

(p 452) IV. Circulation in mammals is in a double loop.

A. Blood moves through a series of vessels.

(msg 6) 1. Arteries have thick, multilayered, muscular walls that maintain blood pressure.

2. Arteries branch into arterioles, which branch again into capillaries.

(tpq 5) a. Red blood cells can pass through capillaries only in single file.

(msg 5) b. Capillary walls are only one cell thick, allowing easy diffusion inward and outward.

c. Capillary beds link arterial and venous blood vessels.

3. Capillaries converge into venules, which merge to become veins.

a. Venous pressure is low.

b. Valves that extend into the lumen prevent backflow.

B. The heart consists of four chambers.

---

tpq 5. Why are capillaries so small and thin-walled?

---

1. The right atrium receives deoxygenated blood from the superior vena cava and inferior vena cava and pumps it into the right ventricle via an atrioventricular valve.
2. The right ventricle pumps blood past the pulmonary semilunar valve and through the pulmonary artery to the lungs.
3. From the lungs, blood flows to the left atrium, which pumps it into the left ventricle.
4. The left ventricle pumps blood through the aortic semilunar valve and into the aorta, which leads to the rest of the body.

(msg 7) C. The heart is made of specialized muscle cells.

(tr 99)
1. Unlike other muscles, the healthy heart is always pumping at its maximum strength.
2. The cells are linked end to end by intercalated discs through which electrical impulses travel to stimulate contraction.

(tpq 6) D. Cardiac muscle contracts automatically, without stimulation from the nervous system.

1. Pacemaker cells touch off the contraction of the entire heart.
2. Pacemaker cells are located in the sinoatrial (SA) node.

---

tpq 6. What is the natural selection advantage of having a heartbeat that doesn't depend on the central nervous system?

---

3. The signal from the SA node causes the atria to contract in unison.
4. The atrioventricular (AV) node delays the signal before it stimulates the contraction of the ventricles to allow blood to pass from the atria to the ventricles.
5. During the systole phase, the atria or ventricles contract. The relaxed phase is diastole. Together these phases make up the cardiac cycle.

E. Blood pressure forces blood through the vessels.

1. Pressure is measured in millimeters of mercury.
2. Low pressure in the veins leads to the accumulation of blood. Veins act as a blood reservoir.
3. Muscular activity forces blood out of the veins; one-way valves prevent backflow.

F. Vasoconstriction and vasodilation shunt blood where it is needed.

(tpq 7) 1. Vasoconstriction causes the precapillary sphincter to constrict, restricting blood flow.
2. Vasodilation allows capillary beds to become engorged.

(msg 8) G. A cascade of events stops the bleeding from a damaged vessel.

---

tpq 7. People with high blood pressure sometimes take drugs that relax smooth muscles. How might this lower blood pressure?

---

1. Smooth muscles in the vessel walls contract.
2. Platelets release serotonin, which maintains muscle contraction, and stick to each other to form a plug.
3. Coagulation forms a clot. Fibrinogen is transformed from a soluble form to fibrin threads that trap red blood cells.
4. Hemophiliacs have mutant, ineffective clotting proteins.

(p 461) (msg 9) (tr 100)

V. The lymphatic system is a second circulatory system, which collects the fluid that leaks out of blood vessels and returns it to veins near the heart. It is also a component of the immune system.

A. This system is not a loop.
B. Lymph contains proteins, dead cells, and invasive microorganisms. Lymph nodes filter the lymph and harbor infection-fighting lymphocytes.
C. Muscle action pumps lymph; valves prevent backflow. Edema results whem lymph accumulates.
D. The thymus plays a role in the immune system. The spleen filters the blood, removing old red blood cells and other debris.

INSTRUCTIONAL AIDS

VIDEOS, FILMS, SLIDES

Circulatory System. Coronet Film and Video, Deerfield, IL. Animation portrays the operation of the human circulatory system. Video.

Circulatory System & Its Function. Educational Images Ltd., Elmira, NY. The basics of human circulation and a more detailed look at heart and vessel anatomy. Transparencies, filmstrip.

Dissection and Anatomy of the Heart. Nebraska Scientific, Omaha, NE. A detailed presentation with excellent close-up photography. Video.

Specialized Connective Tissue: Blood. Carolina Biological Supply Company, Burlington, NC. Twenty transparencies that present the histology of blood.

REFERENCES

Cantin, Marc, and Jacques Genest. 1986. The heart as an endocrine gland. *Scientific American* 254 (2):76–81. Discusses atrial production of a recently discovered hormone that fine-tunes blood pressure and volume.

Goldstein, Gary W., and A. L. Betz. 1986. The blood-brain barrier. *Scientific American* 255 (3):74–83. Compares brain capillaries with other vessels and discusses their special properties.

Gunstream, Stanley E. 1986. *Human Biology: Laboratory Explorations*. Minneapolis, MN: Burgess. Twenty-six lab exercises on human anatomy, physiology, and other topics.

Moffat, Anne Simon. 1991. Three li'l pigs and the hunt for blood substitutes. *Science* 253:32–34. An analysis of the use of bioengineered pigs that produce human hemoglobin.

KEY TERMS

aorta	page 452	lymphocyte	449
arteriole	452	monocyte	449
artery	452	neutrophil	449
atrioventricular (AV) node	455	pacemaker cell	455
atrium	451	plasma	448
basophil	449	platelet	448
capillary	452	precapillary sphincter	458
capillary bed	453	pulmonary artery	454
cardiac cycle	456	pulmonary circulation	452
circulatory system	446	sinoatrial (SA) node	455
coagulation	459	spleen	461
diastole	456	superior vena cava	454
eosinophil	449	systemic circulation	452
erythrocyte	448	systole	456
fibrin	459	thrombocyte	450
fibrinogen	459	thymus	461
hemolymph	450	valve	454
inferior vena cava	454	vasoconstriction	458
leukocyte	449	vasodilation	458
lumen	454	vein	453
lymph	461	ventricle	451
lymph node	461	venule	453
lymphatic system	461		

FUN FACTS

ARTIFICIAL ERYTHROCYTES

Because blood transfusions can transmit diseases such as AIDS and hepatitis, and because blood of the correct type is not always available, scientists are searching for blood substitutes. Research has followed two paths – one involves the formulation of oxygen-carrying chemicals; the other includes modification of natural hemoglobin.

In the 1960s, a researcher experimented with silicon oil as a blood substitute. He filled a beaker with the oil, bubbled oxygen through it, and dropped in a lab rat. The rat survived for several minutes because the silicon oil held the same amount of oxygen as air does, about 20 percent. Unfortunately, the rat died from lung inflammation a few days later.

Other research focused on the use of hemoglobin, either free in the bloodstream or encapsulated in artificial "cells." Microcapsule technology, used to package odors in scratch-and-sniff ads, has been used to package hemoglobin, but the body quickly attacks and breaks down the fake cells. Free hemoglobin molecules introduced into the circulatory system do not work well. They are so small that they are easily lost from the system. Without certain accessory proteins, they bind oxygen too tightly and won't surrender it to tissues where it's needed. Bacterial toxins stick to hemoglobin and can cause tissue damage, and the problem of contamination with pathogens accompanies any extraction of hemoglobin from humans.

See: Weiss, Rick. 1987. Sanguine substitutes. *Science News* 132:200 – 202.

DEEP DIVERS

While most humans can stay underwater for no more than three minutes, diving mammals like Weddell seals can remain submerged for as much as an hour. Their circulatory system adaptations include an enormous amount of blood, fine control over circulation, and a spleen that acts like a scuba tank.

Humans keep about a third of their oxygen supply in the lungs, half in the blood, and the remainder in muscles. Seals store only 5 percent in the lungs, 70 percent in the blood, and 25 percent in the muscles. Blood constitutes 14 percent of a seal's body weight, compared with 7 percent in humans. Red blood cells make up 60 percent of the volume of a seal's blood, compared with 35 to 40 percent in humans. Like other mammals, including humans, a seal's circulation patterns change when it submerges (the "diving reflex"). Its heart rate slows and vasoconstriction restricts the flow of blood to many muscles and organs, with normal flow supplied to the retina, brain, and spinal cord. The restriction of blood flow to the muscles not only conserves oxygen, it reduces the buildup of lactic acid and carbonic acid in the blood, which could alter the blood pH drastically. The spleen stores oxygenated red blood cells and releases them when a seal dives, increasing the red-blood-cell-content of the blood by 50 percent. Estimates indicate that seals store 60 percent of their red blood cells in the spleen, compared with 10 percent in humans.

See: Zapol, Warren M. 1987. Diving adaptations of the Weddell seal. *Scientific American* 256 (6):100 – 105.

CHAPTER 21 CIRCULATION: TRANSPORTING GASES AND MATERIALS

MESSAGES

1. The circulatory system moves body fluids, maintains homeostasis, delivers materials to each body cell, and removes metabolic wastes.

2. Blood plasma contains dissolved proteins and nutrients and carries red blood cells, white blood cells, and platelets.

3. Open circulatory systems, which are relatively inefficient, occur in arthropods and mollusks. Closed circulatory systems allow greater pressure and efficiency.

4. The evolution of vertebrate hearts has led from a two-chambered heart to four chambers, reducing the mixing of oxygenated and deoxygenated blood.

5. Capillaries are thin-walled sites of nutrient and gas exchange.

6. Arteries carry blood away from the heart. They are thick-walled, elastic, and pressurized. Veins carry blood back to the heart. They are low-pressure blood reservoirs.

7. Cardiac muscle fibers generate and transmit impulses that stimulate heart contraction.

8. Clotting is a multi-step process that stops the flow of blood from a wound.

9. The lymphatic system drains the body of extracellular fluid and is a component of the immune system.

22

THE IMMUNE SYSTEM AND THE BODY'S DEFENSES

PERSPECTIVE

This chapter presents the immune system, the vertebrate body's main defense against pathogens and cancers. While immunity is the product of a distinct set of organs, cells, and molecules, it relies on the circulation of blood and lymph, detailed in Chapter 21, and is partially controlled by the body's communications systems, presented in Chapters 26 through 28.

The themes and concepts of this chapter include:

- The immune system recognizes an invader, communicates between cells regarding the invader, and eliminates the invader.
- The specific immune response includes recognition of an antigen, selective proliferation of lymphocytes, and differentiation of some lymphocytes into plasma cells that produce antibodies.
- Antibodies include variable regions that bind to specific antigens and constant regions that mark the antigens for destruction.
- Gene shuffling leads to the millions of different antibodies. Each B cell is studded with one kind of antibody.
- Humoral immune responses are carried out by B cells. T cells carry out cell-mediated responses.
- Killer T cells eliminate foreign cells. Helper T cells and suppressor T cells regulate the action of killer T cells. HIV attacks helper T cells.
- Balanced regulation results in immunity to foreign antigens and tolerance of one's own cells.
- Passive immunization provides fast, but short-lived, protection. Vaccination is a slower, but longer-lasting, process that stimulates the production of memory cells.

CHAPTER OBJECTIVES

Students who master the material in this chapter will gain an overview of the immune system and will understand how this complicated system protects them against disease. They will be able to distinguish between nonspecific and specific responses and between active and passive immunity. They will understand how antibodies are produced, how they act, and how clonal selection of B cells results from exposure to specific antigens. They will see the roles of helper and killer T cells and how AIDS damages the immune system.

LECTURE OUTLINE

(p 464) **I. AIDS, the new plague, introduces the <u>immune system</u>.**

(tpq 1) (msg 1) **A. <u>Immunity</u> is the work of white blood cells and associated organs.**

1. White blood cells attack foreign cells and molecules directly.

2. White blood cells produce antibodies that mark intruders for destruction.

B. The cells and molecules of the immune system recognize foreign invaders.

C. Cells of the immune system communicate with one another.

D. Elements of the immune system selectively eliminate foreign invaders.

---

tpq 1. The AIDS virus takes years to kill its host. Other diseases kill rapidly. How has delayed lethality enabled the AIDS virus to be highly successful?

(p 466) **II. Immune system components may be specific or nonspecific.**

A. Nonspecific components are the first line of defense.

1. **The skin and mucous membranes are barriers to invasion.**
2. **At the site of injury, inflammation occurs. The temperature rises, blood flow increases, and phagocytes aggregate.**
3. **Natural killer cells recognize bacteria, certain cancer cells, and virus-infected cells, and kill them.**

(msg 2) **B. The specific immune response is the body's most powerful defense.**

(tr 101) 1. **Several organs and types of cells are involved.**

   a. **Lymphocytes include two main types of cells.**

      i. **B cells (B lymphocytes) produce antibodies.**

      ii. **T cells (T lymphocytes) kill foreign cells directly.**

   b. **Macrophages consume debris and stimulate lymphocytes to attack invaders.**

   c. **Lymphoid organs include the spleen, thymus, and lymph nodes.**

(msg 3) 2. **Antibodies are made by B cells. The binding site of an antibody attaches to a specific antigen.**

3. **The initiation of the specific response is a multistep process.**
    a. **Debris binds to antibodies on the membranes of circulating B cells.**
    b. **B cells carrying antibody-antigen complexes move to the nearest lymph node and interact with helper T cells.**
    c. **The helper T cells secrete interleukins, stimulating the division of the B cells, which produce antibodies against the specific antigen.**
    d. **Antibodies circulate in the lymph and bloodstream and attach to the bacterial antigen.**
    e. **Tagged bacteria can be devoured by macrophages or killed by complement proteins.**

(p 469) **III. Antibodies are immunoglobins that are directed against specific targets.**

(tpq 2) **A. Newborn infants do not have a fully functioning immune system and must rely on antibodies they get from their mothers.**

B. Each antibody consists of two identical heavy amino acid chains and two identical light chains, all arranged in a Y shape.

---

tpq 2. What benefits do infants get from nursing instead of bottle-feeding?

---

1. The tips of the two branches bind to antigens.
    a. The amino acid sequence is variable.
    b. Engineered abzymes may allow the creation of specific catalysts.
(tpq 3) 2. The stem of the Y stimulates the attack by macrophages or complement. It has a relatively constant sequence.

C. Variations in the amino acid sequence allow great diversity in binding sites.
    1. More than 100 million distinct antibodies may be found in one human.
    2. Tests for antibodies against the AIDS virus reveal whether a person has been exposed.

D. Different classes of antibodies exist, each with a distinct stem.
    1. IgG, or immunoglobin G, attaches to viruses and bacteria.
        a. It is the only class of antibodies that can cross the placenta.
        b. Gamma globulin is rich in IgG.
    2. IgA antibodies can make their way into bodily secretions, such as milk.

---

tpq 3. Would antibodies work as well if they were I-shaped instead of Y-shaped?

---

3. IgE binds to specialized cells in the skin and intestinal lining, fighting parasites and causing allergies.

(p 471) IV. B cells mobilize the immune response and retain a memory of antigens.

(msg 4) A. Each B cell carries on its membrane 100,000 copies of the one antibody it will secrete when stimulated.

(tr 102) B. When an antigen enters the body, it selects from the entire population of B cells by binding only to those few that carry the best-fitting antibodies.

1. The binding of the antigen stimulates the B cell to divide rapidly into a clone of cells, the clonal selection mechanism.
2. Some of the clonal cells differentiate into plasma cells, which produce circulating antibodies.
3. Others become memory cells that can divide and redivide when stimulated by the antigen.

(tpq 4) a. This allows rapid response (the secondary response) in subsequent exposures to the antigen.

---

tpq 4. What is the purpose of the booster shots that follow some vaccinations?

---

b. This <u>immunological memory</u> provides immunity to diseases to which one has previously been exposed.

(msg 5) C. Antibody genes in three or four gene groups are shuffled to produce the many different antibodies.

1. <u>Humoral immune response</u> describes the synthesis and secretion of antibodies.
2. Antibodies cannot act on microbes that have entered cells.

(p 474) V. <u>Cell-mediated immune response</u> is the action of T cells directly against invaders.

(msg 6) A. T cells mature in the thymus. They can kill infected, cancerous, or transplanted body cells.

B. T-cell receptors can bind antigen only when a macrophage presents antigen to them along with <u>histocompatibility proteins</u>, which distinguish self from nonself.

(msg 7) C. Most effector T cells are cytotoxic; they poke holes in the membranes of invaders or mutant body cells.

(tpq 5) 1. Organ donors and recipients must have similar or identical histocompatibility markers.

---

tpq 5. Why must physicians be sure that an organ recipient has the same blood type as the donor?

---

2. Tissue transplants (except between identical twins) require suppression of the T cells to avoid tissue rejection.

D. Half of the T cells function as regulators.

1. Helper T cells stimulate the activities of B cells and effector T cells.
2. Suppressor T cells prevent these activities.
3. T cells communicate with other cells by secreting molecules such as interferon and interleukin.

(msg 8) E. The AIDS virus kills a type of helper T cell,
(tr 103) reducing the body's ability to fight off disease.

(msg 9) F. Self-tolerance may be a function of macrophages.

1. Macrophages digest antigens and display parts of the antigenic molecules on their membrane along with histocompatibility proteins.
2. They do not process or present intact substances that originate in the same body they do; thus they avoid the stimulation of autoimmune diseases.
3. B cells carrying antibodies that react with self may be killed during fetal development.
4. Pregnancy is a special case of immune suppression.

(p 477) **VI. The immune system may be manipulated to induce immunity, provide passive protection, or to mass-produce antibodies.**

(msg 10) **A. Passive immunization, the injection of specific antibodies, provides fast-acting (but short-lived) protection against pathogens and toxins.**

B. Active immunity is the body's response to vaccination or other exposure to antigens.

1. Vaccines are made of microbes and toxins that have been killed or modified.

2. The first vaccination stimulates proliferation of specific B cells.

3. A booster shot induces the formation of many memory cells.

(tpq 6) **4. AIDS viruses attack the lymphocytes that promote protection, and they readily change their surface antigens, which makes vaccine development difficult.**

---

tpq 6. Would a high or low mutation rate be more advantageous to a virus?

INSTRUCTIONAL AIDS

VIDEOS, FILMS, SLIDES

Death of a Disease. Time-Life Video, Paramus, NJ. Discusses the eradication of smallpox. Film, video.

The Fight Against Microbes. Pennsylvania State University AV Services, University Park, PA. Presents the history of the fight against disease. Film.

Our Immune System. National Geographic Society, Washington, DC. Reviews the basics of the immune system, including the sequence of events that occurs following invasion by microorganisms.

Viruses: The Mysterious Enemy. Human Relations Media, Pleasantville, NY. Describes the life cycle and health effects of viruses. Filmstrip, video.

REFERENCES

Piel, Jonathan, ed. 1988. What science knows about AIDS. *Scientific American* 259 (4):40–135. A single-topic edition that summarizes current knowledge about AIDS.

Russo, A. J. 1988. Immunological assays for the classroom II–Hybridoma technology: Production of monoclonal antibodies. *American Biology Teacher* 50:48–51. A simple hybridoma assay that can be used to demonstrate the production of monoclonal antibodies.

Tonegawa, Susumu. 1985. The molecules of the immune system. *Scientific American* 253 (4):122–131. A survey of the diverse battery of proteins that recognize and fight foreign invaders.

Waldmann, Thomas A. 1991. Monoclonal antibodies in diagnosis and therapy. *Science* 252:1657–1662. Discusses advances in molecular biology and genetic engineering that enhance the effectiveness of monoclonal antibodies in medicine.

KEY TERMS

antibody	page 467	immunity	465
antigen	468	immunological memory	472
B cell (B lymphocyte)	467	macrophage	467
cell-mediated immune response	474	self-tolerance	476
histocompatibility protein	474	specific immune response	467
immune system	464	T cell (T lymphocyte)	467

FUN FACTS

SELF-DESTRUCTION AND SELF-RECOGNITION

Autoimmune disorders include multiple sclerosis, in which the immune system attacks the fatty sheath around nerves; rheumatoid arthritis, which affects joints between bones; type I diabetes mellitus, which results from immune-system destruction of the glandular cells that produce insulin; and a wide variety of other acute and minor diseases. Autoimmunity often follows exposure to a foreign antigen that closely resembles surface markers on one's own cells. Rheumatic fever, for example, is preceded by an acute infection by a bacterium known as streptococcus. Antibodies that bind to streptococcus also bind to human heart tissue, causing the immune system to attack the heart as if it were a foreign invader.

In many cases, mammals recover from autoimmune diseases. This observation led scientists to conclude that the recognition of self is an ongoing process, not one that was completed early in life, as some researchers had hypothesized. In the 1970s, immunologist Neils Kaj Jerne deduced that the recognition site of antibodies could itself be an antigen for other antibodies, which would therefore attack the initial antibodies. He predicted that this interlocking, self-recognizing network regulates the behavior of the immune system.

Although Jerne's picture of the immune system seems correct in many ways, it has not been conclusively demonstrated. Scientists have used it as a basis for constructing new vaccines that target the specific immune-system cells that produce anti-self antibodies. Preliminary results indicate that this technique may find widespread application in the treatment of autoimmune diseases.

To study multiple sclerosis (MS), immunologists injected a component of the myelin sheath from neurons into rats. The protein triggered an immune attack on neurons, causing symptoms like those of MS. The researchers found that they could induce MS symptoms in other rats by injecting them with T cells from rats immunized by the protein. They also discovered that the same T cells, when inactivated and incapable of stimulating the immune response, actually protected the rats against a subsequent challenge with active T cells or the protein. Apparently the rats were immunized against myelin-reactive T cells. These trials suggest that MS may be prevented by immunization against the body's own defenses.

See: Cohen, Irun R. 1988. The self, the world, and autoimmunity. *Scientific American* 258 (4):52–60.
Marx, Jean. 1990. Taming rogue immune reactions. *Science* 249:246–247.

CHAPTER 22 THE IMMUNE SYSTEM AND THE BODY'S DEFENSES

MESSAGES

1. The immune system recognizes an invader, communicates between cells regarding the invader, and eliminates the invader. Immunity depends on nonspecific and specific lines of defense.

2. The specific immune response includes recognition of an antigen, selective proliferation of lymphocytes, and differentiation of some lymphocytes into plasma cells that produce antibodies.

3. Antibodies include variable regions that bind to specific antigens and constant regions that mark the antigens for destruction.

4. Each B cell is studded with one kind of antibody.

5. Gene shuffling leads to the millions of different antibodies.

6. Humoral immune responses are carried out by B cells. T cells carry out cell-mediated responses.

7. Killer T cells eliminate foreign cells. Helper T cells and suppressor T cells regulate the action of killer T cells.

8. HIV, the AIDS virus, attacks helper T cells.

9. Balanced regulation results in immunity to foreign antigens and tolerance of one's own cells.

10. Passive immunization provides fast, but short-lived, protection. Vaccination is a slower, but longer-lasting, process that stimulates the production of memory cells.

23

RESPIRATION: GAS EXCHANGE IN ANIMALS

PERSPECTIVE

This chapter presents another of the systems that serve the needs of animal cells. The respiratory system is intimately linked to the circulatory system (discussed in Chapter 21) and to cellular respiration (covered in Chapters 4 and 5). Material presented in this chapter will reappear in Chapter 24, Animal Nutrition and Digestion.

The themes and concepts of this chapter include:

- Respiration involves specialized exchange surfaces that are often housed in protective structures.
- Countercurrent flow in gills maximizes the diffusion between capillaries and the environment.
- Air-breathing animals have internal respiratory surfaces that are kept moist and may have substances that carry gases throughout the body. Insects have tracheae, while vertebrates have lungs.
- In humans, breathing rate and depth are mostly involuntary and are controlled by nerves that detect the amount of carbon dioxide in the blood.
- In vertebrates, respiratory and circulatory systems are closely interdependent. Hemoglobin carries oxygen.
- Differences in the partial pressures of gases lead to their diffusion.
- Most of the carbon dioxide in the bloodstream is converted to bicarbonate ions.

CHAPTER OBJECTIVES

Students who master the material in this chapter will see that different groups of animals have evolved different means of exchanging gases with their environments. These students will understand that diffusion and bulk flow underlie all gas exchanges. They will be able to describe the structures that make up the human respiratory system and how those structures function.

LECTURE OUTLINE

(p 482) **I. The deep-diving Weddell seal introduces respiration.**

- **A. The blood and muscles of the seal act as a major oxygen reservoir.**
- (tpq 1) **B. The dive reflex shunts blood away from most organs and toward the central nervous system. The muscles switch to anaerobic respiration.**
- **C. During shorter dives, the muscles remain aerobic. The spleen releases great quantities of red blood cells.**
- **D. Two themes emerge in this chapter:**
  1. **The architecture of an animal's respiratory system is constrained by diffusion.**
  2. **An animal's activities are limited by the quantity of oxygen available.**

---

tpq 1. Do humans have the dive reflex?

(484) **II. A wide variety of structures and mechanisms have evolved to carry out gas exchange in animals.**

(msg 1)

- **A. Diffusion is the underlying mechanism and simplest form of gas exchange.**
    - **1. Oxygen and carbon dioxide enter and leave all cells by diffusion.**
        - **a. Gases must diffuse through a watery medium.**
        - **b. Diffusion is much more rapid in air than in water.**
    - **2. Most large animals rely on a circulating internal fluid to transport gases throughout the body.**

(tpq 2)

- **B. Gills extract oxygen from water.**

(tr 104)

- **1. Gills provide large surface areas through which gases can be exchanged between the environment and the bloodstream. Gills are thin-walled, allowing diffusion to and from underlying capillaries.**
- **2. External gills are exposed. Most fishes have internal gills.**
    - **a. <u>Opercular flaps</u> cover the openings.**
    - **b. The gill is divided into <u>gill filaments</u> which are composed of plate-like <u>lamellae</u>.**

---

tpq 2. Why don't whales have gills instead of lungs?

---

c. **In each lamella is a meshwork of capillaries just one cell layer away from water.**

(msg 2) d. **Water moves past the gills in the opposite direction to blood flow within the capillaries, known as countercurrent flow.**

3. **Land animals inhabit an environment with higher oxygen content but face dessication.**

(msg 3) C. **Adaptations to life in air include tracheae and lungs.**

1. **Insects and some other arthropods have tracheae linked to the outside world through spiracles.**

   a. **The tracheae branch into air capillaries.**

   b. **Oxygen dissolves in fluid droplets at the end of the air capillaries and diffuses into adjacent cells.**

(tr 105) 2. **Air-breathing vertebrates have lungs connected to the outside by hollow tubes.**

(tpq 3) a. **Some fishes that live in oxygen-depleted water have lungs. In many fishes, the lung pouch serves as a swim bladder.**

   b. **Lung walls are lined with a thin, moist membrane through which gases diffuse.**

---

tpq 3. Were freshwater or oceanic fishes more likely to be the ancestors of land vertebrates?

---

c. **In most vertebrates, the lungs are blind-ended and the air flows in and out via the same path, in a process called <u>tidal ventilation</u>.**

d. **In birds, air sacs form an interconnected system with the lungs that enables air to flow in only one direction, thus sustaining a high gas exchange rate.**

(p 487) **III. All mammals share a basic set of respiratory structures and mechanisms.**

(tr 106) **A. Respiratory plumbing includes structures that condition and distribute air.**

(tpq 4) **1. The mouth cavity or <u>nasal cavities</u> warm and humidify the air.**

2. The pharynx branches into the <u>esophagus</u> and the <u>trachea</u>.

a. **The <u>larynx</u>, housing the vocal cords, lies at the anterior end of the trachea.**

b. **The <u>epiglottis</u>, at the opening to the larynx, acts as a valve.**

c. **The trachea branches into <u>bronchi</u>, which branch into <u>bronchioles</u>.**

d. **Gas exchange takes place in <u>alveoli</u>.**

---

tpq 4. Why is it better for you to breathe through your nose than through your mouth?

---

3. **Airway cells secrete mucus, which traps debris. Cilia sweep the mucus toward the pharynx.**

   a. **Cystic fibrosis is the most common lethal genetic disease of Caucasians and is characterized by heavy mucus production.**

   b. **Tobacco smoke damages cells lining the alveoli and paralyzes cilia.**

B. **Ventilation in mammals is accomplished by the actions of the diaphragm and rib cage.**

1. **The lungs are enclosed in fluid-filled pleural sacs and suspended in the thoracic cavity.**

(tpq 5) 2. **Expansion of the thoracic cavity by contraction of the intercostal muscles or the diaphragm lowers the pressure in the cavity.**

   a. **Ambient air pressure forces air into the lungs.**

   b. **Exhalation is the reverse.**

(msg 4) C. **Respiratory centers in the medulla and elsewhere in the brain control when and how we breathe. The brain is better at sensing a rise in blood carbon dioxide than it is at sensing a decrease in blood oxygen.**

(p 490) **IV. Gas exchange must continue uninterrupted.**

(msg 6) A. **Diffusion is based on gas partial pressure.**

---

tpq 5. How does an "iron lung" work?

---

1. The partial pressure of a gas equals its concentration multiplied by the total pressure of the gas mixture.
2. Gases diffuse from areas of higher partial pressure to areas of lower partial pressure.

(msg 5) B. Hemoglobin transports oxygen.

1. A hemoglobin molecule contains four iron-containing heme groups that hold oxygen molecules.
2. Oxyhemoglobin is bright red; deoxyhemoglobin is bluish.
3. Blood can carry 70 times the oxygen that an equivalent volume of water could absorb.

C. Carbon dioxide may be transported in several forms.

(msg 7) 1. Carbon dioxide may be dissolved in the plasma, combined with deoxyhemoglobin, or, largely, combined with water to form carbonic acid, which dissociates to carbonate ions.

2. Buildup of carbonic acid and lactic acid in hard-working muscles leads to easier release of oxygen from hemoglobin, a phenomenon known as the Bohr effect.

INSTRUCTIONAL AIDS

VIDEOS, FILMS, SLIDES

Breath of Life. Films for the Humanities and Sciences, Princeton, NJ. Explains why the body needs air and how oxygen is distributed. Video.

Partners for Life: The Human Heart and Lungs. Human Relations Media, Pleasantville, NY. Examines the natural engineering that makes the heart and lung function well; emphasizes health maintenance. Filmstrip, video.

The Respiratory System and Its Function. Connecticut Valley Biological Suppply, Southampton, MA. Presents the respiratory tract, blood oxygenation, and the release of carbon dioxide. Transparencies, filmstrip.

REFERENCES

Eckert, R., and D. Randall. 1983. *Animal Physiology*, Second Edition. New York: W. H. Freeman. Includes an excellent discussion of respiration and circulation.

Rahn, Hermann, et al. 1979. How bird eggs breathe. *Scientific American* 240 (2):46–55. Discusses how bird embryos exchange gases by diffusion through pores in the eggshell.

West, J. 1985. *Respiratory Physiology: The Essentials*, Third Edition. Baltimore: Williams & Wilkins. A well-written, brief introduction to the physiology of respiration.

KEY TERMS

air capillary	page 486
alveolus	487
Bohr effect	492
bronchiole	487
bronchus	487
countercurrent flow	485
diaphragm	489
epiglottis	487
esophagus	487
gill filament	485
lamella	485
larynx	487
nasal cavity	487
opercular flap	484
partial pressure	491
pleural sac	489
respiration	483
spiracle	486
thoracic cavity	489
tidal ventilation	487
trachea	487
ventilation	489

FUN FACTS

LUNGS, SKIN, AND GAS EXCHANGE

Many amphibious and terrestrial vertebrates don't rely on lung ventilation for gas exchange. In fact, some salamanders don't even have lungs. Instead, respiration occurs through their skin.

The functional similarities between lungs and skin are great. Wherever a membrane separates the environment from cells or blood, gases can pass through. The thinner the membrane, the greater its blood supply, and the larger its surface area, the faster gases will diffuse. The alveoli of lungs are thin-walled, richly endowed with capillaries, and have a large surface area. Skin-breathers show similar adaptations that maximize gas exchange.

The male hairy frog, *Astylosternus robustus,* seems to have fur on its hindquarters. A closer examination reveals that the "hair" is actually tiny skin outgrowths that increase the skin's surface area to maximize gas exchange. Another frog, *Telmatobius culeus,* looks as if it is wearing a suit designed for a much larger animal; its skin hangs down in extensive folds. Some male newts develop enlarged tail fins and dorsal crests during the mating season. These structures may increase the newt's sexual attractiveness but they also serve to increase the skin area for respiration.

Blood flow in skin capillaries of skin-breathing amphibians varies as the need for oxygen changes. When a frog is submerged in water with little oxygen, blood flow to the skin increases. In air, its skin capillaries constrict, reducing the blood flow because oxygen is more available.

Skin breathing allows amphibians to survive for long periods buried in mud, permits turtles to survive in ice-covered lakes, and enables lungless salamanders to thrive in the canopy of tropical forests. Skin breathing occurs in mammals, but it can't provide the great volumes of oxygen necessary for mammalian metabolism. We must respire lest we expire.

See: Feder, Martin E., and W. W. Burggren. 1985. Skin breathing in vertebrates. *Scientific American* 253 (5):126 – 140.

HUMAN RESPIRATION

At rest, a typical adult human inhales and exhales 10 to 14 times a minute, moving 5 to 7 liters of air per minute. During heavy exertion, the respiration rate can jump to 60 per minute; the tidal volume may be 90 liters in a minute. During a normal day, an adult breathes about 15,000 liters of air. During a lifetime, a human may breathe 368 million liters, or twice the capacity of the dirigible *Hindenburg.*

Because air enters and exits from the lungs through the same opening, fresh air and stale air mix. Only one-sixth of the air in the lungs is evacuated with each breath, leading to the speculation that some of the molecules of air drawn in during the first breath after birth may remain there until death. In compensation for this poor ventilation system, the surface area of the alveoli is huge. There are hundreds of millions of alveoli in an adult's lungs; together, their area adds up to 90 square meters, some 40 to 50 times greater than the surface area of the skin.

See: Smith, Anthony. 1986. *The Body*. New York: Viking Penguin, Inc.

CHAPTER 23 RESPIRATION: GAS EXCHANGE IN ANIMALS

MESSAGES

1. Respiration involves specialized exchange surfaces that are often housed in protective structures.

2. Countercurrent flow in gills maximizes the diffusion between capillaries and the environment.

3. Air-breathing animals have internal respiratory surfaces that are kept moist; they may have substances that carry gases throughout the body. Insects have tracheae, while vertebrates have lungs.

4. In humans, breathing rate and depth are mostly involuntary and are controlled by nerves that detect the amount of carbon dioxide in the blood.

5. In vertebrates, respiratory and circulatory systems are closely interdependent. Hemoglobin carries oxygen.

6. Differences in the partial pressures of gases lead to their diffusion.

7. Most of the carbon dioxide in the bloodstream is converted to bicarbonate ions.

24

ANIMAL NUTRITION AND DIGESTION: ENERGY AND MATERIALS FOR EVERY CELL

PERSPECTIVE

This chapter reveals how animals break food down into particles that can be absorbed and utilized by cells. It builds on concepts presented in Chapter 5, which discusses how cells harvest energy from nutrients, and in Chapter 21, Circulation. Dietary needs and adaptations discussed in this chapter will be relevant to Part IV, Interactions: Organisms and Environment.

The themes and concepts of this chapter include:

- Animals must take in sufficient amounts of specific nutrients or they develop diseases and die.
- An animal's anatomy and physiology are closely tied to its diet and nutritional needs.
- The four basic steps of digestion include the grinding of large food particles, the breakdown of macromolecules, the absorption and transport of materials, and the elimination of waste.
- Essential amino acids must be eaten in the right combinations and ratios. Vitamins and minerals are required in small quantities.
- Peristalsis moves food through the digestive system. Digestive activities are controlled and coordinated by feedback nervous and hormonal systems.
- The stomach stores and breaks down food and regulates the flow of chyme into the small intestine. The liver produces bile salts and detoxifies the blood.
- The lining of the small intestine has a huge surface area as a result of folds and extensions. The colon absorbs water and processes waste for elimination.

CHAPTER OBJECTIVES

Students who master the material in this chapter will understand how food is broken down and absorbed by the body during digestion. They will see the necessity of a balanced dietary intake of essential amino acids, vitamins, and minerals. They will be able to describe the activities that occur in the various digestive organs and how these activities are integrated and controlled.

LECTURE OUTLINE

(p 494) **I. The eucalyptus-eating koala introduces nutrition and digestion.**

(tpq 1) **A. The koala's digestive tract has evolved to digest toxic, low-nutrient eucalyptus leaves.**

B. Three themes emerge.

(msg 1) **1. Animals must take in sufficient amounts of energy and materials or they develop diseases and die.**

(msg 2) **2. An animal's anatomy and physiology are tied closely to its diet and its nutrient needs.**

3. In most animals, digestion occurs in steps.

(p 496) **II. Nutrients supply energy and materials to sustain life.**

A. The science of nutrition is concerned with the amounts of nutrients and calories an animal must consume to stay healthy.

B. Carbohydrates contribute carbon and energy.

tpq 1. What are the strangest animal diets you know of? What diets are the hardest to utilize?

1. Sugars, starches, polysaccharides, and cellulose are common carbohydrates.
2. Glucose is the main supplier of energy for glycolysis and aerobic respiration. Cells in the brain and nerves are particularly sensitive to fluctuations in blood glucose levels.
3. (tpq 2) Humans cannot digest cellulose. Some herbivores have cellulose-digesting microbes in their guts.

C. Lipids are highly compact energy-storage nutrients.
1. Most of the lipids in the human body are triglycerides; others are phospholipids or sterols.
2. Because carbohydrates bind to water, storing energy in the form of carbohydrates instead of lipids would take more space and add more weight.
3. Fats also cushion internal organs, provide insulation, and store fat-soluble vitamins.
4. Linoleic acid is an essential fatty acid.

D. Proteins are basic to the structure and function of cells.
1. Proteins are constantly built and broken down, requiring a continuous supply of protein in food.

---

tpq 2. Does cellulose have any value in the human diet?

(msg 3)

2. The body cannot manufacture the eight <u>essential amino acids</u>: lysine, leucine, phenylalanine, isoleucine, tryptophan, valine, threonine, and methionine.
   a. They must be included in the same meal because the does not store free amino acids.
   b. If an amino acid is missing, some proteins cannot be synthesized.

(tpq 3)

   c. Vegetarians must be especially careful to balance their dietary amino acids.

E. Vitamins and minerals are absolutely necessary but only in small amounts.
   1. <u>Vitamins</u> are organic compounds needed in small amounts but which higher animals cannot manufacture.
      a. They often serve as coenzymes or activators.
      b. High levels of fat-soluble enzymes may be toxic; excess water-soluble vitamins (B and C) are rapidly excreted in the urine.
   2. <u>Minerals</u> are inorganic elements that contribute to structure, as in bone, and to function, as in hemoglobin.

---

tpq 3. What health advantages or disadvantages might there be to a vegetarian diet?

---

3. Most people in affluent countries get an adequate supply of vitamins and minerals from a balanced diet. Deficiencies can lead to serious consequences.
4. Many teenagers and adults in this country are overweight or obese.
5. Anorexia nervosa and bulimia are eating disorders.

(tpq 4) F. Calories indicate the energy value in food. Regulation of caloric intake is necessary for weight control.

(p 504) III. Digestion breaks food down into units that cells can use.

A. Intracellular digestion circumvents the need for a gut but limits the animal's size and structure.

B. Extracellular digestion breaks food down into small particles and constituent molecules.

1. In simple organisms with only one gut opening, food particles cannot be broken apart mechanically and areas of the gut cannot become specialized.

---

tpq 4. Are diets effective at controlling weight? Are fad diets safe?

2. In animals with one-way flow in the gut, mechanical food-processing organs such as teeth and gizzards grind food. Areas of the gut can become specialized.

C. The alimentary canal in vertebrates includes the mouth, esophagus, stomach, small intestine, and large intestine.

1. Accessory organs produce enzymes, bile, and other material that aids digestion.

(msg 4)

2. In the first step of the digestive process, the food is ground into smaller particles, increasing its surface area.

   a. In the second step, enzymes break macromolecules into smaller molecules.

   b. The third step is the absorption of monomers into the lymphatic system or bloodstream.

   c. The last step is the elimination of waste.

(msg 5)

3. The layers of the gut wall include the mucosa, covered with mucus; the submucosa, richly supplied with blood vessels and nerves; a muscular layer that moves food by peristalsis; and the serosa, which attaches the gut to the inner wall of the body cavity.

(tpq 5) **4. Evolution has resulted in a number of variations in digestive systems that allow animals to exploit very different foods. Many herbivores, such as the koala, have a cecum that ferments cellulose.**

(p 507) (tr 107) **IV. The human digestive system processes all the food a person eats.**

(msg 4) **A. The teeth cut and grind the food and form a <u>bolus</u> that can be easily swallowed.**

1. The <u>tongue</u> moves the food into the teeth and is a main organ of taste and speech.

2. Each bite is mixed with <u>saliva</u>, which is secreted by three pairs of <u>salivary glands</u>. Saliva includes amylase.

3. When food is swallowed, the epiglottis and <u>soft palate</u> rise to prevent food from entering the nasal cavities and trachea.

4. Muscle contractions in the throat push food into the <u>esophagus</u>. The cardiac <u>sphincter</u> keeps the stomach contents from rising back into the esophagus.

(msg 6) **B. The <u>stomach</u> stores food and starts the chemical breakdown.**

---

tpq 5. How are the digestive systems of carnivores, herbivores, and detritivores likely to differ?

---

1. Peristalsis mixes the stomach's contents with an acid bath of gastric juices secreted by glands in the stomach wall.
2. Gastric juices are a mixture of water, hydrochloric acid, mucus, and pepsinogen, a precursor to the protein-cleaving enzyme pepsin.
3. The resulting chyme moves through the pyloric sphincter into the small intestine in carefully regulated doses.

C. Most of the chemical digestion of food and some absorption takes place in the small intestine.
   1. The pancreas is situated near the stomach.
      a. It produces digestive enzymes, including proteases, lipases, and amylase, and secretes them into the small intestine.
      b. It produces bicarbonate ions that neutralize the stomach acids entering the intestine.
      c. Some pancreatic cells secrete insulin and glucagon.

(msg 7)

   2. The liver is a multipurpose organ.
      a. It produces bile salts, modified cholesterol molecules that act like detergents.
      b. Bile salts are stored as bile in the gall bladder.

c. Bile salts emulsify fat in chyme and aid in its absorption across the lining of the small intestine and into the lymph.

d. The liver stores compounds, including amino acids, glucose, glycogen, and vitamins, and detoxifies poisons.

e. Cirrhosis is irreversible scarring of the liver.

(msg 8) (tr 108) 3. The lining of the small intestine is pleated into folds covered by villi, which are in turn covered by microvilli, structures that maximize surface area.

4. The upper section of the small intestine is the duodenum, the central section is the jejunum, and the remainder is the ileum.

(msg 9) D. The colon absorbs water, ions, and vitamins.

1. It stores the feces until they are excreted through the rectum and out the anus.

2. It houses bacteria that produce vitamins.

(p 511) V. Digestive activities must be coordinated.

A. Nerves and hormones regulate the speed of food passage through the gut.

B. Food in the stomach lowers the overall acidity, triggering the production of the hormone gastrin, which stimulates the production of more hydrochloric acid.

1. **A negative feedback loop regulates stomach acidity. Gastrin production is suppressed by acidity.**
2. **Ulcers are sores in the mucosa resulting from decreased secretion of mucus or excessive gastric secretions.**

C. **Stomach acid entering the small intestine causes the secretion of secretin, a hormone that causes the pancreas to secrete bicarbonate.**

D. **Partially digested proteins cause the small intestine to release cholecystokinin, which stimulates the release of digestive enzymes and signals "full" to the brain.**

INSTRUCTIONAL AIDS

VIDEOS, FILMS, SLIDES

Digestion: A Tough Dirty Job; It Takes a Lot of Guts. University of Wisconsin, Milwaukee. A videotaped lecture that presents an imaginative scenario, with visual aids and props.

Digestive System. Coronet Film & Video, Deerfield, IL. Animation and X-ray photography demonstrate the path and chemistry of digestion.

The Digestive System and Its Function. Connecticut Valley Biological Supply, Southampton, MA. A tour of the anatomy and processes of the human alimentary tract. Transparencies, filmstrip.

Eating to Live. Films for the Humanities and Sciences, Princeton, NJ. A look at appetite, hunger, salivation, swallowing, and digestion. Video.

REFERENCES

Cohen, Leonard A. 1987. Diet and cancer. *Scientific American* 257 (5):42–48. Surveys the studies and experiments that led to dietary recommendations against carcinogens.

Frye, B. L., and R. L. Neill. 1987. A laboratory exercise in human nutrition. *American Biology Teacher* 49:370–373. Relates caloric intake, weight gain, and activity levels.

Gershoff, Stanley. 1990. *The Tufts University Guide to Total Nutrition*. Scranton, PA: Harper and Row. A compilation of up-to-date nutrition information, covering topics as diverse as deciphering food labels and dealing with food allergies.

Russo, Salvatore F., and E. T. Wahl. 1987. Autocatalytic activation of trypsinogen by trypsin. *Journal of Chemical Education* 64:83–84. A lab exercise that demonstrates zymogen activation.

KEY TERMS

accessory organ	page 505	nutrient	495
alimentary canal	505	nutrition	496
bile	509	obese	501
bolus	508	overweight	501
cholecystokinin	513	pancreas	509
chyme	509	pepsin	508
colon	511	pepsinogen	508
digestion	495	peristalsis	506
esophagus	508	rectum	511
essential amino acid	498	saliva	508
essential fatty acid	497	salivary gland	508
extracellular digestion	504	secretin	513
feces	511	serosa	506
gallbladder	509	soft palate	508
gastrin	513	sphincter	508
intracellular digestion	504	stomach	508
liver	509	submucosa	506
microvillus	510	tongue	508
mineral	500	villus	510
mucosa	506	vitamin	498

FUN FACTS

TERMITE DIGESTION

Termites have strong jaws that can tear wood fibers apart, but they lack the enzymes necessary to digest the long-chain molecules that constitute wood. A termite relies on a complex community of microbes in its gut to break down macromolecules and to produce the vitamins and proteins the wood lacks. There can be dozens of species of bacteria and protists in a termite's gut.

The relationship between the microbes and the termite is mutualistic; neither can live without the other. To avoid being expelled by defecation, microbes may attach firmly to the gut walls, swim upstream, or burrow into their host's cells. Termites lose their gut microbes when they molt and shed the gut lining. A newly-molted termite will eat the feces of a nestmate and reinoculate its gut.

See: Margulis, Lynn, et al. 1986. Microbial communities. *BioScience* 36 (3):160–170.

A NEWLY DISCOVERED ROLE FOR VITAMIN A

The importance of vitamin A in vision has been known for decades, but scientists have recently discovered that it plays essential roles in fetal development and in the regulation of cell proliferation and differentiation throughout life. The vitamin A–derivative retinoic acid is a morphogen ("shape producer") that affects cell and tissue specialization in developing embryos. It binds to receptor proteins in cell nuclei; the complex becomes attached to specific locations on DNA to regulate gene expression. Different receptor proteins appear in different tissues. Concentration gradients of retinoic acid determine, for example, whether limb tissue will become forearm, hand bones, or fingers.

See: Hoffman, Michelle. 1990. The embryo takes its vitamins. *Science* 250:372–373.

CHAPTER 24 ANIMAL NUTRITION AND DIGESTION: ENERGY AND MATERIALS FOR EVERY CELL

MESSAGES

1. Animals must take in sufficient amounts of specific nutrients or they develop diseases and die.

2. An animal's anatomy and physiology are closely tied to its diet and nutritional needs.

3. Essential amino acids must be eaten in the right combinations and ratios. Vitamins and minerals are required in small quantities.

4. The four basic steps of digestion include the grinding of large food particles, the breakdown of macromolecules, the absorption and transport of materials, and the elimination of waste.

5. Peristalsis moves food through the digestive system. Digestive activities are controlled and coordinated by feedback nervous and hormonal systems.

6. The stomach stores and breaks down food and regulates the flow of chyme into the small intestine.

7. The liver produces bile salts and detoxifies the blood.

8. The lining of the small intestine has a huge surface area as a result of folds and extensions.

9. The colon absorbs water and processes waste for elimination.

25

EXCRETION AND THE BALANCING OF WATER AND SALT

PERSPECTIVE

This chapter discusses the vertebrate kidney, the organ responsible for osmoregulation and excretion. The material in this chapter builds on concepts of diffusion and osmosis introduced in Chapter 2, Atoms, Molecules, and Life, and Chapter 4, The Dynamic Cell. This chapter continues the discussion of animal physiology and homeostasis (Chapter 20) and circulation (Chapter 21). Osmoregulation and excretion in invertebrates (Chapter 18) are addressed.

The themes and concepts of this chapter include:

- The vertebrate kidney excretes wastes and maintains salt and water balance in body fluids.
- Waste removal is an active process that requires energy.
- Kidney structures and functions are adaptations to particular environments.
- Animals excrete nitrogenous waste as ammonia, uric acid, or urea.
- Nephrons are the kidney's functional units, with specialized parts involved in filtration, reabsorption, and secretion.
- Active transport establishes osmotic gradients that conserve water.
- In mammals, the hypothalamus is the thirst center. Hormones regulated by negative feedback systems control diuresis.
- The kidney helps regulate blood pressure and volume.
- Gills and salt glands can help osmoregulation. Some animals are osmoconformers.
- Land animals conserve water by physiological and behavioral means.

CHAPTER OBJECTIVES

Students who master the material in this chapter will understand the importance of osmoregulation. They will be able to describe the fine structures of the kidney and how those parts filter the blood and recycle water and important plasma solutes. These students will appreciate the importance of excretory adaptations that allow animals to live in a wide variety of habitats.

LECTURE OUTLINE

(p 516) **I. The kangaroo rat, master of water conservation, illustrates the importance of water.**

A. This rat does not drink, but retains the water produced in cellular respiration.

B. Behavioral tactics and physiological adaptations allow the rat to retain virtually all its water.

(msg 1) **C. The kidney cleanses the blood of nitrogen-containing waste molecules, excreting uric acid or urea, and regulates the balance of salt and water in vertebrates.**

D. Three unifying themes occur.

1. The critical issue for survival is the maintenance of constant conditions inside body cells.

(msg 2) **2. Natural selection has varied the shape and function of the kidneys and other excretory organs to fit the animal to its environment.**

(msg 3) **3. The kidneys are the body's major blood-filtering organs, and their activities require a large expenditure of energy.**

(p 518) **II. <u>Excretion</u> is the process of removing excess water, salts, and the by-products of metabolism.**

A. When proteins are broken down to release energy, the amino portion cannot be oxidized. The waste is in the form of ammonia.

(msg 4) **B. Ammonia is very toxic to cells. It may be excreted directly to the environment or converted to <u>uric acid</u> or <u>urea</u>.**

(tpq 1) **1. Fish and aquatic invertebrates excrete ammonia.**

2. Birds, reptiles, and most land arthropods excrete uric acid.

a. The liver converts ammonia to uric acid.

b. Uric acid is insoluble and remains suspended (not dissolved) in the blood until removed by the kidney.

c. Using uric acid as the excretory product conserves water and reduces waste buildup inside eggs with developing embryos.

3. Mammals and some bony fishes excrete urea.

a. Ammonia and carbon dioxide are combined in the liver to form urea.

b. Urea is soluble and less toxic than ammonia.

---

tpq 1. Why don't aquatic invertebrates have to convert ammonia to urea or uric acid?

---

c. Gout results from the accumulation of uric acid that results from the breakdown of nucleic acids.

(p 520) III. The kidney is the master organ of waste removal, water recycling, and salt balance.

(tr 109) A. Blood flows to the kidney through renal arteries and leaves via renal veins.

B. The excess water and wastes flow down the ureters to the bladder.

C. During urination, the urine exits the body through the urethra.

(tr 110) D. The kidney has three distinct zones.

1. The outer renal cortex is the site of initial blood filtration.
2. The central medulla does much of the work of conserving water and valuable solutes.
3. The hollow pelvis is where urine is stored.
4. The nephrons reach from the cortex to the medulla.

(msg 5) E. The nephron is the working unit of the kidney.

(tr 111) 1. Blood flow follows a complex pathway.

a. The incoming arteriole brances into the glomerular tuft of capillaries.

b. These capillaries rejoin into an arteriole that carries blood into the <u>peritubular capillaries</u> that twist around the nephron's looped portion.

c. The capillaries finally merge into venous capillaries and then into a larger venule that connects to the renal vein.

2. The nephron loop includes different parts.

    a. The <u>proximal tubule</u> meanders through the kidney's cortex.

    b. At the inner edge of the cortex, the tubule forms the <u>loop of Henle</u>. It becomes thin-walled, straightens, dips into the medulla (the <u>descending limb</u>), makes a hairpin curve, and the <u>ascending limb</u> heads back into the cortex.

    c. The ascending limb leads to the <u>distal tubule</u>, which winds about and connects to the <u>collecting ducts</u>.

(tr 112)

3. During <u>filtration</u>, <u>Bowman's capsule</u> removes small molecules from the plasma.

    a. The capsule surrounds a bundle of capillaries, the <u>glomerulus</u>.

    b. Podocytes lining the capsule have projections that enclasp the capillaries completely.

    c. A fine-meshed filter strains the plasma between the capillary and the podocytes.

d. Blood pressure forces plasma out of the capillaries and into the capsule. The filtrate is like plasma but contains no blood cells or large proteins.

(tpq 2) 4. During tubular reabsorption, useful solutes are actively transported back into the blood in the peritubular capillaries. Water follows by osmosis.

a. In the proximal tubule, most of the water, sodium and chloride ions, sugars, and amino acids from the filtrate are returned to the blood.

(msg 6) b. Cells of the ascending arm pump chloride ions into the extracellular fluid and sodium ions follow passively. This salinity causes water to diffuse out of the descending limb by osmosis.

c. The ascending limb is not permeable to water, but the collecting duct is, and water diffuses out.

5. During tubular secretion, residual wastes are removed from the blood.

a. The pH of the blood is regulated by excretion of hydrogen ions.

---

tpq 2. Why do the kidneys require so much energy?

---

b. Tubular secretion makes <u>drug testing</u> possible.

F. The structures and physiological processes of the kidney are adapted to the animal's environment.

(p 526) IV. <u>Thirst</u> and <u>hormones</u> regulate the body's water content.

(msg 7) A. Rising solute concentrations trigger nerve cells in the brain and stimulate thirst.

1. Drinking water leads to distension of the stomach walls, setting up nerve impulses that inhibit the thirst center.
2. Restoration of osmotic balance inhibits the thirst center.

(tr 113) B. Hormones act on nephrons, controlling the flow of water.

1. <u>Antidiuretic hormone</u> (ADH or vasopressin) reduces diuresis, the production of urine.
   a. ADH is produced in the hypothalamus and stored in the pituitary gland.
   b. ADH controls the reabsorption of water.
   c. Release of ADH is controlled by negative feedback.
2. <u>Aldosterone</u> is secreted by the adrenal glands when blood concentration of sodium ions is low and of potassium ions is high.

a. **Aldosterone stimulates the cells in the distal tubule to reabsorb more sodium ions and return them to the blood, increasing the osmosis of water into the peritubular capillaries.**

b. **Aldosterone causes potassium ions to enter the tubules and be excreted.**

(tpq 3) 3. **High sodium levels in the blood can lead to elevation of blood pressure because the total volume of the blood increases.**

(msg 8) 4. **When blood pressure falls, the enzyme <u>renin</u>, which forms the hormone <u>angiotensin</u> from blood proteins, is released. This hormone causes blood vessels to constrict, increasing blood pressure.**

5. **The heart secretes atrial natriuretic factor (ANF), a hormone that increases the volume of urine and decreases blood pressure.**

(p 529) **V. <u>Osmoregulation</u> relies on different strategies in different environments.**

(msg 9) **A. Terrestrial animals lose water rapidly; adaptations conserve water or enhance water intake.**

---

tpq 3. Why should people with high blood pressure avoid salty foods?

(msg 10) **B. Aquatic animals face the problem of osmoregulation in an environment with salinity that differs from the salinity of body fluids.**

1. **Aquatic invertebrates may be osmoconformers with fluids that are isotonic to surrounding seawater.**
2. **Many invertebrates are osmoregulators; they maintain a constant internal salinity by excreting or taking up salt and/or water.**
3. **Aquatic vertebrates maintain a constant internal solute level. Gills, skin, or salt glands may be the site of salt excretion or uptake.**

INSTRUCTIONAL AIDS

VIDEOS, FILMS, SLIDES

Excretory System. Coronet Film & Video, Deerfield, IL. Animation portrays the roles of blood, lungs, kidneys, and skin as excretory organs. Video.

The Mammalian Kidney. Educational Images Ltd., Elmira, NY. Details the anatomy and function of the kidney during osmoregulation and excretion. Transparencies, video.

The Vertebrate Kidney. Pennsylvania State University AV Services, University Park, PA. Surveys the structure and function of the mammalian kidney. Film.

REFERENCES

Despopoulos, A., and S. Silbernagl. 1986. *Color Atlas of Physiology*, Third Edition. New York: Thieme. A compact book containing many color charts and drawings that illustrate physiological processes.

Morel, Francois, and Alain Doucet. 1986. Hormonal control of kidney functions at the cell level. *Physiological Reviews* 66 (2):377–468. A survey of the sites and mechanisms whereby hormones control functions in their target cells along the nephron. For advanced students.

Stricker, Edward M., and J. G. Verbalis. 1988. Hormones and behavior: the biology of thirst and sodium appetite. *American Scientist* 76 (3):261–267. Discusses the interrelated behavioral and biological mechanisms that control the amounts of salt and water in the body fluid.

KEY TERMS

aldosterone	page 527	loop of Henle	524
angiotensin	528	medulla	521
antidiuretic hormone	527	nephron	521
ascending limb	524	osmoregulation	529
bladder	521	pelvis	522
Bowman's capsule	523	peritubular capillary	524
collecting duct	524	proximal tubule	524
cortex	521	renal artery	521
descending limb	524	renal vein	521
distal tubule	524	renin	528
drug testing	525	tubular reabsorption	522
excretion	518	tubular secretion	522
filtrate	524	urea	518
filtration	522	ureter	521
glomerulus	523	urethra	521
gout	520	uric acid	518
hormone	527	urination	521
kidney	521	urine	517

FUN FACTS

A TASTE FOR SALT

Sodium is necessary for many biological functions, from osmotic balance to nerve conduction. Many behavioral and physiological adaptations have evolved to insure that salt levels are maintained properly.

Carnivores are not as salt-hungry as are herbivores because their diet of flesh includes relatively large amounts of sodium. Natural selection among herbivores has led to neural, hormonal, and behavioral systems that ensure the intake of adequate sodium. Porcupines have been known to chew through glass jars to get at salt. Deer and cattle will walk miles to get to salt licks.

Our tongues have the ability to sense only a few basic tastes, including salt. Experiments have shown that the less sodium there is in the diet, the more sensitive the taste buds become. Many animals like the taste of salt and will consume it in excess, a trait that can lead to health problems in humans.

In the U.S., the average daily consumption of salt is estimated to be 10 to 20 grams, at least ten times as much as the body needs. Salt has been implicated in heart disease, high blood pressure, and other maladies. But we continue to pour it on, perhaps because our omnivorous ancestors didn't have enough.

See: Beauchamp, Gary K. 1987. The human preference for excess salt. *American Scientist* 75:27–33.

FLOWS IN HUMAN KIDNEYS

Each human kidney contains an estimated 1,250,000 nephrons. Together, the two kidneys in an adult filter about 1900 liters of blood per day, removing more than 140 liters of fluid and restoring 99 percent of it to the blood. The remainder goes to the urine. A normal day's output of urine is between one and two liters, a quantity that can be increased by disease, nervousness, and diuretics such as alcohol, coffee, and tea. A day's urine typically contains 50 grams of solids, including a wide array of organic and inorganic chemicals.

The composition of urine has been used as a diagnostic tool for centuries. Diabetes, for example, produces high glucose concentrations in the urine. Starvation is indicated by increased nitrogenous wastes that result from consuming protein for energy after the body's fat stores are exhausted. Urine color may be symptomatic of diseases such as alkaptonuria, cholera, and typhus.

See: Smith, Anthony. 1986. *The Body*. New York: Viking Penguin, Inc.

CHAPTER 25 EXCRETION AND THE BALANCING OF WATER AND SALT

MESSAGES

1. The vertebrate kidney excretes wastes and maintains salt and water balance in body fluids.

2. Kidney structures and functions are adaptations to particular environments.

3. Waste removal is an active process that requires energy.

4. Animals excrete nitrogenous waste as ammonia, uric acid, or urea.

5. Nephrons are the kidney's functional units, with specialized parts involved in filtration, reabsorption, and secretion.

6. Active transport establishes osmotic gradients that conserve water.

7. In mammals, the hypothalamus is the thirst center. Hormones regulated by negative feedback systems control diuresis.

8. The kidney helps regulate blood pressure and volume.

9. Land animals conserve water by physiological and behavioral means.

10. Gills and salt glands can help osmoregulation. Some aquatic animals are osmoconformers.

26

HORMONES AND OTHER MOLECULAR MESSENGERS

PERSPECTIVE

This chapter presents hormones, which make up one of the two major control systems in the animal body. The other, the nervous system, is addressed in Chapter 27. Some concepts discussed in this chapter are based on material presented in Chapters 3 and 4, which cover cellular activities. The processes of maturation detailed in Chapter 14, The Human Life Cycle, are largely controlled by hormones. The organismic processes presented in Part IV, How Animals Survive, are also regulated by hormones.

The themes and concepts of this chapter include:

- Molecular messengers, including hormones, are secreted by regulator cells and trigger activity in target cells by binding to receptor molecules.
- Molecular messengers differ in the organization of their regulatory cells, in their transport fluids, and in their target cells.
- Hydrophobic messengers can pass through a cell membrane and bind to intracellular receptors.
- Hydrophilic messengers bind to cell-surface receptors and trigger the release of second messengers.
- The hypothalamus and the pituitary act as master endocrine glands.
- The thyroid gland regulates the rates of metabolism, growth, and calcium deposition. Parathyroid hormone causes bones to release calcium, keeping blood calcium levels above critical limits.
- Adrenal glands secrete hormones that prepare the body for fight or flight, that are involved in stress reactions, and that regulate sodium, potassium, and water balance.
- Hormones help maintain homeostasis via feedback loops and govern seasonal and developmental changes.
- The molecular messenger strategy has been conserved throughout evolution.

CHAPTER OBJECTIVES

Students who master the material in this chapter will understand that molecular messengers maintain homeostasis and regulate maturation and seasonal cycles. They will see that all molecular messengers act in the same way, by binding to receptors and affecting cell activity. These students will appreciate the master roles of the hypothalamus and pituitary gland and be impressed by the long evolutionary history of hormones.

LECTURE OUTLINE

(p 534) **I. The maturation and metamorphosis of the cecropia moth introduces hormones.**

(tpq 1) **A. Hormones are produced by one set of cells and transported in body fluids to other parts of the body, where they bind to a target cell and change cell activity.**

B. Molecular messengers pass through one of several fluid or gaseous media and regulate activities of other cells.

C. Three themes will emerge in this chapter:

1. All molecular messengers bind to a specific protein in or on a target cell and trigger some activity.

2. Molecular messengers regulate physiological systems by preventing or provoking change.

tpq 1. Are the effects of hormones psychological as well as physiological?

3. Very similar molecular messengers occur throughout the living world.

(p 536) II. Molecular messengers are central to survival and function.

(msg 1) A. Molecular messengers work by responding to a change in the cell's environment and by triggering some new physiological activity.

B. The molecular messenger is secreted by a regulator cell in response to a perturbation in the environment.

C. The molecular messenger diffuses to target cells that respond by carrying out the cellular activity needed to adjust to the original perturbation.

1. Each target cell includes protein receptors with shapes complementary to specific molecular messengers.

(tr 114) 2. When a molecular messenger binds to a receptor, the receptor changes shape and triggers a cell activity.

(tpq 2) a. A messenger that enters the cytoplasm may interact with its receptor to act on nuclear DNA.

---

tpq 2. How can a hormone trigger changes in cell activity by binding to DNA?

---

b. A receptor on the cell surface may trigger the release of a second messenger in the cytoplasm.

D. Most molecular messengers are produced in glands.

(msg 2) 1. Endocrine glands do not have ducts; they dump their molecular messengers directly into the extracellular fluid.

2. Exocrine glands empty into ducts that generally lead out of the body.

E. There are five types of molecular messengers that may be classified by the organization of the secretory cells, the transport fluid, and the location of the target cells.

1. Paracrine hormones act on cells adjacent to the secretory cells.

2. Neurotransmitters relay signals between nerve cells.

3. Neurohormones are hormones secreted by nerve cells.

4. True hormones are nonneural hormones.

(tpq 3) 5. Pheromones act on other organisms.

F. Molecular messengers differ in their sites of activity.

---

tpq 3. How might pheromones be used to control insect pests?

(msg 3)

1. **Hydrophobic messengers can enter through the cell membrane and act within the cell.**
   a. **The messenger binds to a receptor molecule.**
   b. **The messenger-receptor complex binds to a stretch of DNA, triggering gene transcription.**
2. **Hydrophilic messengers tend to bind to cell surface receptors.**
   a. **When a messenger binds to a cell surface receptor, a relay triggers the release of an intracellular, second messenger.**
   b. **The second messenger changes the activity of a specific protein which in turn alters the activity of the cell.**

(p 539)
(tr 115)

III. Hormones are produced by the mammalian endocrine system.

A. **Most mammalian hormones belong to four molecular groups.**
   1. **Polypeptide hormones include oxytocin and luteinizing homone.**
   2. **Steroid hormones include estrogen, testosterone, and ecdysone.**
   3. **Amine hormones include thyroid hormones.**
   4. **Fatty acid hormones include prostaglandins.**

(msg 4)

B. **The hypothalamus controls the pituitary gland.**

1. **The posterior lobe of the pituitary stores oxytocin and ADH, which are made in the hypothalamus and transported to the pituitary in axons.**
2. **The anterior lobe of the pituitary produces more than six hormones.**
3. **The hypothalamus controls the secretion of these pituitary hormones by secreting hypothalamic releasing hormones and hypothalamic release-inhibiting hormones.**
4. **The hypothalamus and the pituitary gland participate in several regulatory feedback loops that govern other glands and organs.**

(msg 5) **C. The thyroid gland, controlled by the hypothalamus and the pituitary, regulates the body's energy use and growth.**

(tpq 4)

1. **Thyroxine, an iodine-containing amine, governs metabolic and growth rates and stimulates nervous system function.**
2. **Underactivity leads to cretinism or myxedema.**
3. **A person without enough iodine might develop a goiter.**
4. **An overactive thyroid can cause weight loss, nervousness, irritability, and emotional instability.**

---

tpq 4. How can exposure to radioactive iodine cause cretinism?

---

5. The thyroid gland and the parathyroid glands regulate the level of calcium in the blood.
   a. Calcitonin, released by the thyroid, causes excess calcium to be deposited in bones.
   b. When blood calcium levels drop, the parathyroid glands release parathryoid hormone, which causes bones to release calcium.

(tpq 5) (tr 116) (msg 6)

D. Adrenal glands, atop the kidneys, produce hormones that enable fight or flight.
   1. The adrenal medulla secretes epinephrine (adrenaline) and norepinephrine, which bring about the stress response.
   2. The adrenal cortex and medulla secrete other hormones that help an animal respond to stress.
      a. Cortisol, a glucocorticoid, speeds up the metabolic breakdown of carbohydrates, proteins, and fats.
      b. ACTH, which triggers the release of cortisol, is produced by the anterior lobe of the pituitary.
      c. Aldosterone, a mineralocorticoid hormone, causes the kidney to conserve sodium and water.

---

tpq 5. What's the value in natural selection of the stress response (or adrenaline rush)?

(p 546) **IV. Hormones control physiological change.**

(msg 7) **A. Hormones maintain homeostasis. For example, diabetes mellitus, the most common hormonal disorder, is the inability to maintain blood glucose levels within a normal range.**

(tr 117) **1. In the pancreas of a normal person, alpha and beta cells in the islets of Langerhans secrete glucagon and insulin.**

(msg 8) **2. Glucagon and insulin work together by means of a feedback loop to contol blood sugar levels.**

a. Glucagon causes liver cells to break down glycogen and release glucose.

b. Insulin causes cells in the liver and other organs to remove glucose from the blood and store it as glycogen.

3. In a diabetic, beta cells fail to produce insulin and body cells don't remove glucose from the blood.

B. Hormones regulate cyclical physiological changes.

1. Cycles are tied to circadian cycles and seasonal day/night cyles.

2. The pineal gland measures day length and controls cycles and moods by secreting melatonin in response to darkness, promoting sleep and inhibiting gonadal activity.

3. Endorphins may regulate the activities of the pineal gland, the hypothalamus, the pituitary, and the gonads.

C. Hormones control permanent developmental changes such as sexual maturation and metamorphosis.

(tpq 6) 1. Ecdysone controls the timing of silkworm molting.

2. Juvenile hormone controls whether the larva will molt to another larval stage or to a pupa.

(p 549) V. The simplest and the most complex organisms have similar molecular messengers.

A. There are similarities between releasing factors that stimulate mammalian gonads and the peptide factor that prepares yeasts for mating.

(msg 9) B. This implies that molecular messengers must have had a very ancient origin.

---

tpq 6. How might hormones be used to control insect pests?

INSTRUCTIONAL AIDS

VIDEOS, FILMS, SLIDES

Endocrine System. Coronet Film & Video, Deerfield, IL. Animation set against the human body shows how the major glands coordinate the body's activities. Film, video.

Homeostasis: Maintaining the Body's Internal Environment. Human Relations Media, Pleasantville, NY. Presents a comprehensive look at homeostasis in the control of temperature, weight, fluids, oxygen, and glucose levels. Filmstrip, video.

Messengers. Films for the Humanities and Sciences, Princeton, NJ. Follows the roles hormones play in response to a sudden emergency. Video.

REFERENCES

Belzer, William R. 1986. A nonlethal laboratory demonstration of antidiuretic hormone's action. *American Biology Teacher* 48:360–361. ADH-induced permeability is demonstrated by changes in the permeability of the skin of frogs injected with ADH.

Berridge, M. 1985. The molecular basis of communication within the cell. *Scientific American* 253 (4):142–152. Discusses the ways signals are transmitted in the cell's interior in response to hormonal stimuli.

Bower, Bruce. 1991. Pumped up and strung out. *Science News* 140:30–31. A discussion of steroid addiction among body builders.

Evans, Ronald M. 1988. The steroid and thyroid hormone receptor superfamily. *Science* 240:889–895. A review of studies that have led to the identification of a superfamily of regulatory proteins.

KEY TERMS

Term	Page
ACTH	page 545
adrenal gland	544
alpha cell	546
amine hormone	539
axon	540
beta-endorphin	545
beta cell	546
calcitonin	544
circadian rhythm	547
cortex	544
cortisol	545
diabetes mellitus	546
ecdysone	548
endocrine gland	536
endocrine system	539
epinephrine	544
exocrine gland	536
fatty acid hormone	539
gland	536
glucagon	546
goiter	544
hormone	535
hypothalamic release-inhibiting hormone	540
hypothalamic releasing hormone	540
hypothalamus	540
insulin	546
juvenile hormone	549
medulla	544
melatonin	548
metamorphosis	535
molecular messenger	535
neurohormone	537
neurotransmitter	537
norepinephrine	544
paracrine hormone	537
parathyroid gland	544
parathyroid hormone	544
pheromone	537
pineal gland	547
pituitary gland	540
polypeptide hormone	539
receptor	536
regulator cell	536
second messenger	536
steroid hormone	539
target cell	536
thyroid gland	544
thyroxine	544
true hormone	537

FUN FACTS

OUR HORMONAL HERITAGE

Biologists have found that the amino acid sequences of many human hormone receptors are virtually identical – except for differences at the hormone binding site. This similarity is strong evidence that these proteins are all divergent forms of one original receptor, modified by natural selection for different signalling tasks. There are many similarities among chemical messengers produced by nerves, endocrine cells, and a variety of other tissues. This suggests that an early ancestor had a simple system of chemical communication, a system that became more elaborate and specialized with time.

Other researchers have found a chemical similar to insulin in a protist, *Tetrahymena,* and in a bacterium, *Escherichia coli*. Brewer's yeast, *Saccharomyces cerevisiae,* produces a chemical messenger much like gonadotropin-releasing hormone, a sex hormone in humans. In yeast, this messenger orchestrates mating between cells.

Some scientists now argue that chemical messengers arose long before plants and animals evolved, and that multicellular life just adapted and elaborated an already successful communications system. The ability of our cells to interact with substances produced by other organisms is the result of descent from shared ancestors.

See: Evans, Ronald M. 1988. The steroid and thyroid hormone receptor superfamily. *Science* 240:889 – 895.

Roth, Jesse, and Derek LeRoith. 1987. Chemical cross talk. *The Sciences* 27 (3):51 – 54.

DIABETES

An autoimmune attack on the body's own insulin-producing cells causes insulin-dependent, or type I, diabetes. The attack is highly specific and is restricted to the insulin-producing beta cells in the pancreas. Antibodies bind to antigens found only on beta cells; they are followed by white blood cells that kill and devour the cells. Exposure to a foreign antigen that closely resembles a component of the beta cell may trigger the autoimmune response. Prevention may involve immune system supression.

See: Atkinson, Mark A., and N. K. Maclaren. 1990. What causes diabetes? *Scientific American* 263 (1):62 – 71.

CHAPTER 26 HORMONES AND OTHER MOLECULAR MESSENGERS

MESSAGES

1. Molecular messengers, including hormones, are secreted by regulator cells and trigger activity in target cells by binding to receptor molecules.

2. Molecular messengers differ in the organization of their regulatory cells, in the transport fluid, and in their target cells.

3. Hydrophobic messengers can pass through a cell membrane and bind to intracellular receptors. Hydrophilic messengers bind to cell-surface receptors and trigger the release of second messengers.

4. The hypothalamus and the pituitary act as master endocrine glands.

5. The thyroid gland regulates the rates of metabolism, growth, and calcium deposition. Parathyroid hormone causes bones to release calcium, keeping blood calcium levels above critical limits.

6. Adrenal glands secrete hormones that prepare the body for fight or flight, that are involved in stress reactions, and that regulate sodium, potassium, and water balance.

7. Hormones help maintain homeostasis and govern seasonal and developmental changes.

8. Hormones act via negative feedback loops when maintaining constant conditions.

9. The molecular messenger strategy has been conserved throughout evolution.

27

HOW NERVE CELLS CONTROL BEHAVIOR

PERSPECTIVE

This chapter introduces the nervous system, the second of the major systems that control the animal body. Chapter 26 presented the other contolling system, the endocrine system. Concepts in this chapter depend on an understanding of chemistry, presented in Chapter 2, and of cell membrane processes, discussed in Chapters 3 and 4. Chapters 27 through 29 will continue the discussion of the central nervous system, senses, and muscle stimulation. Material from this chapter will appear in Chapter 38, Animal Behavior.

The themes and concepts of this chapter include:

- The nervous and endocrine systems share several characteristics and work together to control and coordinate body processes. Nerve cells are arranged in complex networks that allow for rich behavioral repertoires.
- Neurons receive stimuli at dendrites, transmit them through axons, and release neurotransmitters at boutons, stimulating other cells.
- A nerve impulse, or action potential, is a unidirectional wave of ion migration through a cell membrane.
- An action potential is initiated when sodium ions leak through a section of the membrane and cause depolarization, which stimulates the opening of ion gates.
- Following an action potential, membrane pumps move sodium out of the cell and potassium in, restoring the resting potential.
- An action potential propagates rapidly in giant axons or axons wrapped in myelin sheaths.
- Neurons may form electrical junctions or chemical synapses. Neurotransmitters are rapidly degraded or recycled after transmission across the synapse.
- A synapse may be excitatory or inhibitory. Learning can involve habituation or sensitization, which change the effectiveness of signals crossing synapses.

CHAPTER OBJECTIVES

Students who master the material in this chapter will understand how nerves conduct messages through axons to other cells. They will understand that the nerve impulse is an all-or-nothing event that is stimulated by the summation of excitatory and inhibitory inputs. These students will understand how a reflex arc works and see that the central nervous system in vertebrates is composed of an elaborate network of interconnected neurons.

LECTURE OUTLINE

(p 552) **I. Irreversible drug-induced damage to a baby's <u>nervous system</u> introduces the topic.**

(msg 1) **A. Nervous systems control and coordinate the activities of the animal's body systems.**

B. Nervous systems receive data about environmental conditions, integrate these sensory data into a meaningful form, and effect a change in behavior, physiology, or information storage.

(tpq 1) **C. The nervous and endocrine systems work together to coordinate and control the body and share several important features.**

D. The central task of the nervous system is communication.

(msg 2) **E. Communication is possible because nerve cells contact each other in specific ways and are organized into highly elaborate networks.**

---

tpq 1. How do the endocrine system and the nervous system differ in the way they control activity?

(p 554) **II. Nerve cells are structured for long-range communication.**

A. The cells of the human brain and nervous system may be classified as neurons or glial cells.

(msg 3) **B. Neurons are characterized by certain structural features.**

1. **Dendrites gather information.**
2. **The soma, the main cell body, produces proteins.**
3. **The signals pass through the axon to the boutons.**
4. **A synapse is a junction where a bouton of a presynaptic cell approaches a postsynaptic cell, which may be a neuron, muscle cell, or secretory cell.**
5. **An axon may branch and rebranch; conversely, many cells might form synapses with the dendrites and soma of a single cell.**

(p 554) **III. Regardless of the stimulus, all neurons produce the same kind of signal: a nerve impulse.**

(tpq 2) (tr 118) **A. By controlling the traffic of ions into and out of the cell, a cell membrane generates and conducts a nerve impulse.**

---

tpq 2. In terms of nerve impulses, how is the sensation of being poked by a pin different from being caressed by a loved one?

---

1. The fluid inside the neuron is relatively rich in potassium ions. The extracellular fluid contains more sodium ions.
2. There are more positive charges surrounding the neuron than occuring inside of it, so the inside is negatively charged with respect to the outside.
3. The resting potential across the membrane is about −70 mV.

B. Three cell membrane proteins are involved in the movement of ions through the membrane.

1. The potassium channel allows potassium ions to pass through the membrane down the concentration gradient, from the inside to the outside, leaving a negative charge inside.
2. The sodium channel closes tightly and allows few sodium ions to leak inward.
3. The sodium-potassium pump uses ATP energy to bring potassium into the cell and pump sodium out.
4. The actions of the three proteins cause the neuron to be polarized.

(msg 4) C. The action potential, or nerve impulse, is a
(tr 119) reversal of the polarity of the resting potential.

(msg 5) 1. An impulse begins when a stimulus causes a region of the membrane to leak sodium ions and become locally depolarized.

2. When a specific threshold (about −50 mV) is reached, the sodium ion channels open wide.

   a. Sodium ions rush in and the inside of the neuron becomes positively charged with respect to the outside, about +50 mV.

(tpq 3)

   b. The sodium channels close rapidly and cannot open again for a while, the <u>refractory period</u>.

3. As the sodium channels close, the potassium channels open fully, allowing potassium to rush out of the cell and restore the original polarity.

(msg 6)

4. The sodium-potassium pump moves sodium ions out and potassium ions in in preparation for the next action potential.

5. Action potentials are all-or-none.

6. Anesthetics may bind to the ion channels.

(tr 120)

D. <u>Propagation</u> of an impulse depends on the action potential passing from one patch of the membrane to an adjacent patch.

   1. The refractory period prevents reverse propagation.

   2. Propagation can be speeded up by increasing the diameter of the axon or by saltatory conduction.

---

tpq 3. What might happen to a person exposed to a poison that kept the sodium channels open?

---

a. In vertebrates, Schwann cells form the myelin sheath that insulates the axon from the extracellular fluid.

b. At the uninsulated nodes of Ranvier, ions can flow across the membrane.

c. The impulse leaps from one node to the next.

(p 559) **IV. Neurons may be electrically coupled to other neurons or may rely on a chemical neurotransmitter to span the synapse.**

(msg 7) A. At electrical synapses, cells are fastened together by gap junctions, allowing ions to flow from one cell to the next.

(tr 121) B. Chemical synapses are slower but are capable of greater modification.

1. Neurotransmitters are stored in synaptic vesicles in the bouton.
2. When the action potential reaches a bouton, it causes calcium ion channels in the cell membrane to open.
3. Calcium ions stimulate the synaptic vesicles to fuse with the cell membrane and release their neurotransmitters.
4. Neurotransmitters diffuse across the synaptic cleft and bind to receptors on the postsynaptic cell, stimulating the postsynaptic cell to react.

(tpq 4) 5. The neurotransmitter is rapidly degraded or reabsorbed.

a. Cocaine blocks the cleanup of dopamine.

b. Addiction results from imbalance of neurotransmitter production and cleanup rates.

C. There are many types of neurotransmitters.

1. Acetylcholine transmits nerve impulses in the brain and to skeletal muscles.

2. Norepinephrine (also known as a hormone) seems central to keeping mood and behavior on an even keel and suppresses the flight-or-fight response.

3. Neuromodulators are longer-lived substances that influence the postsynaptic cell's response to neurotransmitters.

(msg 8) D. Synapses may be excitatory or inhibitory.

E. The activity in a cell results from the summation of all the molecular messages.

(p 564) V. Networks of neurons control behavior.

(tr 122) A. The reflex arc underlies simple behaviors that are rapid, involuntary, and stereotyped.

---

tpq 4. What might happen if a drug blocked or impeded the cleanup of neurotransmitters?

B. **Vertebrate nerves fall into three general categories:**

   1. **Sensory neurons receive information from the environment and send it toward the central nervous system (CNS).**
   2. **Interneurons relay messages between nerve cells and integrate and coordinate messages.**
   3. **Motor neurons send messages from the CNS to muscles or glands.**

C. **Learning enables organisms to adapt to changing conditions.**

   1. **Habituation is a progressive decrease in the strength of a behavioral response to a constant or repeated stimulus that is followed by trivial consequences.**
   2. **Sensitization is an exaggerated reaction that occurs in response to a repeated mild stimulus followed by an intensely threatening stimulus.**
   3. **Learning is due to changes in the effectiveness with which signals cross synapses.**

INSTRUCTIONAL AIDS

VIDEOS, FILMS, SLIDES

The Nature of the Nerve Impulse. Films for the Humanities and Sciences, Princeton, NJ. Experiments demonstrate a locust neuron's response to movement in the visual field. Video.

The Nerve Impulse. Brittanica Films and Video, Chicago. Laboratory demonstrations ranging from simple frog nerve-muscle preparations to squid axon experiments. Film, video.

The Nervous System and its Function. Connecticut Valley Biological Supply, Southampton, MA. Presents information on neural transmission and major brain areas. Transparencies, filmstrip.

REFERENCES

Alkon, Daniel L. 1989. Memory storage and neural systems. *Scientific American* 261 (1):42–50. Discusses the changes in molecular and electrical properties of nerve cells that accompany learning.

Koshland, Daniel E., ed. 1988. Frontiers in neuroscience. *Science* 242:692–745. This special issue surveys the leading edge of research in neuroscience.

Miller, Christopher. 1991. 1990: Annus mirabilus of potassium channels. *Science* 252:1092–1096. A review of research on voltage-gated potassium channel proteins found in cell membranes.

Veca, A., and J. H. Dreisbach. 1988. Classical neurotransmitters and their significance within the nervous system. *Journal of Chemical Education* 65:108–111. A clearly written summary of neurotransmitters. Includes molecular diagrams.

KEY TERMS

action potential	page 557	neuron	554
addiction	561	polarized	557
axon	554	postsynaptic cell	554
bouton	554	potassium channel	555
dendrite	554	presynaptic cell	554
depolarized	557	propagation	557
glial cell	554	reflex arc	564
habituation	564	refractory period	557
interneuron	564	sensitization	565
motor neuron	564	sensory neuron	564
myelin sheath	558	sodium channel	555
nerve	554	soma	554
nerve impulse	555	summation	564
nervous system	552	synapse	554
neuromodulator	563	synaptic vesicle	560

FUN FACTS

SHOCKING BEHAVIOR

Many types of fish, from catfish to rays, can perceive electrical currents in water. Animals produce electricity as nerves conduct signals and muscles contract, and many predators have evolved the ability to detect these minute currents.

Some fish have electric organs that can produce electricity. The current is often generated by modified muscle cells that are oriented in series so the voltage is cumulative. The best known electric fish is the electric eel, which can put out 600 volts, enough to stun a human. Like the glass knife fish, *Eigenmannia virescens,* most electric fish produce fields on the order of one volt.

Glass knife fish emit regular pulses of electricity at 300 to 1000 cycles per second interrupted, like a cricket's call, by brief silences a few times a second. They use the electrical currents for navigation and communication. Researchers who put electrodes in the water and translate the pulses into sound hear "chirps," much like bird chirps. Like birds, glass knife fish vary the tone and frequency to suit the circumstances. A dominant male will threaten adversaries with repeated chirps. Females and males communicate their state of sexual readiness with chirps. Young glass knife fish produce electrical currents by the time they are ten days old. Glass knife fish recognize each other as individuals by their electrical pulses. If two fish are put in close proximity, they alter their frequencies so their signals don't overlap. Their nervous systems are so finely tuned that they can detect electrical impulses that are just 400 billionths of a second apart. By contrast, our brains can detect differences in arrival time of signals from our ears only if they are more than 15 millionths of a second apart.

See: Anonymous. 1986. A fish that makes its own current to find its way. *Science 86* 7 (2):7.

Hagedorn, Mary, and Walter Heiligenberg. 1985. Court and spark: Electric signals in the courtship and mating of gymnotoid fish. *Animal Behavior* 33:254–265.

CHAPTER 27 HOW NERVE CELLS CONTROL BEHAVIOR

MESSAGES

1. The nervous and endocrine systems share several characteristics and work together to control and coordinate the body.

2. Nerve cells are arranged in complex networks that allow for rich behavioral repertoires.

3. Neurons receive stimuli at dendrites, transmit them through axons, and release neurotransmitters at boutons, stimulating other cells.

4. A nerve impulse, or action potential, is a unidirectional wave of ion migration through a cell membrane.

5. An action potential is initiated when sodium ions leak through a section of the membrane and cause depolarization, which stimulates the opening of ion gates.

6. Following an action potential, membrane pumps move sodium out of the cell and potassium in, restoring the resting potential.

7. Neurons may form electrical junctions or chemical synapses. Neurotransmitters are rapidly degraded or recycled after transmission across the synapse.

8. A synapse may be excitatory or inhibitory. Learning can involve habituation or sensitization, which change the effectiveness of signals crossing synapses.

28

THE SENSES AND THE BRAIN

PERSPECTIVE

This chapter continues the discussion of the nervous system with an overview of sensation and the highest levels of integration. This chapter builds on concepts presented in Chapter 27, How Nerve Cells Work, and leads into the material in Chapter 29, The Dynamic Animal: The Body in Motion.

The themes and concepts of this chapter include:

- The nervous system's major role is to monitor internal needs and external conditions, process the information, and initiate the appropriate responses.
- Sense organs contain bare nerve cell endings modified in ways that increase their sensitivity to one physical aspect of the environment.
- Nervous systems become more complex in more complicated animals.
- Different senses and different behaviors can be localized to specific regions or groups of regions in the brain.
- The cochlea of the inner ear converts sound waves into nerve impulses. Semicircular canals are involved in balance and detecting motion.
- Rhodopsin molecules change shape when struck by light, initiating a cascade of events that culminates in the sensation of vision.
- In humans, the peripheral nervous system includes the somatic and autonomic nervous systems. The autonomic nervous system includes the sympathetic and parasympathetic nervous systems.
- The human brain includes the brain stem, which helps control autonomic functions; the cerebellum, involved in coordination of fine movement; and the cerebrum.
- The human cerebrum contains structures that regulate homeostasis and basic drives as well as areas that are involved in personality, intelligence, memory, and learning.

CHAPTER OBJECTIVES

Students who master the material in this chapter will gain an overview of the vertebrate central nervous system, the senses that gather information, and the systems that process, integrate, and act on that information. They will understand how the eye perceives light, the ear distinguishes sounds, and the senses of taste and smell operate. They will see that the evolution of the nervous system includes increasing complexity and cephalization. These students will be able to describe the major subdivisions of the human brain and their functions.

LECTURE OUTLINE

(p 568) **I. The night-hunting tawny owl introduces sense organs and the vertebrate brain.**

- **A. The bird's ability to capture prey depends on the integration of sensory information and coordinated muscle action.**
- **B. Four main themes recur.**
  - (msg 1) **1. The nervous system's major role is to monitor internal needs and external conditions, process the information, and initiate appropriate behavioral responses.**
  - (msg 2) **2. Sense organs contain excitable cells that are modified in ways that increase their sensitivity to one physical aspect of the environment.**
  - (msg 3) **3. Nervous systems become more complex in more complicated animals.**
  - (msg 4) **4. Different senses and different behaviors can be localized to specific regions or groups of regions in the brain.**

(p 570) **II. Sense organs are an organism's windows on the world.**

(tpq 1) **A. Sense organs are groups of specialized neurons that receive stimulus energy and convert it into a kind of energy that can trigger a nerve impulse.**

(msg 2) **B. All sense organs have bare dendrites not covered by a protective myelin sheath.**

(tr 123) **C. The ear is the body's most complex mechanical device.**

(msg 5) **1. Cells in the cochlea detect sound by sensing subtle movements in the fluid that surrounds them.**

2. The stimulated cells send nerve impulses to the brain, which interprets the signal as a specific sound.

3. A sound is a compression wave that strikes the eardrum, setting in motion the three small bones of the middle ear that in turn vibrate the cochlea.

a. The cochlea is divided into three fluid-filled chambers.

b. The basilar membrane bears hair cells, each crowned by stereocilia.

---

tpq 1. Are there more than five senses?

---

c. **Deflection of the basilar membrane moves the hair cells relative to the tectorial membrane, bending the stereocilia, which stimulates action potentials in the hair cells.**

d. **High pitches vibrate the wide part of the basilar membrane; low pitches cause movement near the narrow tip. Each region of the membrane is connected to a different region of the brain, allowing the brain to distinguish pitches.**

(tpq 2) **e.** **Loud sounds cause more neurons to fire, and to fire more frequently, than do quieter sounds.**

4. **Semicircular canals serve as organs of balance and motion detection. Similar hair cells in a fish's lateral line organ detect changes in water pressure.**

(tr 124) **D.** **The eye is an outpost of the brain.**

1. **The eye converts electromagnetic radiation in the form of light energy into neural energy.**

2. **Light passes through the cornea, the pupil (the opening in the iris), the lens, and onto the retina, which contains photoreceptor cells.**

---

tpq 2. How do loud noises cause permanent hearing loss?

---

a. **Receptor cells are rods, which are sensitive to low light levels but cannot detect color, and cones, which can detect color but need more light.**

b. **Axons leading to the brain are bundled in the optic nerve.**

(tpq 3) 3. **A rod contains stacks of membranous discs in which millions of molecules of rhodopsin are embedded.**

(msg 6) a. **The protein portion cradles the pigment portion, known as retinal, which changes shape when struck by light.**

b. **This shape change initiates a cascade of reactions that we eventually perceive as light.**

E. **Chemical senses include taste and smell.**

1. **The tongue is studded with papillae, which house the taste buds.**
2. **Taste bud receptors can distinguish only four classes of flavors: sweet, salty, bitter, or sour.**
3. **Volatile aroma molecules bind to receptors in the olfactory epithelium.**
4. **Neurons from the olfactory epithelium connect with the olfactory bulb on the underside of the brain.**

---

tpq 3. Rhodopsin is found in some photosynthetic bacteria. What does this indicate about the evolutionary history of the sense of vision?

F. **In summary, a brain can decipher the stimulus it receives as light, sound, taste, or smell because each sense organ is tuned to a different physical or chemical stimulus and sends a signal to a different part of the brain. The strength of the signal reflects the number of sensory neurons activated.**

(p 576) **III. The variety of nervous systems ranges from nerve nets to the human brain.**

A. **Two general evolutionary trends are apparent:**

(tpq 4) 1. **Animals that move in a forward direction have sense organs concentrated in the part of the body that first meets the environment.**

a. **Cnidarians have nerve nets but lack brains.**

b. **Cephalization includes a concentration of nerves in the animal's anterior end.**

c. **Ganglia are distinct clumps of nerve cell bodies that act like a primitive brain.**

(msg 3) 2. **The more complex the animal, the more interneurons it will have.**

(tr 125) B. **A vertebrate's nervous system is organized into a peripheral nervous system and a central nervous system (CNS).**

---

tpq 4. Do all animals have brains? If not, how can they function?

(p 578) **IV. Peripheral neurons sense changes and cause actions, but do little information processing.**

A. Sensory neurons conduct impulses to the CNS.

(tr 126) **1. Thirty-one pairs of spinal nerves are attached to the spinal cord.**

2. Receptors send action potentials to the cell body in the dorsal root ganglion.

3. Interneurons in the spinal cord carry a message to the brain.

4. Motor neurons leave the CNS via the ventral root of the spinal nerve.

(msg 7) **B. Motor neurons form two systems.**

1. The somatic nervous system activates muscles under voluntary control.

(tr 127) **2. The autonomic nervous system regulates the body's internal environment.**

a. Parasympathetic nerves form a housekeeping system that regulates normal functioning of internal organs.

b. The sympathetic system dominates during emergencies.

c. Parasympathetic and sympathetic nerves often act antagonistically.

(p 580) **V. The CNS acts as the information processor.**

(tpq 5) **A. The spinal cord is responsible for preliminary integration of signals and for conducting signals to and from the brain.**

1. Gray matter includes neuron cell bodies, unmyelinated nerve fibers, and many synapses.

2. White matter consists of myelinated axons that transmit information long distances.

(msg 8) **B. The human brain has three interconnected parts:**

(tr 128) **1. The brain stem integrates sensory and motor systems, regulates body homeostasis, and controls arousal.**

a. The medulla oblongata receives information and regulates many subconscious body activities.

b. The pons and midbrain relay sensory information to other parts of the brain.

c. The hypothalamus links the endocrine and nervous systems and plays important roles in maintaining homeostasis.

d. The thalamus relays information to the cerebrum.

---

tpq 5. Why is the spinal cord considered part of the CNS instead of part of the peripheral nervous system?

---

e. The <u>reticular formation</u> is a network of tracts that reaches into the cerebrum and cerebellum and is involved in sleep and wakefulness.

2. The <u>cerebellum</u> compares outgoing commands with incoming information about posture and muscle activity and refines motor commands.

(msg 9) 3. The <u>cerebrum</u> is the seat of perception, thought, and humanness.

a. During evolution, the <u>cerebral cortex</u> increased greatly in size.

(tr 129) b. All our basic cerebral functions can be traced to specific regions of the cerebrum.

i. The <u>motor cortex</u> controls voluntary muscles.

ii. The <u>sensory cortex</u> registers and integrates sensations from body parts.

iii. The right side of the brain controls the left side of the body and vice versa.

iv. Pain helps keep us from damaging ourselves.

(tpq 6) c. Higher cerebral functions may be mapped to specific regions.

d. The <u>hippocampus</u> plays a crucial role in the formation of long-term memory.

---

tpq 6. How can scientists gain understanding of the roles of the different parts of the brain by studying people with brain damage?

---

e. Other areas perform specific high-level mental activities such as speech and recognition of faces.

4. The brain is asymmetrical.

a. In most people, language abilities, analytical thought, and fine motor control take place on the left side of the brain.

b. The right hemisphere is mainly responsible for intuition, musical aptitude, recognition of visual patterns, and emotion.

c. The corpus callosum bridges the hemispheres. If this bridge were severed, the visual cortex of the right hemisphere would not be able to communicate with the verbal cortex of the left brain.

d. The main cerebral lobes include regions called the association cortex.

INSTRUCTIONAL AIDS

VIDEOS, FILMS, SLIDES

The Brain. Films Incorporated, Chicago. Eight programs covering matters as diverse as vision, memory, madness, and states of mind. Film, video.

The Ear as a Sensory Organ. International Film Bureau, Chicago. An explanation of the physical basis of sound, the anatomy and physiology of the ear, nerve impulses, and the organs of balance. Film, video.

REFERENCES

Barinaga, Marcia. 1991. How the nose knows: Olfactory receptor cloned. *Science* 252:209–210. Discusses the discovery and cloning of olfactory receptor molecules.

Ezzell, Carol. 1991. Memories might be made of this. *Science News* 139:328–330. A discussion of recent discoveries in the biochemistry of learning.

Freeman, Walter J. 1991. The physiology of perception. *Scientific American* 264 (2):78–85. An argument that chaotic, collective activity of millions of neurons is essential for transforming sensory messages into conscious perceptions.

Hubel, David H. 1988. *Eye, Brain, and Vision*. New York: Scientific American Books. An eminent scientist's eloquent description of a fascinating field, intended for a lay audience.

Sachs, Frederick. 1988. The intimate senses. *The Sciences* 28 (1):28–34. An easy-to-read discussion of the sense of touch.

Weiss, Rick. 1991. Toward a future with memory. *Science News* 137:120–123. A review of the suspected causes and proposed treatments for Alzheimer's disease.

KEY TERMS

FUN FACTS

THE THIRD EYE

The pineal gland is in many ways a third eye. It can be seen in some lizards as a tiny lens sitting in an opening between the bones at the top of the skull. It has a retina, lens, and cornea, although it may not be able to focus an image. In fish, reptiles, and birds, the pineal may lie just under the skin, still exposed to (and sensitive to) light. In humans, it is buried by the great expansion of the cerebral hemispheres and is never exposed to light, but it receives day-night signals from the brain. Scientists have discovered that the pineal gland of many vertebrates has proteins once thought to be unique to the retina of the eye, including rhodopsin, the pigment that absorbs light, and a host of other molecules normally involved in photoreception.

At night, the pineal gland produces melatonin, a hormone that regulates sexual maturation and other seasonal activites. Melatonin inhibits gonadal development in many species of vertebrates, including birds. Some birds never mature if their heads are covered; their pineal glands produced melatonin continuously, inhibiting development of the gonads. In humans, melatonin levels are highest in children under the age of five and decrease before puberty. Destruction of the pineal gland may result in premature sexual maturation.
See: Miller, Julie Ann. 1985. Eye to (Third) Eye. *Science News* 128:298 – 299.

BIRDSONG AND BRAIN CELLS

Many birds, such as zebra finches, are able to learn songs only during a short period as they mature, while other finches can learn throughout life. Neurobiologists have discovered that areas of the brain that control singing continue to produce new neurons during the period of song development. Neuron growth in this brain region does not occur as dramatically in females, which do not sing. The generation of new neurons and new dendritic connections seems linked with the ability to learn.
See: Marler, Peter. 1988. Birdsong and neurogenesis. *Nature* 334:106 – 107.

CHAPTER 28 THE SENSES AND THE BRAIN

MESSAGES

1. **The nervous system monitors needs and conditions, processes information, and initiates responses.**

2. **Sense organs contain nerve cells sensitive to one physical aspect of the environment.**

3. **Nervous systems become more complex in more complicated animals.**

4. **Different senses and different behaviors can be localized to specific regions or groups of regions in the brain.**

5. **The cochlea of the inner ear converts sound waves into nerve impulses. Semicircular canals are involved in balance and detecting motion.**

6. **Rhodopsin molecules change shape when struck by light, initiating a cascade of events that culminates in the sensation of vision.**

7. **In humans, the peripheral nervous system includes the somatic and autonomic nervous systems. The autonomic nervous system includes the sympathetic and parasympathetic nervous systems.**

8. **The human brain includes the brain stem, which helps control autonomic functions; the cerebellum, involved in coordination of fine movement, and the cerebrum.**

9. **The human cerebrum contains structures that regulate homeostasis and basic drives as well as areas that are involved in personality, intelligence, memory, and learning.**

29

THE DYNAMIC ANIMAL: THE BODY IN MOTION

PERSPECTIVE

This chapter completes the survey of animal organ systems and their control. It builds on and integrates concepts introduced in Chapters 20 through 28 and refers to material presented in Chapter 4, The Dynamic Cell; Chapter 5, How Living Things Harvest Energy; and Chapter 18, Invertebrates, the Quiet Majority.

The themes and concepts of this chapter include:

- Muscles and skeletons adapt to stresses both over the lifetime of an individual and over the evolution of the species.
- Fibers made of protein provide the strength needed for building skeletons and muscles.
- Rigorous physical activity involves all of the body's physiological systems.
- There are three main types of skeletons: hydroskeletons, exoskeletons, and endoskeletons.
- Bone structure includes attachment of muscles, ligaments, and tendons; compact bone for strength; spongy bone near joints; and marrow for blood-cell production.
- Cells deposit or reabsorb bone minerals, maintaining homeostasis in blood calcium levels and remodeling and repairing bone.
- To initiate muscle contraction, membranes in muscle fibers release calcium ions. Muscles contract as myosin fibers use ATP energy to slide past actin fibers.
- ATP for muscle contraction comes from three sources: an immediate system, a glycolytic system, and an oxidative system.
- Muscle fibers may be fast-twitch or slow-twitch, differing in their energy sources, myoglobin content, and duration of exertion.
- During a fight-or-flight response, the hypothalamus directs the body to prepare for action.

CHAPTER OBJECTIVES

Students who master the material in this chapter will understand how skeletons and muscles work together to enable an animal to move. They will see that muscles function by the sliding filament mechanism under control of membranes that release calcium ions. These students will appreciate the parallels between the mechanisms of nerve activity and muscle contraction. They will understand how muscle cells acquire energy and how exercise increases one's health and athletic ability.

LECTURE OUTLINE

(p 588) **I. The struggle for dominance between two wild stallions introduces how muscles work and how physiological systems are integrated to maintain homeostasis during heavy exercise.**

(msg 1) **A. Muscles and skeletons adapt to stresses over the lifetime of the individual and the evolution of the species.**

(msg 2) **B. Protein fibers provide the strength needed for building skeletons and muscles.**

(msg 3) **C. Rigorous physical activity involves all of the body's physiological systems so homeostasis can be maintained.**

(p 590) **II. The skeleton provides a scaffold for support and movement.**

(msg 4) **A. Muscles pull against bones or other hard structures. The rigid body support to which muscles attach and apply force is called a <u>skeleton</u>.**

(tpq 1) **B. Many invertebrates have <u>hydroskeletons</u>.**

1. A force exerted against an incompressible fluid in one region can be transmitted to other regions.

2. Muscles are arranged in <u>antagonistic muscle pairs</u>.

a. Circumferential muscles encircle the body wall.

b. Longitudinal muscles extend the length of the animal.

C. Braced framework skeletons include <u>exoskeletons</u> and <u>endoskeletons</u>.

(tr 130) **D. Mammalian endoskeletons are made of <u>bone</u> and are divided into an <u>axial skeleton</u> and an <u>appendicular skeleton</u>.**

1. The axial skeleton includes the <u>skull</u>, <u>vertebral column</u>, ribs, and tailbone.

a. The skull includes the <u>cranium</u>, hyoid bone, and several bones of the face and middle ear.

b. The vertebral column is made of 33 <u>vertebrae</u>.

2. The appendicular skeleton includes the <u>pectoral (shoulder) girdle</u> and <u>pelvic (hip) girdle</u>.

---

tpq 1. A caterpillar has a hydroskeleton while a beetle has an exoskeleton. Which moves more efficiently? What is the selective advantage of a caterpillar's having a hydroskeleton instead of an exoskeleton?

(msg 1) 3. **Natural selection acting over many generations has molded bones to perform specific functions.**

4. **The anatomy of bones reflects their functions.**

   a. **A long bone such as a femur has a shaft, or diaphysis, with an expanded epiphysis at each end. The epiphysis makes a joint, or articulation, with other bones.**

   b. **Processes are projections that serve as attachment sites for ligaments and tendons.**

   c. **The outer layer of compact bone provides support.**

   d. **Spongy bone is crisscrossed by girders to provide strength.**

   e. **Marrow is the site of blood cell production.**

   f. **Cells within the bone secrete or reabsorb proteins and minerals.**

   g. **Bones elongate at epiphysial plates.**

(msg 5) h. **Deposition and removal of protein and minerals goes on continuously.**

(tpq 2) i. **Osteoporosis results from calcium loss.**

5. **Joints are areas of contact between bones.**

   a. **Depending on the mobility of the joint, it may have connective tissue, cartilage pads, and/or a sac (bursa) filled with synovial fluid.**

---

tpq 2. How can osteoporosis be prevented?

(msg 4)

b. Bursitis is inflammation of a synovial sac. Arthritis is long-term inflammation of a joint.

6. Bones act as levers to transmit force.
   a. The origin of a muscle attaches to a bone that remains stationary.
   b. The insertion attaches to the bone that moves.
   c. One muscle in an antagonistic pair usually works against the other. (tpq 3)

(p 594) (tr 131)

III. Skeletal muscles, made of muscle fibers, are the motors of the body.

(msg 2) (tr 132)

A. Myofibrils are threads that lie within each muscle cell.
   1. Sarcomeres are repeating units of light and dark bands.
   2. Each sarcomere consists of interdigitating actin and myosin filaments.
   3. The sarcomeres shorten during muscle contraction.

(msg 6)

B. The myosin fibers slide past each other, a process termed the sliding filament mechanism.
   1. Each myosin molecule has a long straight tail and a head that reaches out like an oar.

---

tpq 3. Why does permanent damage to a muscle often result in atrophy of the antagonistic muscle?

---

2. The heads bind to actin molecules and rotate, pulling the actin and shortening the sarcomere.
3. To release its hold, the myosin binds an ATP molecule, cleaves it, and uses the energy to swing the head forward and bind again to actin.
4. This process is repeated many times as the sarcomere shortens.
5. Without ATP, myosin would not release actin and rigor mortis would result.

C. Muscle contraction is controlled by cell membranes.

(tpq 4) 1. A motor neuron releases acetylcholine onto the sarcolemma, initiating an action potential.
2. Transverse tubules propagate the action potential into the cell's interior.
(msg 6) 3. A signal from the transverse tubules spreads to the sarcoplasmic reticulum, causing the sarcoplasmic reticulum to release calcium ions.
4. Calcium ions cause regulatory proteins, troponin and tropomyosin, to release their hold on actin, allowing myosin heads to bind.
5. Membrane pumps actively transport calcium ions back into the sarcoplasmic reticulum, preparing the muscle for the next contraction.

---

tpq 4. Some insecticides poison the enzyme that cleans up acetylcholine. What effects might they cause?

(msg 7) D. Cardiac muscle and smooth muscle both differ from skeletal muscles.

1. Cardiac muscle cells are electrically connected to each other via intercalated discs and can propagate impulses. They are striated.
2. Smooth muscle cells lack striations and communicate with each other by gap junctions.

(msg 8) E. Three energy systems supply muscles with ATP.

1. The duration of activity determines which system is used.
2. The immediate energy system is instantly available for one brief action. Stored ATP and creatine phosphate are the energy sources.
3. The intermediate system depends on glycolysis and fuels activities lasting from 1 to 3 minutes. Lactic acid is the waste product of anaerobic glycolysis.
4. The oxidative energy system supplies energy for a longer duration. The Krebs cycle and oxidative phosphorylation generate ATP.

(msg 9) F. Some skeletal muscle types rely more on one of these systems than another.

1. Slow oxidative muscle fibers obtain most of their ATP from the oxidative system. They have abundant myoglobin and resist fatigue.

2. Fast glycolytic fibers react faster and fatigue faster. They derive most of their ATP from glycolysis and contain little myoglobin.
3. Fast oxidative-glycolytic fiber has characteristics midway between fast- and slow-twitch fibers.

(p 600) **IV. Exercise physiology is linked to survival.**

(msg 10) A. The fight-or-flight response galvanizes nearly every system in the body into action.

1. This response is managed largely by the hypothalamus.

(tpq 5)

2. In times of stress, the hypothalamus sends signals to various body tissues via the sympathetic nervous system.
3. Responses include an increase in the supply of energy-rich molecules in the blood, elevation of heart rate and blood pressure, increased ventilation, release of epinephrine, and diversion of blood to the muscles.
4. The hypothalamus causes the pituitary to release ACTH.
   a. ACTH stimulates the adrenal gland to secrete more cortisol.
   b. Cortisol results in gluconeogenesis and the liberation of fatty acids and amino acids.

---

tpq 5. Is stress good or bad for you as a student?

---

5. The pituitary secretes antidiuretic hormone.

B. The body adapts to repeated stress.

1. Athletic training increases the amount of oxygen that can be delivered to muscles (the maximal oxygen uptake).
   a. The amount of blood pumped per heartbeat increases.
   b. Exercise increases the muscles' ability to extract oxygen from the blood.
2. Exercise stimulates the release of endorphins.
3. Hypokinetic diseases are caused by inactivity.
4. Regular exercise decreases the risk of heart disease.

INSTRUCTIONAL AIDS

VIDEOS, FILMS, SLIDES

Bones and Muscles. International Film Bureau, Chicago. Portrays how human skeletal and muscular systems work together and presents the anatomy of joints, vertebrae, and the skull. Film, video.

The Mammalian Skeleton and Teeth. Educational Images Ltd., Elmira, NY. A rabbit skeleton is examined in detail and compared with different mammals. Transparencies.

Muscle: Chemistry of Contraction. Pennsylvania State University AV Services, University Park, PA. Describes internal muscle structure, actin and myosin, and the roles of ATP and calcium ions. Film.

REFERENCES

Casey, Timothy M. 1991. Energetics of caterpillar locomotion: Biomechanical constraints of a hydraulic skeleton. *Science* 252:112–114. An analysis of the energetic costs of locomotion in caterpillars and a comparison with animals with jointed skeletons.

Dodson, Edward O., and Peter Dodson. 1985. *Evolution: Process and Product*, Third Edition. Boston: PWS. Includes a concise and clearly-written description of the evolutionary history of the vertebrate skeleton.

Hanegan, James L., and H. R. McKean. 1989. Muscle activities that stretch the mind. *American Biology Teacher* 51 (8):489–493. How to build a simple model made of wood and rubber bands that students can use to understand concepts of muscle contraction and limb control.

Ruben, John E., and A. A. Bennett. 1987. The evolution of bone. *Evolution* 41 (6):1187–1197. A technical discussion of the origin of the calcium phosphate skeleton, arguing that vertebrate physiology makes calcium phosphate a better material than calcium carbonate.

Weisburd, Stefi. 1988. The muscular machinery of tentacles, trunks and tongues. *Science News* 133:204–205. Discusses how cephalopods, which have no hard skeleton or hydroskeleton, manage to move.

KEY TERMS

antagonistic muscle pair	page 590
appendicular skeleton	590
axial skeleton	590
bone	590
cardiac muscle	598
cranium	590
endoskeleton	590
exoskeleton	590
hydroskeleton	590
joint	593
ligament	592
muscle fiber	594
myofibril	594
pectoral (shoulder) girdle	591
pelvic (hip) girdle	591
process	592
sarcomere	594
skeletal muscle	594
skeleton	590
skull	590
smooth muscle	598
tendon	592
vertebra	590
vertebral column	590

FUN FACTS

HORMONES AND BONES

Osteoporosis is especially common in inactive women past menopause, which often occurs around the age of fifty. Through processes that are poorly understood, low estrogen levels lead to an imbalance between bone mineral deposition and reabsorption.

There is no evidence of estrogen receptors in skeletal tissues, so scientists believe that other hormones may play an intermediary role, perhaps by suppressing the cells that break down bone. It has been estimated that 1.2 million fractures occur in the United States each year as a consequence of osteoporosis. Typical injuries include hip fractures and vertebral collapse, which happens twelve times as often among older women than among older men.

Physicians often administer estrogen to postmenopausal women to prevent or treat osteoporosis. Women with bone loss may be advised to take large doses of calcium, but researchers are not entirely convinced that additional calcium intake can counter osteoporosis unless it is accompanied by hormone therapy. One effective way to prevent or treat osteoporosis is through physical exercise, especially in activities that stress the skeleton. Prolonged bed rest in males and females results in a net loss of bone, as does weightlessness, a significant problem for astronauts who spend much time in space. Scientists don't understand the mechanism by which mechanical stress affects bone, but have suggested that electricity may be important.

When bone is bent, electrical fields are produced. The electricity may stimulate production of some hormone-like skeletal growth factor. The external application of electrical fields promotes bone repair and the magnitude of the fields produced in a bone increases as the stress on the bone increases. Women who remain active, especially in weight-bearing sports such as running, reduce their chances of osteoporosis even without hormone therapy.

See: Johnson, C. Conrad, and C. Slemenda. 1987. Osteoporosis: an overview. *Physician and Sportsmedicine* 15 (11):65 – 68.

Kolata, Gina. 1986. How important is dietary calcium in preventing osteoporosis? *Science* 233:519 – 520.

Lindsay, Robert. 1987. Estrogen and osteoporosis. *Physician and Sportsmedicine* 15 (11):105 – 108.

Smith, Everett L., and Catherine Gilligan. 1987. Effects of inactivity and exercise on bone. *Physician and Sportsmedicine* 15 (11):91 – 100.

CALCIUM'S ROLE AS A MESSENGER

In a wide variety of animal cells, calcium conveys signals received at the cell surface to the inside of the cell. Muscle cells, for example, rely on intracellular calcium cycling to regulate contraction. Calcium levels within cells are typically 0.0001 the levels found in the plasma; membrane pumps actively export calcium ions from cells. When a cell is stimulated by an extracellular signal, membrane channels allow calcium to enter, prompting responses within the cell. Complex reactions among ATP, protein kinases, cyclic adenosine monophosphate, and calcium ions are involved in transient and sustained responses.

See: Rasmussen, Howard. 1989. The cycling of calcium as an intracellular messenger. *Scientific American* 261 (4):66 – 73.

CHAPTER 29 THE DYNAMIC ANIMAL: THE BODY IN MOTION

MESSAGES

1. **Muscles and skeletons adapt to stresses over the lifetime of an individual and the evolution of the species.**

2. **Fibers made of protein provide the strength needed for building skeletons and muscles.**

3. **Rigorous physical activity involves all of the body's physiological systems.**

4. **Bones typically articulate with other bones, serve as attachment sites for ligaments and tendons, and are levers that transmit force applied by antagonistic muscle pairs.**

5. **Cells deposit or reabsorb bone minerals, maintaining homeostasis in blood calcium levels and remodeling and repairing bone.**

6. **To initiate muscle contraction, membranes in muscle fibers release calcium ions. Muscles contract as myosin fibers use ATP energy to slide past actin fibers.**

7. **Vertebrates have three types of muscle: cardiac, skeletal, and smooth muscle.**

8. **ATP for muscle contraction comes from three sources: an immediate system, a glycolytic system, and an oxidative system.**

9. **Muscle fibers may be fast-twitch or slow-twitch, differing in their energy sources, myoglobin content, and duration of exertion.**

10. **During a fight-or-flight response, the hypothalamus directs the body to prepare for action.**

30

PLANT ARCHITECTURE AND FUNCTION

PERSPECTIVE

This chapter begins Part V, How Plants Survive, by describing the anatomy and life cycle of vascular plants. It continues the discussion of plant life cycles started in Chapter 17 and presents the plant structures associated with photosynthesis (Chapter 6) and the transport of materials, which will be covered in greater depth in Chapter 32, Water, Nutrients, and Plant Survival.

The themes and concepts of this chapter include:

- Plant form and function represent compromises between conflicting needs.
- Anatomical parts that we examine separately make up tissues, organs, and systems that function collectively as whole plants.
- Plants have open growth based on perpetual embryonic centers known as meristems.
- Vascular plants have two major parts, the root and shoot, with three major tissue systems: dermal, ground, and internal transport tissue systems.
- Flower parts serve to unite pollen and egg and to support the developing seed.
- A seed is an embryo, with nutritive tissue, surrounded by a protective seed coat. A ripened ovary forms the fruit.
- Roots and shoots have primary and secondary meristems, an epidermis, a cortex, xylem, phloem, and waxy layers that prevent uncontrolled water gain or loss.
- Secondary growth in woody plants occurs in the vascular cambium. Bark is formed by layers of dead cells from the periderm.

CHAPTER OBJECTIVES

Students who master the material in this chapter will understand that a plant's form and function represent a compromise between conflicting needs. They will see that vascular bundles and other structural features extend throughout the plant body and unite leaves, stems, and roots. These students will understand how and why patterns of growth and reproduction differ between plants and animals and will be able to recount the life cycle of plants from the germination of a seed through the production of new seeds. They will be able to describe the anatomy of roots, stems, leaves, flowers, and seeds.

LECTURE OUTLINE

(p 606) **I. The life cycle of a pear tree introduces plant architecture and function. Three themes emerge:**

(tpq 1) (msg 1) **A. A plant's form and function strike a compromise between conflicting needs.**

B. Anatomical parts that we examine separately make up tissues, organs, and systems that function collectively as whole plants.

(msg 2) **C. Plants have open growth based on perpetual embryonic centers that continually produce new organs and larger body size throughout life.**

(p 608) **II. The anatomy of vascular plants is characterized by internal transport tubes.**

(msg 3) **A. The plant's main axis includes roots and shoot.**

1. The shoot includes the stem, branches, leaves, flowers, and fruit.

tpq 1. How do plants and animals differ in the ways they make their living? How are the hazards of life different?

2. Roots support the plant physically and nutritionally.

   a. Taproot systems develop directly from the radicle and grow straight down into the soil. They can act as storage organs.

   b. Fibrous root systems grow downward and outward from the plant stem and branch repeatedly.

   c. Adventitious roots arise from aboveground structures.

(tpq 2) B. Plants have tissues composed of different cell types.

(msg 3) 1. The dermal tissue system provides protection and reduces water loss.

   a. In young seedlings, stems, and leaves, the dermal tissue consists of an epidermis.

   b. In woody plants, the periderm replaces the epidermis.

2. The ground tissue system provides support and stores starch.

   a. Parenchyma, which makes up the majority of ground tissue, is made up of loosely packed, thin-walled parenchyma cells.

---

tpq 2. Do plants have as many cell types and tissue types as animals?

---

i. Most cells have only a thin primary cell wall and are often unspecialized, able to give rise to new cells and new cell types.

ii. Some are specialized, e.g., for photosynthesis or starch storage.

b. Collenchyma cells help hold the plant body together; they often have a cylindrical shape and thick primary walls, and connect end-to-end to form fibers.

c. Sclerenchyma tissue is hard and often surrounds and reinforces vascular tubes.

i. Sclerenchyma cells die at maturity and leave behind a thick secondary cell wall hardened with lignin.

ii. Sclereids, or stone cells, are hard and crystalline.

iii. Hemp and flax fibers are another type of sclerenchyma tissue.

C. Two kinds of vascular tissues extend from the roots to the leaves.

(tpq 3) 1. Xylem transports water and minerals from the roots upward. Wood consists of xylem vessels.

---

tpq 3. Must cells be alive to function?

---

a. Xylem includes tracheids and vessel members, which die before they transport water.

b. Tracheids are long, narrow cells that overlap at their ends, and which have pits to allow water to pass between cells.

c. Vessel members occur mainly in flowering plants. They have perforation plates between cells.

2. Phloem transports dissolved sugars and proteins from source to sink.

a. Phloem cells, known as sieve tube members, are thin-walled and alive at maturity, but lack a nucleus.

b. Sieve plates allow fluids and solutes to pass in and out.

c. Companion cells have nuclei that direct the activities of nearby sieve tube members.

(tpq 4) (msg 2)

D. Open growth is characterized by the addition of new organs throughout life.

1. Since plants cannot migrate, they grow toward light, water, and nutrients and away from harmful situations.

(tr 133)

2. Continual growth is based on meristems, perpetually embryonic tissue.

---

tpq 4. Are there any animals with an open growth body plan?

a. Apical meristems are at the tips of roots and stems and give rise to primary growth.

b. Lateral meristems produce secondary growth in stems or roots.

i. Vascular cambium produces wood.

ii. Cork cambium generates cork.

c. Woody plants have secondary growth; herbaceous plants have only primary growth.

(tpq 5) 3. Annual plants complete their life in a single year. Biennials have a two-year life cycle. Perennials live for many years.

(p 614) III. Flowers, fruits, seeds, and embryos perpetuate the species.

(msg 4) A. Flowers are modified shoots that contain the reproductive organs.

(tr 134) 1. The various parts of the flower help ensure the transfer of pollen to egg.

a. Colored petals and sepals attract pollinators.

b. A stamen produces pollen in its anther, which sits atop a filament.

c. Carpels include ovaries, which house the ovules. The style supports the stigma, on which pollen grains germinate.

---

tpq 5. What are the evolutionary advantages of an annual life cycle? Of a perennial life cycle?

---

2. **The numbers, sizes, and shapes of flower parts vary from species to species.**
    a. **Perfect flowers have both stamens and carpels.**
    b. **Imperfect flowers lack either male or female parts.**
    c. **Monoecious plants have both female and male flowers.**
    d. **Dioecious plants are either male or female.** (tpq 6)
3. **Pollination involves the transfer and germination of pollen.**
    a. **One of the three haploid nuclei in each pollen grain directs the growth of the pollen tube.**
    b. **The other two cells, the sperm, travel through the tube to the ovule and embryo sac.**
    c. **In double fertilization, one sperm nucleus fuses with the egg nucleus and forms the diploid embryo.**
    d. **The other fuses with two polar nuclei, forming the endosperm.**

(msg 5) B. **In the pear, the wall of the ovary and associated structures enlarge to form the fruit.**
1. **The ovule's integument gives rise to the seed coat.**

---

tpq 6. What is the evolutionary advantage of being monoecious? Of being dioecious?

---

2. The embryo produces a radicle (to become the main root) at one end and a plumule (which will develop into the shoot) at the other.
3. The embryo's first leaves are called cotyledons. Flowering plants may be dicotyledons or monocotyledons.
4. The hypocotyl is the stem between the radicle and cotyledons.
5. The epicotyl is the stem between the cotyledons and the next leaf, and will give rise to the entire shoot.
6. Various mechanisms disperse seeds.

C. During germination, the embryo grows, breaks free of the seed coat, and begins independent life.

D. As a pear seedling grows, primordia differentiate into new shoots, leaves, and flowers.
1. Nodes form at the level of each leaf primordium.
2. Bud primordia form in the upper angle between the leaf and stem and develop into buds, which can mature into flowers or shoots.

(p 618) IV. Roots anchor the plant, absorb water and minerals, and often store starch.

A. Roots are zoned along their length.
1. The root cap pushes through the soil. Cells are damaged and scraped off.

(tpq 7) 2. The apical meristem produces cells that replace the root cap and others that enlarge in the zone of elongation, pushing the root cap through the soil.

3. In the zone of maturation, these cells mature and specialize.

(msg 6) B. Roots are composed of layers.

1. The root hairs of the epidermis absorb water and minerals from the soil.

2. The cortex surrounds the central core of the vascular system and often stores starch.

3. The endodermis isolates the cortex from the vascular system.

a. The walls of endodermal cells are impregnated with water-resistant suberin.

b. The Casparian strip encircles each cell, preventing the diffusion of water through the cell wall into the vascular system.

c. Water and minerals can reach the vascular system only by passing through the cytoplasm of the endodermal cells.

C. The vascular system includes xylem, phloem, and other tissues.

1. The pericycle gives rise to lateral roots.

---

tpq 7. How can plants, which do not have muscles, push their roots through the soil?

---

2. In monocots, pith is a central tissue consisting of large, thin-walled parenchyma storage cells.

(tr 135) D. Secondary growth occurs in the roots of perennial plants.

1. The pericycle gives rise to the lateral meristems, the vascular cambium, and the cork cambium.
2. Cork and cork cambium make up the periderm.

(p 622) V. Stems are the major structural elements of the shoot.

A. Stems support the plant vertically and transport water, minerals, sugars, and other substances.

(msg 6) B. The primary growth of stems resembles the primary growth in roots.

1. The dermal system consists of a layer of epidermal tissue one cell thick. The cuticle resists dessication.
2. The ground system is primarily cortex made of parenchyma cells. It may include strands of collenchyma and pith.
3. Xylem and phloem are organized into vascular bundles.
   a. In most dicots, vascular bundles form a ring around the central core of pith.
      i. Cambium from each bundle extends to adjacent bundles, forming a ring.

ii. In conifers and woody dicots, this cambium gives rise to secondary growth.

(tpq 8) b. In monocots and some dicots, the bundles are scattered throughout the cortex, there is no vascular cambium, and stems are incapable of secondary growth.

(msg 7) (tr 136) C. Secondary stem growth in woody plants produces secondary xylem and secondary phloem.

1. The cambium makes far more secondary xylem than phloem. Most of the stem tissue is made of dead xylem cells (or wood).
2. Trees growing in temperate regions will have growth rings with thicknesses influenced by environmental factors.
3. Heartwood is composed of xylem cells that have ceased to conduct water and have become infiltrated with oils, gums, resins, and tannins.
4. The sapwood continues to conduct water.

D. The periderm, phloem, and cork cells make up the bark.

---

tpq 8. If monocots are incapable of secondary growth, how can monocots like palms become trees?

(p 625) **VI. Leaves serve to expose a photosynthetic surface area, to obtain carbon dioxide, and to help draw water and nutrients up through the vascular system.**

A. Most leaves have a <u>blade</u> and a <u>petiole</u> or sheath.

1. **Blades are often flat and thin, maximizing surface area, but must be able to minimize water loss.**
2. **Leaves may be reduced in size, absent, or modified as spines, holdfasts, flower petal substitutes, or to serve other functions.**
3. **Leaves may be simple or compound.**

B. A leaf's anatomy is associated with its functions.

(msg 1)
1. **The surface is covered with a waxy cuticle perforated by <u>stomata</u>, which open and close, allowing water vapor and gases to enter or exit.**

(msg 8)
2. **The <u>mesophyll</u> layer is made of two types of parenchyma surrounding transport vessels.**
   a. **Palisade parenchyma is the main photosynthetic tissue.**
   b. **Spongy parenchyma absorbs carbon dioxide.**
3. **Vascular tissue in leaves is continuous with vascular tissues in stems.**
   a. **Monocots have parallel <u>veins</u> (called parallel venation). Dicots have netted venation.**
   b. **Bundle-sheath cells surround veins and protect them from exposure to air.**

INSTRUCTIONAL AIDS

VIDEOS, FILMS, SLIDES

How Pine Trees Reproduce: Pine Cone Biology. Britannica Films and Video, Chicago. Time-lapse and close-up photography reveals fertilization, dissemination, and germination of seeds. Film, video.

The Kingdom of Plants. Human Relations Media, Pleasantville, NY. This three-part discussion of botany covers anatomical and functional diversity, physiology, reproduction, and genetics. Filmstrip, video.

Plant Life Cycles. Connecticut Valley Biological Supply, Southampton, MA. Diagrams and photos illustrate the major types of plant life cycles. Transparencies.

Plants: Parts and Processes. National Geographic Society, Washington, DC. Describes the functions of and differences between plant parts, compares woody and herbaceous plants, presents flowers, seeds, and germination. Filmstrips.

REFERENCES

Bold, Harold C., and J. W. LaClaire II. 1987. *The Plant Kingdom*, Fifth Edition. Englewood Cliffs, NJ: Prentice-Hall. A small text covering botany from cyanobacteria to angiosperms. Many black-and-white photos and drawings.

Hafner, Robert. 1990. Fast plants: Rapid-cycling brassicas. *American Biology Teacher* 52 (1):40 – 46. Varieties selected for fast growth and maturation allow the whole life cycle to be observed in 35 to 40 days. Suggests experiments and lab activities.

Kaufman, Peter B., et al. 1983. *Practical Botany*. Reston, VA: Reston Publishing. Covers a wide variety of botanical topics with a focus on propagation, culture, and care of plants.

Pietraface, William J. 1988. Plant regeneration. *American Biology Teacher* 50:234 – 235. Laboratory exercises in plant tissue culture.

KEY TERMS

adventitious root	page 608	perennial	614
annual plant	614	pericycle	621
apical meristem	613	petiole	627
bark	624	phloem	610
biennial	614	pith	621
blade	627	plumule	616
bud	618	primary growth	613
Casparian strip	620	primordium	617
collenchyma	610	radicle	616
companion cell	613	root	608
cork	622	root cap	618
cork cambium	613	root hair	620
cortex	620	sclerenchyma	610
cotyledon	617	secondary growth	613
dermal tissue system	608	shoot	608
endodermis	620	sieve plate	613
epicotyl	617	sieve tube member	613
fibrous root	608	stoma	627
ground tissue system	609	taproot	608
hypocotyl	617	tracheid	610
lateral meristem	613	vascular cambium	613
meristem	613	vascular tissue system	609
mesophyll	627	vein	627
node	618	vessel member	610
open growth	613	wood	624
parenchyma	610	xylem	610

FUN FACTS

LAND OF THE GREEN GIANTS

Some of the largest organisms that have ever lived are conifers that thrive in the narrow coastal zone west of the Cascade Mountains, from northern California to Alaska. These coniferous forests hold the world's record for biomass accumulation; redwood forests can contain more than 3,300 tons of organic material per hectare (a square 100 meters on a side), several times the biomass of a tropical rain forest.

Conifers have adaptations that enable them to outcompete hardwoods (broad-leaved trees)in the Pacific Northwest, where 90 percent of the year's precipitation, mostly rain, falls between late autumn and early spring. Summers are characterized by extended droughts. Being evergreens, conifers can continue photosynthesis throughout the mild winters. Their massive trunks allow them to store water and nutrients through summer droughts. Hardwoods cannot store as much water as conifers can, and hot, dry weather reduces their transpiration and inhibits photosynthesis.

Hardwoods need rain throughout the growing season, not only to maximize transpiration and photosynthesis, but to keep the litter decomposers active. Hardwoods shed their leaves every year and are able to withdraw only about a third of the nutrients before losing them. They rely on decomposers in the litter layer to recycle the nutrients. But when the litter is dry, as it usually is during the Pacific Northwest summer, microbial activity ceases and hardwoods face nutrient starvation. During the winter, the soil remains unfrozen, decomposers are active, and nutrients are available, but hardwoods have lost their leaves and thus cannot transpire and draw in soil water and nutrients. Conifers continue transpiring and photosynthesizing throughout the winter, even when the temperature is a few degrees below freezing. They can make half of their annual production of fixed carbon during the winter.

The spire-like shape and drooping branches are adaptations that allow conifers to shed snow and to maximize the leaf area exposed to the low winter sun. Their needle-like leaves shed rain quickly, lowering the humidity at the leaf surface so transpiration can occur.

See: Waring, Richard H. 1982. Land of the giant conifers. *Natural History* 91 (10):55–62.

CHAPTER 30 PLANT ARCHITECTURE AND FUNCTION

MESSAGES

1. Plant form and function represent compromises between conflicting needs.

2. Plants have open growth based on perpetual embryonic centers known as meristems.

3. Vascular plants have two major parts, the root and shoot, with three major tissue systems: dermal, ground, and internal transport tissue systems.

4. Flower parts serve to unite pollen and egg and to support the developing seed.

5. A seed is an embryo, with nutritive tissue, surrounded by a protective seed coat. A ripened ovary forms the fruit.

6. Roots and shoots have primary and secondary meristems, an epidermis, a cortex, xylem, phloem, and waxy layers that prevent uncontrolled water gain or loss.

7. Secondary growth in woody plants occurs in the vascular cambium. Bark is formed by layers of dead cells from the periderm.

8. Leaves have two kinds of parenchyma cells for photosynthesis and carbon dioxide absorption. Stomata control the movement of gases into and out of the leaf.

31

REGULATORS OF PLANT GROWTH AND DEVELOPMENT

PERSPECTIVE

This chapter details the mechanisms that regulate the activities presented in Chapter 30, Plant Architecture and Function. In many ways, plant control systems parallel the hormonal systems of animals, the topic of Chapter 26. The material in this chapter provides a basis for the discussion of water, nutrients, and plant survival presented in Chapter 32.

The themes and concepts of this chapter include:

- Plants generally adjust to changing environmental conditions via growth.
- Changes in plant morphology or physiology are often regulated by plant hormones produced in response to environmental factors.
- Plant hormones interact in complex ways and may promote or inhibit growth.
- Phytochrome allows seeds and plants to detect changes of season and to respond to shading.
- Plants become oriented in space by means of tropisms.
- Apical dominance results from auxins produced at the apical meristem. Cytokinins from the roots act in opposition.
- The regulation of germination and flowering often depends on external cues such as temperature and photoperiod.
- Fruit development may be regulated by ethylene or by hormones produced by viable seeds.
- Leaf senescence and dormancy allow plants to survive adverse conditions.
- Plants protect themselves from herbivores and disease by producing toxic compounds and walling off injured areas.

CHAPTER OBJECTIVES

Students who master the material in this chapter will understand how hormones and environmental cues interact to regulate plant growth and development. They will see how plants respond by oriented growth, seasonal growth or senescence, flowering, and dormancy. These students will understand how phytochrome allows plants to perceive photoperiod and how plants protect themselves from damage, herbivores, and disease.

LECTURE OUTLINE

(p 630) **I. Foolish seedling disease, which causes rice plants to grow abnormally tall and spindly, introduces the topic of plant growth regulators.**

A. Foolish seedling disease is caused by a fungus that releases plant growth hormones known as gibberellins.

B. External regulators include temperature, the amount of light, gravity, and the duration of daylight.

C. Unlike animals, which have nervous systems as well as hormonal systems, plants rely largely on hormones to respond to environmental cues. Individual plant hormones often have widespread effects.

D. Four principles unify this chapter:

(tpq 1) (msg 1) **1. Plants generally adjust to changing environmental conditions via growth.**

tpq 1. Do plants need homeostatic mechanisms like those of animals?

(msg 2) 2. **Changes in plant morphology or physiology are often regulated by hormones produced in response to environmental factors.**

(msg 3) 3. **Plant hormones interact in complex ways.**

4. **Plant hormones act at the level of cells to induce cell division, enlargement, or maturation.**

(p 632) **II. There are five major kinds of plant hormones.**

A. Gibberellins include several related compounds, are produced in a variety of organs, and are primarily involved in regulating plant height.

B. Auxins are generally growth promoters.

1. **The most common natural auxin is indoleacetic acid, or IAA.**
2. **Auxins are generally made in shoot apical meristems and developing leaves.**
3. **They promote growth through cell elongation by triggering enzymes that soften cell walls.**
4. **Auxins also prevent senescence and inhibit lateral growth.**

(tpq 2) 5. **The herbicides 2,4-D and 2,4,5-T are synthetic auxins.**

C. Cytokinins generally stimulate cell division, especially in lateral buds, and appear to move from root to shoot.

---

tpq 2. For what purposes might artificial plant hormones be used?

D. Abscisic acid (ABA) is a growth inhibitor that apparently functions by inhibiting translation of mRNA.

1. ABA moves only a short distance from the site of production.
2. ABA stimulates the closure of stomata, decreases photosynthesis, and promotes dormancy and abscission.

E. Ethylene is a gas that is produced by ripening fruits and stimulates the ripening of adjacent fruits, senescence, and abscission.

(p 634) III. Germination is regulated by internal and external factors.

(tpq 3) A. Some seeds, especially in the tropics, germinate as soon as they imbibe enough water.

(msg 4) B. Many temperate-zone plants do not germinate until spring, thereby avoiding harsh winter conditions.

C. Moisture stimulates germination by leaching away growth-inhibiting hormones such as ABA.

D. Light is a determining factor in germination of some seeds.

E. Seeds of some species can lie dormant for thousands of years.

---

tpq 3. Why would tropical seeds be less likely to remain dormant than temperate-zone seeds?

F. Water-soaked corn kernels demonstrate the activities of hormones produced by the embryo during germination.

(p 635) **IV. Environment and hormones influence growth and development.**

(msg 5) A. Tropisms are directional growth or orientation in response to external or internal stimuli.

1. Tropisms include hydrotropism, thigmotropism, phototropism, and gravitropism.
2. The plant root is positively gravitropic; the shoot is negatively gravitropic.

B. Phototropism is plant growth toward light.

1. Light causes auxin to move from the tip to the stem.
2. Auxin softens cell walls on the side away from light, allowing cells to elongate.

C. Not all plant movements are tropisms.

1. Nastic responses are plant movements in response to stimuli but not oriented with respect to the stimuli. They may be reversible.
2. (tpq 4) The leaves of some plants droop at night and rise during the day. Even if kept in the dark, they continue periodic movement, evidence of a circadian clock.

---

tpq 4. Can plants tell time?

D. External factors can influence plant growth and shape in ways that adapt the plant to its environment.

1. **Plants may elongate in the direction of light. Plants that have grown long and pale in darkness are etiolated.**

(msg 6)
(tr 137)

2. **Plants have a pigment called <u>phytochrome</u> that allows them to detect the quality and quantity of light.**

   a. **Phytochrome does not diffuse from cell to cell, and is therefore not a hormone.**

   b. **Phytochrome regulates germination, flowering, and other activities linked to the light cycle.**

   c. **Phytochrome interconverts between two forms, depending on the relative intensity of red light (absorbed by P_r) and far-red light (absorbed by P_{fr}).**

   d. **P_{fr}, the active form of the molecule, spontaneously changes to P_r in darkness.**

   e. **If phytochrome is largely in the P_r form, the plant loses chlorophyll and becomes pale, spindly, and elongated.**

(msg 7)

3. **There are two major growth forms in plants, conical or spherical, depending on <u>apical dominance</u>.**

a. Auxins promote apical growth but suppress lateral growth.

i. The further from the apex, the lower the concentration, leading to more lateral growth on lower branches.

ii. Cytokinins come up from the roots, counteract auxins, and cause lower branches to grow more strongly.

b. Removal of the apical bud releases lower buds from apical dominance.

(p 639) V. Environmental cues and internal hormonal triggers control flowering and fruit formation.

(msg 8) A. Some plants flower only after a period of exposure to cold, a process known as vernalization.

B. Many plants track seasons by photoperiod rather than daily temperature.

(tpq 5) 1. Depending on the timing of flowering, plants may be called short-day plants, long-day plants, or day-neutral plants.

a. Long-day plants flower in spring or early summer, when days are becoming longer.

b. Short-day plants bloom in late summer or fall.

(tr 138) 2. Plants measure night length (instead of day length) by accumulating P_r. Nocturnal exposure to light resets the clock by converting P_r to P_{fr}.

---

tpq 5. What is the evolutionary value of precise timing of flowering?

---

a. In short-day plants, P_{fr} blocks flowering.

b. In long-day plants, P_{fr} promotes flowering.

C. Hormones generated in leaves apparently control flowering. Florigen is the name given to a putative flower-inducing hormone.

(msg 9) D. Fruit development may be controlled by hormones or the presence of viable seeds.

1. Seeds may produce auxins, promoting fruit development.

(tpq 6) 2. Ethylene accelerates ripening.

(p 641) VI. Plants often respond to the onset of winter with senescence or abscission.

A. In annuals, the life cycle ends with seed production and dispersal; the adult plants are programmed to die.

(msg 10) B. Many perennials are deciduous.

1. Deciduous leaves may serve as organs of excretion.

2. In the autumn, the deciduous plant withdraws nutrients and chlorophyll from the leaves, leaving behind other pigments that are the sources of fall colors.

C. After the leaves have fallen, the plant enters dormancy.

---

tpq 6. How is it true that one bad apple can spoil a barrel full of apples?

(p 643) **VII. Plants protect themselves from environmental threats.**

(msg 11) **A. Plants produce secondary compounds that repel, kill, or interfere with the normal activities of herbivores.**

1. **Some compounds deter insects from eating plant parts or laying eggs on them.**
2. **Some plants produce compounds that disrupt insect metamorphosis.**
3. **Biologists recommend avoiding damaged fruits and vegetables to reduce one's risk of cancer.**

B. Plants can wall off damaged tissues to prevent invaders from gaining access to healthy tissues, a process called compartmentalization.

INSTRUCTIONAL AIDS

VIDEOS, FILMS, SLIDES

The Growth of Plants. Britannica Films and Video, Chicago. Illustrates the dynamics of plant growth and explains how materials are dispersed through the plant and how various stimuli affect plant growth. Film, video.

How Plants Move. International Film Bureau, Chicago. Demonstrates phototropism, thigmotropism, and gravitropism. Video, film.

Plant Growth Regulators. Carolina Biological Supply Company, Burlington, NC. Explains the physiology of plant growth hormones. Filmstrip, video.

REFERENCES

Bohnsack, Charles W. 1989. Cytokinin induces cell division and differentiation using intact plants. *American Biology Teacher* 51:106–108. A demonstration/lab that shows the effects of plant hormones while avoiding the necessity of tissue culture and aseptic technique.

Emmeluth, Donald S., and D. E. Brott. 1985. Some plant hormone investigations that work. *American Biology Teacher* 47:480–482. Presents lab exercises that demonstrate hormonal regulation of plant development.

Friend, Douglas J. C. 1987. Measuring circadian rhythms in plants. *American Biology Teacher* 49:46–47. A simple exercise demonstrating daily activity cycles.

Meidner, Hans. 1984. *Class Experiments in Plant Physiology*. Winchester, MA: Allen and Unwin. A source book for teachers who conduct class experiments and demonstrations in botany.

Moses, Phyllis B., and Nam-Hai Chua. 1988. Light switches for plant genes. *Scientific American* 258 (4):88–93. Discusses segments of DNA that respond to light energy by activating gene transcription.

KEY TERMS

abscisic acid	page 633	ethylene	633
abscission	633	florigen	641
apical dominance	639	gibberellin	630
auxin	632	photoperiod	640
cytokinin	633	phytochrome	637
deciduous	642	senescence	641
dormancy	633	tropism	635

FUN FACTS

TASTY TOXINS

Plants produce many poisons to protect themselves against herbivores, fungi, bacteria, and other plants. Humans eat many plants despite their toxicity, and in fact we treasure some plants specifically because of their toxins. Coffee is prized for its caffeine, a toxin that causes nerve damage, hyperactivity, tremors, and stunted growth among insects that graze on the coffee plant. Caffeine is an alkaloid, a class of substances that includes other toxic plant compounds: cocaine, nicotine, morphine, and coniine, the toxin in the hemlock that poisoned Socrates.

Vanilla is an extract from the seedpods of an orchid, but this phenol and closely related compounds are produced by a wide variety of plants. They inhibit the germination of seeds and the growth of competing plants. Black pepper contains safrole, which has been demonstrated to cause cancer in laboratory animals. The common grocery-store mushroom contains agaritine and related compounds that are carcinogens. Plants of the Umbelliferae family, including celery, poison hemlock, carrots, and parsley, have psoralens, which are potent mutagens when exposed to ultraviolet light. Potatoes produce solanine and chaconine, which interfere with neurotransmitters and may cause birth defects. Alfalfa sprouts contain canavanine, a toxic molecule that resembles the amino acid arginine and may be incorporated into proteins in place of arginine. Canavanine seems to cause lupus erythematosus, a disease characterized by damage to the immune system and the production of antibodies that attack the body's own DNA and tissues.

Biochemist Bruce Ames estimates that the human dietary intake of nature's pesticides is at least 10,000 times greater that our intake of manufactured pesticides. But he points out that the diet includes many substances that counteract toxins, including vitamin E, carotene (which gives carrots their orange color), selenium (an element that occurs as a soil mineral), and a variety of other beneficial compounds. Research has demonstrated again and again that a balanced, diverse diet helps guard against disease.

See: Ames, Bruce N. 1983. Dietary carcinogens and anticarcinogens. *Science* 221:1256 – 1264.

Nathanson, James A. 1984. Caffeine and related methylxanthines: Possible naturally occurring pesticides. *Science* 226:184 – 187.

Rice, Elroy. 1974. *Allelopathy*. New York: Academic Press.

CHAPTER 31 REGULATORS OF PLANT GROWTH AND DEVELOPMENT

MESSAGES

1. Plants generally adjust to changing environmental conditions via growth.
2. Changes in plant morphology or physiology are often regulated by plant hormones produced in response to environmental factors.
3. Plant hormones interact in complex ways and may promote or inhibit growth.
4. Seeds of many plants remain dormant until conditions are suitable for germination and growth.
5. Plants become oriented in space by means of tropisms.
6. Phytochrome allows seeds and plants to detect changes of season and to respond to shading.
7. Apical dominance results from auxins produced at the apical meristem. Cytokinins from the roots act in opposition.
8. The regulation of germination and flowering often depends on external cues such as temperature and photoperiod.
9. Fruit development may be regulated by ethylene or by hormones produced by viable seeds.
10. Leaf senescence and dormancy allow plants to survive adverse seasons.
11. Plants protect themselves from herbivores and disease by producing toxic compounds and walling off injured areas.

32

THE DYNAMIC PLANT: TRANSPORTING WATER AND NUTRIENTS

PERSPECTIVE

This chapter completes the discussion of plant anatomy and physiology by presenting the mechanisms that plants use to obtain and transport water and nutrients. The material in this chapter integrates material presented earlier: water and ions, Chapter 2; nutrients and building blocks, Chapter 5; photosynthesis, Chapter 6; mycorrhizae, Chapter 17; plant anatomy and function, Chapter 30; and plant growth regulation, Chapter 31. Topics covered in this chapter will reappear in the discussion of global mineral cycles in Chapter 36.

The themes and concepts of this chapter include:

- Plants depend on the physical properties of water to obtain and transport materials.
- Plants have mechanisms for maintaining homeostasis. The vascular system transports materials and integrates the parts of a plant into a smoothly functioning unit.
- Soil quality is crucial to the growth and survival of plants.
- Water inside xylem is under tension as transpiration removes water from leaves.
- When the water supply is adequate and the plant needs carbon dioxide, guard cells inflate, opening stomata.
- Plants are made primarily of carbon, oxygen, and hydrogen. Other essential minerals come from soil water. Plants cannot use atmospheric nitrogen, which must be fixed to ammonium or nitrate ions before uptake by plants.
- Sugar and other substances move through the phloem by source-to-sink flow. Some minerals can be redistributed in the plant body.
- Tissue culture and genetic engineering supplement traditional breeding techniques.

CHAPTER OBJECTIVES

Students who master the material in this chapter will understand how and why water and dissolved substances move from soil into roots and throughout the plant body. They will see that the special physical properties of water molecules make this flow possible. These students will be able to describe the roles of the major nutrients and how source-to-sink flow moves sugars and other substances. They will appreciate the importance of new biological techniques in plant breeding.

LECTURE OUTLINE

(p 646) **I. Tomatoes, once feared to be toxic, introduce the discussion of how plants obtain the energy, water, and materials they need for growth, self-protection, and reproduction. Four unifying themes are presented:**

(msg 1) **A. Plants depend on the special physical properties of water to obtain and transport materials.**

B. The vascular system transports materials and integrates the parts of a plant into a smoothly functioning unit.

C. Plants have mechanisms for maintaining homeostasis.

D. Soil quality is crucial to the growth and survival of plants.

(p 648) **II. Plants depend on a continuous supply of water to carry out photosynthesis.**

A. Unlike animals, with their closed circulatory systems, the water flow in plants is one-way.

(tpq 1) (tr 139)

B. Waterproof coatings reduce water loss.

C. Stomata open and close to control transpiration and gas exchange. Guard cells operate the stomata. (tpq 1)

D. Root hairs and mycorrhizae absorb water from soil. (tr 139)

1. Water enters by passive diffusion.
2. Casparian strips prevent water from passing through the cell walls of the endodermis.

E. Several mechanisms conduct water from roots into shoots.

1. Guttation, the exudation of water from leaves, is caused by root pressure.
    a. Membrane proteins in the root's parenchyma cells expend energy pumping ions into the xylem; water follows osmotically, generating pressure.
    b. Some plants develop no root pressure.
    c. A pierced stem sucks in air rather than spewing out water, demonstrating that root pressure does not pump water to the tops of plants.
2. Capillary action draws water to a height of about half a meter.
3. Transpiration pulls water upward. (msg 2)

---

tpq 1. How does a plant's need for carbon dioxide conflict with its need to conserve water?

(tpq 2) (tr 140)

F. The transpiration pull theory (also known as the cohesion-adhesion-tension theory) says that as water molecules evaporate from an open stoma, other water molecules replace them from below.

1. **Hydrogen bonds link water molecules in an unbroken chain from the roots.**
2. **Water molecules cohere to one another strongly and adhere to other molecules.**
3. **If a water column breaks, an air bubble forms that blocks the flow of water in that vessel.**

(msg 3)

G. When guard cells swell with water, they arch away from each other and open the stoma.

1. **When they lose water, they slump together and the stoma closes.**
2. **Carbon dioxide, light, temperature, and daily rhythms of the plant affect guard cell function.**

(p 651)

III. Mineral nutrients compose only a small part of a plant's mass. Plants require at least 16 essential elements.

(msg 4)

A. Macronutrients are a plant's fundamental constituents.

1. **Most of a plant's dry weight is made up of carbon, oxygen, and hydrogen.**

(tr 141)

2. **Nitrogen is the next most important element.**

---

tpq 2. How would plant growth be affected if hydrogen bonds were weaker or stronger?

(msg 5)

a. Nitrogen fixation converts gaseous nitrogen to ammonium or nitrate ions.
    i. Bacteria fix nitrogen.
    ii. Legumes and other plants house nitrogen-fixing bacteria in root nodules.

b. Nitrifying bacteria convert ammonia into nitrate ions, which can be taken up by plants.
    i. Nitrifying bacteria cannot survive in bogs.
    ii. Carnivorous bog plants obtain nitrogen from the bodies of animals they trap.

3. Calcium serves as an intracellular messenger that controls cell membrane permeability, serves as a component of pectin, and serves in other roles.
4. Potassium regulates osmosis and helps activate enzymes.
5. Magnesium is a component of chlorophyll molecules and a cofactor in enzymes.
6. Phosphorus occurs in DNA and RNA, ATP, and membrane phospholipids.
7. Sulfur is found in two amino acids and is required for building some fats and coenzymes.

B. Micronutrients are required in minute quantities.
1. Iron is involved in the synthesis of chlorophyll and it occurs in electron acceptors.
2. Copper is present in chloroplasts and enzymes.

3. **Deficiencies of chlorine and zinc result in stunted plants.**

(msg 6) C. **Soil is the source of all nutrients except carbon, oxygen, and hydrogen.**

1. **Soil is a mixture of organic and inorganic particles. Humus is decomposing organic matter.**
2. **Air spaces between particles are crucial to plant life.**

D. **Root hairs and mycorrhize absorb dissolved minerals from soil.**

1. **Mycorrhizal hyphae greatly increase the absorption area of roots.**
2. **Mycorrhizal fungi produce plant growth hormones.**

(tpq 3) 3. **Mycorrhizae allow plants to parasitize other plants.**

(p 656) IV. **Plants must move sugars and other nutrients throughout their bodies.**

(msg 7) A. **Translocation moves sugars from source to sink.**

(tr 142) 1. **The mass flow theory states that companion cells load sugar into phloem vessels. Osmotic pressure increases locally as water moves out of the xylem into the phloem in response to the osmotic gradient.**

---

tpq 3. Are all plants autotrophic?

---

2. **Consumption of sugar at a sink lowers the osmotic pressure locally.**
3. **Water and solutes move through the phloem from areas of high pressure to areas of low pressure.**

(msg 7) **B. Some minerals are deposited permanently when they leave the xylem; others can be redistributed.**

1. **Calcium moves only in the xylem. Symptoms of deficiencies appear first in the newest leaves because calcium cannot be mobilized from other tissues.**
2. **Sulfur and phosphorus can be transferred from one tissue to another via the phloem. Deficiencies show up in older leaves as these minerals are withdrawn.**

(p 658) **V. New biological techniques can be used to improve plants.**

(msg 8) **A. <u>Tissue culture</u> entails the production of new plants from somatic cells.**

1. **In culture, cells dedifferentiate and form a callus.**

(tpq 4) 2. **Callus cells are placed in a culture medium that causes differentiation into new plants.**

---

tpq 4. Why is it that a complete plant can be grown from a single cell of an adult plant but a complete animal cannot be grown from a single cell of an adult animal?

---

3. **Many plants grown in tissue culture contain mutations, providing somaclonal variants that can be selected.**

B. **Genetic engineering allows the transfer of genes from one organism to another.**

1. **Insect-killing genes from bacteria may be introduced into plants to provide resistance to herbivores.**
2. **Protein content of crop plants may be increased.**
3. **Genetic engineering techniques supplement, but do not replace, traditional breeding techniques.**

INSTRUCTIONAL AIDS

VIDEOS, FILMS, SLIDES

An Investigation of Photosynthesis and Assimilate Transport. Films for the Humanities and Sciences, Princeton, NJ. Presents experiments using carbon 14 to demonstrate photosynthesis and translocation of carbohydrates. Video.

Phloem. The Media Guild, San Diego, CA. Reports current research on phloem, sieve tubes, and water flow using radioactive labels, computers, and advanced techniques in microscopy. Video, film.

Plants Through the Seasons: Structure, Growth, and Change. Educational Images Ltd., Elmira, NY. Discusses plant adaptations to environmental conditions and seasonal change. Filmstrip.

REFERENCES

Cocking, Edward C. and Micheal R. Davey. 1987. Gene transfer in cereals. *Science* 236:1259 – 1262. Reviews and summarizes techniques used to transfer genes into grain crops.

Culp, Mary. 1988. An easy method to demonstrate transpiration. *American Biology Teacher* 50 (1):46 – 47. A lab exercise demonstrating and measuring water flow through a plant.

Pool, Robert. 1989. In search of the plastic potato. *Science* 245:1187 – 1189. A review of genetic engineering applied to crops to produce plastics.

Rosenthal, Gerald A. 1986. The chemical defenses of higher plants. *Scientific American* 254 (1):94 – 99. Some plant-produced chemicals poison herbivores or repel them; others reduce a plant's nutritive value or impede an insect's growth.

Scagel, R. F., et al. 1984. *Plants – An Evolutionary Survey*. Belmont, CA: Wadsworth. A comprehensive overview of the evolution of plants from cyanobacteria through angiosperms and fungi. For the advanced student.

KEY TERMS

cohesion-adhesion-tension theory	page 650	nodule	653
legume	653	soil	655
macronutrient	652	tissue culture	659
mass flow theory	657	translocation	656
micronutrient	652	transpiration	650
nitrogen fixation	652	transpiration pull theory	650

FUN FACTS

PITCHER PLANTS

In a bog, the soil is often waterlogged, acidic and nutrient-poor. Anaerobic soil bacteria consume fixed nitrogen and liberate it as nitrogen gas. Few plants can tolerate these conditions, but pitcher plants thrive there because they have alternate sources of nutrients: the decaying flesh of animals.

The cobra lily, in the genus *Darlingtonia,* looks like a green cobra lifting its head, spreading its hood, and flicking out its tongue. The opening of the pitcher is where the snake's mouth would be; two spreading leaflets take the place of the snake's tongue. Insects are attracted to the odor of the leaf and its decaying contents. They enter and find themselves trapped. The top of the "head" of the cobra lily has numerous transparent windows to which insects fly as they try to escape. Eventually the insect falls into the fetid water in the lower part of the pitcher. If it tries to climb out, it will be stopped by downward-pointing hairs. In the tiny pool, a number of bacteria and predatory insects reside, subsisting on the plant's catch.

Most pitcher plants do not produce digestive enzymes and cannot absorb macromolecules. They assimilate the nitrogen-rich waste products of decomposition produced by the resident decomposers and detritivores. In a remarkably self-contained symbiosis, the plant provides its pool-dwellers with oxygen, water, food, and shelter.

See: Bradshaw, William E., and R. A. Creelman. 1984. Mutualism between the carnivorous purple pitcher plant and its inhabitants. *American Midland Naturalist* 112 (2):294 – 304.

Kitching, Roger, and Chris Schofield. 1986. Every pitcher tells a story. *New Scientist* January 23:48 – 50.

LIGHTING UP TOBACCO

To study how gene expression is controlled, scientists light up tobacco. They isolate a firefly gene for the enzyme luciferase ("light bearer") and transplant it into tobacco plants. When the plant lights up, they know the gene has been expressed.

Luciferase catalyzes a reaction of ATP, oxygen, and luciferin (a small organic molecule), producing a cold, green light. The luciferase gene is spliced into DNA from a plant virus and introduced into a bacterium, *Agrobacterium tumefaciens*. The bacterium infects plants and transfers some of its own DNA, including the recombinant gene, into its host's cells. The scientists then grow whole plants from infected cells, water them with a solution containing luciferin, and watch the plants glow.

Luciferase genes make ideal "reporters" on gene activities. Scientists can attach the luciferase gene to the promoter of a gene they are interested in. When the promoter is activated, the gene of interest and the accompanying luciferase gene will be transcribed, and luciferase will be produced. Gene activity can be assessed by feeding the plant luciferin and seeing where and when cells light up.

See: Ow, David W., et. al. 1986. Transient and stable expression of the firefly luciferase gene in plant cells and transgenic plants. *Science* 234:856 – 859.

CHAPTER 32 THE DYNAMIC PLANT: TRANSPORTING WATER AND NUTRIENTS

MESSAGES

1. Plants depend on the physical properties of water to obtain and transport materials.

2. Water inside xylem is under tension as transpiration removes water from leaves. Water molecules stick to one another, pulling the entire chain upward from the roots.

3. When the water supply is adequate and the plant needs carbon dioxide, guard cells inflate, opening stomata.

4. Plants are made primarily of carbon, oxygen, and hydrogen, which they obtain from carbon dioxide and water. Other essential minerals come from soil water.

5. Plants cannot use atmospheric nitrogen, which must be fixed to ammonium or nitrate ions before uptake by plants. Prokaryotes are the only organisms that can fix nitrogen.

6. Soil is composed of inorganic and organic particles. Plants absorb nutrients through root hairs or mycorrhizae.

7. Sugar and other substances move through the phloem by source-to-sink flow. Some minerals can be redistributed in the plant body.

8. Tissue culture and genetic engineering supplement traditional breeding techniques.

33

THE GENETIC BASIS FOR EVOLUTION

PERSPECTIVE

This chapter provides an overview of evolution, tying together genetics (presented in Part II, Perpetuation of Life) and life's variety (the topic of Part III). Material in this chapter builds on the concepts of mutation, presented in Chapters 10 and 12; the origin of life, presented in Chapter 15; and the coevolution of flowering plants and pollinators, presented in Chapter 18.

The themes and concepts of this chapter include:

- A population of interbreeding organisms evolves as its gene frequencies change.
- Most populations have a remarkable amount of genetic diversity. Genetic diversity is a precondition for evolution.
- Evolution occurs because certain individuals are selected by environmental pressures or chance events to be parents for the next generation.
- Sources of new genes are mutation, exon shuffling, and gene duplication and divergence.
- The Hardy-Weinberg principle predicts constant allele frequencies in the absence of natural selection, mutation, gene flow, nonrandom mating, or small population size.
- Selection may be stabilizing, directional, or disruptive. Genetic drift changes gene frequencies nonselectively in small populations.
- Neutralists hold that most alleles confer no disadvantages or advantages for survival.
- Speciation occurs as a result of genetic isolation of a subpopulation.
- Phyletic gradualism describes slow, progressive evolutionary changes; punctuated equilibrium includes long periods of stability interrupted by sudden evolutionary change.

CHAPTER OBJECTIVES

Students who master the material in this chapter will gain an overview of genetic variability and the processes of microevolution and macroevolution. They will understand how allele frequencies change and how species arise. They will see that the fossil record suggests that macroevolution works by processes different from microevolutionary processes.

LECTURE OUTLINE

(p 664) **I. Cheetahs introduce the genetic basis of evolution.**

(tpq 1) **A. A line of great cats gave rise to different species adapted to different ways of life.**

B. Cheetahs have been shaped by natural selection to be sprinters.

C. Cheetahs are threatened by extinction due to a lack of genetic diversity. All cheetahs are descendants of a very few individuals.

D. Four themes emerge:

(msg 1) **1. A group of interbreeding organisms evolves as its constellation of genes changes over the generations.**

(msg 2) **2. Most populations possess a remarkable amount of genetic diversity.**

3. Genetic diversity is a precondition for evolution.

---

tpq 1. Saber tooth cats evolved with fangs of different lengths. How might natural selection have led to different fang lengths?

(msg 3) **4. Evolution occurs because certain individuals are selected by environmental pressures or chance events to be parents of the next generation.**

(p 666) **II. Genetic variation is the raw material of evolution.**

A. The synthetic theory of evolution, or modern synthesis, combines Darwin's theory with genetics. Population genetics includes principles that describe what happens at the genetic level in populations.

B. The gene pool is the sum total of all alleles carried in all members of the population. The genetic makeup of the gene pool changes over time.

C. There is a great amount of genetic variation in most populations.

1. Most of the variation is hidden in the form of recessive alleles.

(tpq 2) **2. Relatively little variation distinguishes human races; nearly all genetic variation exists between individuals of the same race.**

3. Populations that fall below a necessary minimum variability have difficulty adjusting to a changing environment.

---

tpq 2. Is there more genetic variation between human races than there is among the individuals in this classroom?

(msg 4) (tr 143)

D. Genetic variation can arise through various means.

1. A point mutation randomly changes the DNA sequence of a single gene.
2. Gene duplication and exon shuffling create new genes.
   a. Gene duplication and divergence produces new genes while leaving copies of the old gene intact.
   b. An exon with one function can be duplicated and positioned by chance next to a different exon to produce new genetic combinations.
3. Random assortment and recombination rearrange alleles further but do nothing to change gene frequencies.

E. In the absence of outside forces, the frequency of each allele in a large population will not change as generations pass.

(msg 5) (tr 144)

1. The <u>Hardy-Weinberg principle</u> is useful as a predictor of gene frequencies in the absence of evolution.
2. Five conditions must be met for Hardy-Weinberg equilibrium to hold:
   a. No selection occurs.
   b. No mutation occurs.
   c. No gene flow occurs.
   d. Individuals mate randomly.

e. **The population is large.**

3. **Evolution can be defined as changes in allele frequencies over time.**

III. Many agents can change allele frequencies.

A. **Mutations alter allele frequencies by changing an allele into a different one. Changes in frequency are less important than the opportunity the new mutation may provide for natural selection.**

(tpq 3) B. **Migration alters gene frequencies, a phenomenon called gene flow.**

(msg 6) C. **Genetic drift results from chance changes in populations.**

1. **A population bottleneck that drastically reduces a population's size might result in the survival of only a portion of the gene pool.**

   a. **Genetic variability becomes reduced by the random loss of alleles.**

   b. **Habitat destruction endangers the majority of species on earth, an extinction crisis of unprecedented proportions.**

   c. **The minimum viable population is the smallest with enough genetic variability to ensure continued species survival.**

---

tpq 3. Why is it important to leave habitat corridors between wildlife refuges?

---

2. Genetic drift can result from the founder effect, where the initial population includes only a fraction of the alleles of the original large population.

(tpq 4) D. Nonrandom mating, such as sexual selection, causes the frequency of phenotypes to vary from Hardy-Weinberg predictions.

E. Scientists debate the relative importances of natural selection and the accumulation of neutral mutations.

1. Selectionists argue that natural selection is the primary agent of evolution.
2. Neutralists maintain that most variation isn't linked to an organism's fitness.
   a. Organisms in different environments accumulate mutations at similar rates.
   b. The molecular clock concept describes the steady tempo of genetic change.

(p 673) IV. Natural selection results from differential reproduction.

(msg 3) A. Adaptation is the result of natural selection.

B. Sickle cell anemia provides an example of selection.

---

tpq 4. Can selection result in a *decrease* in a bird's ability to survive?

---

1. **A mutant gene for hemoglobin protein causes sickle cell.**
2. **Heterozygotes resist malaria; thus they have a selective advantage.**
   a. **Homozygotes with the normal allele for hemoglobin suffer from malaria.**
   b. **Homozygotes with the sickle allele die of sickle-cell anemia.**
   c. **The greater reproductive success of heterozygotes has led to an increase in gene frequency of the sickle allele.**

(msg 6) **C. Natural selection can affect populations in three ways.**

1. **Directional selection favors one extreme form of a trait over all other forms.**
   a. **Lighter, and hence faster, cheetahs were selected over heavier, slower cats.**
   b. **Dark peppered moths survived after pollution darkened the bark on trees while light moths were eaten by birds.**
   (tpq 5) c. **Drug resistance in pathogens or pesticide resistance in insects spreads as resistant individuals survive to reproduce.**
2. **Stabilizing selection favors intermediate phenotypes.**

---

tpq 5. How do insects develop resistance to pesticides?

---

3. Disruptive selection favors two extremes.

(p 676) V. Speciation occurs as a result of genetic isolation of portions of a population.

A. Species are groups of populations that interbreed with each other in nature to produce healthy and fertile offspring.

1. Each species is reproductively isolated from all other species.

(tpq 6) 2. This definition does not apply to asexual organisms.

(msg 7) B. Reproductive isolating mechanisms prevent interbreeding between species.

1. Different species may breed at different times or sites.

2. Physical structures can hinder mating.

3. The gametes of two species may be so different that fertilization is impossible.

4. Hybrid inviability may follow fertilization of the egg of one species by sperm of another.

5. Hybrid sterility may occur even though the hybrid is robust and healthy (hybrid vigor).

C. Most speciation probably occurs as populations become geographically isolated and evolve in separate ways.

---

tpq 6. Can you suggest how to modify the species definition so it applies to asexually-reproducing organisms?

---

1. Allopatric speciation occurs where geographical barriers prevent gene flow.
2. Parapatric speciation occurs when there is limited gene flow between slightly overlapping populations.
3. Sympatric speciation occasionally occurs between subgroups of a population. Polyploidy leads to sympatric speciation in plants.

(p 679) VI. Macroevolution may operate by mechanisms different from those that influence microevolution.

(tpq 7) A. Life's fossil record has gaps but shows overall consistency.

B. Molecular biology provides strong evidence of descent with variation.

C. Homologous organs, vestigial organs, and comparative embryology provide evidence.

1. Homology describes structures in related organisms with different functions but a common origin.
2. Vestigial organs are rudimentary structures with no apparent utility but a strong resemblance to structures in probable ancestors.

---

tpq 7. Why does the fossil record have gaps? Is the presence of gaps evidence that evolution by descent with modification did not occur?

D. Similarities between organisms in different geographical areas provided clues to Darwin.

1. Convergent evolution occurs when two or more dissimilar and distantly related lineages evolve to become more similar.
2. Biogeography has revealed that many species inhabit only particular islands and yet are closely related to species living on the nearby mainland.

(p 683) VII. A study of phylogeny reveals recurrent patterns.

(tr 145) A. There are several patterns of evolutionary descent.

1. Gradual changes in allele frequencies result in progressive change.
2. Divergent evolution results in the split of one species into two.
3. (tpq 8) Divergence may occur simultaneously among a number of populations, leading to a branching pattern called adaptive radiation.
4. Parallel evolution involves two or more similar and related lineages independently changing in the same direction.

---

tpq 8. Why have the great mass extinctions been followed by great periods of adaptive radiation?

---

5. In coevolution, two species interact so closely that each one's evolutionary fitness depends on the other.

(msg 8) B. Evolution may proceed gradually or in jumps.

(tr 146)
1. Phyletic gradualism occurs as divergent evolution gradually changes a species' characteristics.
2. Punctuated equilibrium is based on the observation that species exist for millions of years without alteration, periods that are punctuated by great phenotypic changes that result in new species. The theory has three tenets:
   a. Alterations in body form evolve very rapidly in evolutionary time.
   b. During speciation, changes in form occur in small populations.
   c. After speciation, species retain much the same form until they become extinct, perhaps after a long period of constancy.

(tpq 9)
3. Both gradualism and punctuated equilibrium probably contribute to the evolutionary process.

C. Macroevolution addresses evolution above the level of species.

---

tpq 9. How would the current mass extinction appear to a paleontologist 100 million years from now?

---

1. **Evolutionary innovations usually follow from a change in size or function in a preexisting body part. Mosaic evolution occurs when some characters evolve without simultaneous changes in other parts.**
2. **Changes in developmental sequences or regulatory genes can cause widespread phenotypic effects.**
3. **Species selection is the survival and diversification of certain species while other species become extinct.**
4. **Extinction is the inevitable fate of all species.**

INSTRUCTIONAL AIDS

VIDEOS, FILMS, SLIDES

Evolution Collection. Carolina Biological Supply Company, Burlington, NC. A comprehensive collection covering natural selection, sources of variety, and speciation. Transparencies.

Fossils and Their Living Kin. Educational Images Ltd., Elmira, NY. Photos of fossil remains and of common living descendants are compared and contrasted. Transparencies.

Natural Selection: Evolution at Work. Pennsylvania State University AV Services, University Park, PA. Explains the role genetic variation plays in the process of natural selection by examining toxin resistance in rats and copper tolerance in grasses. Film.

REFERENCES

Barton, N. H., and B. Charlesworth. 1984. Genetic revolutions, founder effects, and speciation. *Annual Review of Ecology and Systematics* 15:133–164. A review (in moderately technical language) of the origin of species.

Futuyma, Douglas J. 1986. *Evolutionary Biology*. Sunderland, MA: Sinauer. A comprehensive text with an excellent bibliography.

Lister, A.M. 1989. Rapid dwarfing of red deer on Jersey in the last interglacial. *Nature* 342:539–542. Demonstrates the rapidity of evolution; deer isolated on an island became reduced to one sixth of their original body weight in less than 6000 years.

Stanley, Steven M. 1981. *The New Evolutionary Timetable.* New York: Basic Books. A discussion of punctuated equilibrium written for the nonspecialist.

KEY TERMS

adaptation	page 673	microevolution	679
adaptive radiation	683	molecular clock	673
biogeography	683	natural selection	673
coevolution	684	parallel evolution	684
convergent evolution	683	population bottleneck	671
directional selection	674	population genetics	666
disruptive selection	674	phyletic gradualism	685
divergent evolution	683	phylogeny	683
evolution	669	punctuated equilibrium	686
founder effect	671	reproductive isolating mechanism	677
gene flow	669	sexual selection	672
gene pool	666	speciation	677
Hardy-Weinberg principle	668	stabilizing selection	674
homology	681	vestigial organ	681
hybrid	677		
macroevolution	679		

FUN FACTS

QUAGGA QUEST

Horses, donkeys, asses, and zebras have a recent ancestor in common. They can interbreed but they have different numbers of chromosomes, so the offspring are usually sterile. The quagga, extinct since 1883, posed special problems. Its front half looked like a zebra's; its rear, a horse's. Was it more a horse than a zebra? Was it a transitional species bridging the evolutionary gap between horses and zebras? Was it a subspecies of the plains zebra (*Equus burchelli*), of the mountain zebra (*E. zebra*), or of Grevyi's zebra (*E. grevyi*)? Because it was extinct, it couldn't be tested by crossbreeding with other species.

Scrapings were taken from preserved quagga skins in a museum. Mitochondrial DNA from dead skin cells was transplanted into bacteria. When the bacteria reproduced, they replicated the quagga's mitochondrial DNA and produced enough for study. When researchers compared mitochondrial DNA of the quagga with that of other living *Equus* species, they found that it was a subspecies of the plains zebra.

Independently, another scientist examined quagga skin proteins. He too demonstrated that the quagga was more like the plains zebra than any other species. The two biochemical approaches, involving different families of macromolecules, agreed remarkably well, demonstrating molecular biology's effectiveness in resolving long-standing disputes in evolution.

See: Lowenstein, Jerold M. 1985. Half-striped quagga was a plains zebra. *New Scientist* July 13:27.

Lowenstein, Jerold M. 1985. Molecular approaches to the identification of species. *American Scientist* 73:541–547.

CHAPTER 33 THE GENETIC BASIS FOR EVOLUTION

MESSAGES

1. A population of interbreeding organisms evolves as its gene frequencies change.

2. Most populations have a remarkable amount of genetic diversity. Genetic diversity is a precondition for evolution.

3. Evolution occurs because certain individuals are selected by environmental pressures or chance events to be parents for the next generation.

4. Sources of new genes are mutation, exon shuffling, and gene duplication and divergence.

5. The Hardy-Weinberg principle predicts constant allele frequencies in the absence of natural selection, mutation, gene flow, nonrandom mating, or small population size.

6. Selection may be stabilizing, directional, or disruptive. Genetic drift changes gene frequencies nonselectively in small populations.

7. Speciation occurs as a result of genetic isolation of a subpopulation.

8. Phyletic gradualism describes slow, progressive evolutionary changes; punctuated equilibrium includes long periods of stability interrupted by sudden evolutionary change.

34

POPULATION ECOLOGY: PATTERNS IN SPACE AND TIME

PERSPECTIVE

This chapter presents patterns of population growth and decline and the processes and events that affect these patterns. In discussing life's ability to reproduce in excess of the carrying capacity of the environment, this chapter logically follows the discussion of evolution in Chapter 33; the pruning of the gene pool by natural selection is the fundamental process of evolutionary change. Material presented in this chapter will reappear in Chapters 35 through 37, which continue the discussion of the interactions between organisms and their environments.

The themes and concepts of this chapter include:

- Closely interacting biological and physical factors govern the abundance and distribution of any species in any area.
- Ecologists can predict a population's growth with formal models.
- The ecology of populations is closely tied to the evolution of species.
- The probability of survival varies with an organism's age. Different species have different survivorship curves.
- When resources are unlimited, growth is exponential. As the population approaches carrying capacity, growth often follows a logistic curve.
- Limiting factors may be density-dependent or density-independent, extrinsic or intrinsic.
- Individuals allocate their limited energy supplies through *r*- and *K*- selected life history strategies.
- The human population has grown exponentially. A demographic transition has occurred in industrialized countries.
- The human population is likely to double in the near future.

CHAPTER OBJECTIVES

Students who master the material in this chapter will see that biotic and abiotic processes limit natural populations. They will understand exponential growth, logistic growth, survivorship curves, and density-independent and density-dependent population-regulating mechanisms. These students will grasp the importance of limiting human reproduction.

LECTURE OUTLINE

(p 690) **I. The rise and fall of the desert-dwelling Hohokam people introduces population ecology. Three themes emerge:**

(msg 1) **A. Closely interacting biological and physical factors govern the abundance and distribution of any species in any area.**

(msg 2) **B. Ecologists can predict a population's growth with formal models.**

(msg 3) **C. The ecology of populations is closely tied to the evolution of species.**

(p 692) **II. Ecology is the scientific study of how organisms interact with their biotic and abiotic environment. Biologists organize ecology into a hierarchy of four levels:**

A. A population is a group of interacting individuals of the same species that inhabit a defined geographical area.

B. A community consists of two or more populations of different species occupying the same geographical area.

C. The ecosystem is made up of interacting organisms together with the physical factors of the environment.

D. The biosphere includes the entire planet with all its living species, its atmosphere, oceans, and soil, and the physical and biological cycles that affect them all.

(p 693) III. Most species have limited distributions.

(msg 4) A. Three general conditions limit the places where an organism might be found:

1. Organisms are adapted to specific ranges of physical factors.
2. Other species may block survival and limit a population's distribution.
3. Geographical barriers may block access to favorable areas.

(tpq 1) B. Within their ranges, organisms may be distributed in three patterns:

1. Organisms have a uniform distribution if they are spaced at regular intervals. This pattern is rare.

---

tpq 1. In what pattern are humans distributed over this continent? This state? A suburb?

2. Organisms will have a random distribution if individuals do not influence each other's spacing and if environmental conditions are uniform. This pattern is rare.

(msg 1) 3. Most commonly, organisms exist in clumped patterns.

a. Clumping occurs because resources are almost always limited to certain habitats.

b. In some species, clumping is tied to social behavior and community activities.

(p 695) IV. Constraints limit population size and density.

(tpq 2) (tr 147) A. Crude density is the number of individuals in a certain amount of space; ecological density counts organisms only where they actually live.

(msg 2) B. Population size changes over time.

1. Delta N = (births + immigration) − (deaths + emigration)

2. When delta N = 0, there is zero population growth.

(msg 5) C. Survival varies with age.

1. A survivorship curve is a plot of the proportion of a population that survives to each age.

---

tpq 2. The average density of people in the United States is about 70 per square mile. Does this number accurately indicate the density most people experience?

a. Late-loss curves are characteristic of organisms with low mortality in early and middle life and increasing death in old age.

b. Constant-loss curves reflect a fairly constant death rate at all ages.

c. Early-loss curves describe populations in which very young individuals have a high probability of death.

(tpq 3) 2. A life table is a table of the numbers from a survivorship curve. It shows the life expectancy for given ages.

D. Fertility varies with age. Fertility curves are graphs that plot reproduction rate versus the age of females.

(msg 6) E. Biotic potential describes the capacity for reproduction under idealized conditions.

1. Exponential growth results in a J-shaped curve.
2. The speed of population growth is determined by the maximum rate of reproduction (r_m) and the initial population size.

F. Organisms in nature cannot sustain continued, limitless growth because resources are limited.

1. An S-shaped curve represents logistic growth.
2. The carrying capacity (K) is the density at which the population levels off.

---

tpq 3. Why are accurate human life tables important for insurance companies?

---

3. Logistic growth differs from exponential growth in two ways.
   a. The growth rate changes to r, the rate of population growth per individual, which equals birth rate minus death rate.
   b. The logistic model includes a measure of the proportion of resources that remains unused.

(msg 7) G. Environmental resistance to growth includes various population-limiting mechanisms.

1. Density-dependent mechanisms, such as diseases, become more influential as density increases.
2. Density-independent mechanisms, such as adverse weather, exert their effects regardless of population density.
3. Extrinsic population-regulating mechanisms originate outside the population.
4. Intrinsic regulating mechanisms originate in an organism's anatomy, physiology, or behavior.
5. Competition among members of the same species is the most important intrinsic regulating mechanism.
   a. Scramble competition occurs when resources are equally available to all individuals, who must rush for their share.
   b. Contest competition involves clashes for social position or territory.

6. Population crashes may follow rapid population growth.

(msg 8) **H. Individuals must allocate their limited energy supplies among growth, survival, and reproduction.**

1. The way an organism allocates its energy is its <u>life history strategy</u>.
2. Species that reproduce rapidly despite risks to survival are products of <u>r-selection</u>.
   a. These organisms typically show a boom-and-crash population cycle.
   b. They frequently inhabit environments that are unoccupied and unpredictable.
3. (tpq 4) <u>K-selection</u> is characterized by slow rates of reproduction and maintenance of density near the carrying capacity. These species inhabit more stable environments.

(p 703) **V. Human populations are affected by ecological rules.**

(msg 9) **A. The global human population has grown steadily.**

1. Before 10,000 years ago, the population grew slowly to about 10 million.
2. The <u>agricultural revolution</u> accelerated growth. By 1750 A.D., the world population had risen to 800 million.

---

tpq 4. Are humans *r*-selected or *K*-selected?

3. The <u>industrial revolution</u> ushered in the third phase of growth.

(tr 148) 4. The overall curve is one of exponential growth.

B. Carrying capacity cannot be increased forever.

C. Birth rates and death rates are affected by industrial development of a country.

1. A drop in death rate precedes a drop in birth rate.
2. <u>Demographic transition</u> describes the change from high birth and death rates to low rates.
3. Many developing countries have reduced death rates but birth rates have not yet declined, causing enormous population growth.

(tr 149) D. A population's <u>age structure</u> is an indicator of growth rate.

1. Rapidly-growing populations have a great proportion of young individuals and a pyramid-shaped age profile.
2. Stable or declining populations tend to be more bullet-shaped.

E. The factors that cause birth rates to decline are controversial.

F. The global human population is likely in the future to double or triple today's 5 billion.

G. Without dramatic changes, the crush of humanity will reduce or destroy species and ecosystems.

INSTRUCTIONAL AIDS

VIDEOS, FILMS, SLIDES

Population Ecology. Carolina Biological Supply Company, Burlington, NC. Examines populations in theory, in the laboratory, and in nature. Transparencies, filmstrip.

Tropical Rainforest. The Media Guild, San Diego, CA. Examines the complex ecosystem of a tropical evergreen forest; demonstrates diversity and interdependence. Video.

REFERENCES

Brown, Lester R. 1987. Analyzing the demographic trap. In: *State of the World 1987.* New York: Norton. A clearly written discussion of human population growth rates.

Keyfitz, Nathan. 1989. The growing human population. *Scientific American* 261 (3):119–126. Presents the impacts of the burgeoning human population on the environment and analyzes the effectiveness of industrialization in slowing population growth.

KEY TERMS

abiotic	page 692	industrial revolution	703
age structure	706	J-shaped curve	697
agricultural revolution	703	*K*-selection	702
biosphere	693	life expectancy	696
biotic	692	life history strategy	702
biotic potential	697	logistic growth	698
carrying capacity	698	population	692
clumped distribution	694	population ecology	691
community	692	random distribution	694
demographic transition	705	*r*-selection	702
density	695	S-shaped curve	698
ecology	692	survivorship curve	696
ecosystem	692	uniform distribution	694
exponential growth	697		

FUN FACTS

FISH ON THE EDGE

Animals and plants that live on the edge of their species' range may face harsh conditions – but they enjoy a refuge from competition. Intertidal fishes provide examples. Adaptations of tide pool fishes include small size, camouflage, flattened body shapes that permit them to wedge into cracks, and fins that are reduced or modified into suckers that cling tightly to rocks. Many can change color to match their surroundings. They can tolerate dehydration and breathe air better than most fishes.

Males of many intertidal species are the primary parents. They defend territories, build nests, and care for the fertilized eggs until they hatch. Some have remarkable ability to find their homes if they are displaced.

See: Horn, Michael, and R. N. Gibson. 1988. Intertidal fishes. *Scientific American* 258 (1):64–70.

INTRODUCED PESTS

Some of the worst pest species have been intentionally introduced. European starlings (*Sturnus vulgaris*) were released at several locations in North America between 1850 and 1900, but they died out. The current plague of starlings became established in 1890 when 160 birds were released in New York City. By 1930, they had spread to the Mississippi River; by 1960, they reached the Pacific coast. They colonized 3 million square miles during the 50 years after their establishment.

Explorers and settlers carried European rabbits all over the world for use as fresh meat and fur. Rabbits reached Australia with the first settlers in 1788 but, despite repeated releases, did not become established until 1860, after a bush fire in Victoria destroyed the fences surrounding one colony. By 1920 they occupied virtually the entire continent, and they became the most serious agricultural pest ever known in Australia. Famous for prolific breeding, they thrived despite poisoning, shooting, and trapping, and they changed the economy of nature as they became the dominant primary consumers.

See: Krebs, Charles J. 1988. *The Message of Ecology*. New York: Harper and Row.

INTRODUCING RARE AND ENDANGERED SPECIES

In efforts to restore populations of native species exterminated locally or to establish satellite populations of rare non-native species, game managers introduce organisms into new habitats. Some 700 introductions or translocations have occurred each year since the early 1970s. Native game species have made up 90 percent of the translocations; threatened, endangered, or sensitive species account for the remainder. More than 80 percent of the game species translocations have established self-sustaining populations; less than half of the endangered species were successfully introduced. Habitat quality, source population characteristics, and founding population size seemed to be critical in one study.

See: Griffith, Brad, et al. 1989. Translocation as a species conservation tool: Status and strategy. *Science* 245:477–480.

CHAPTER 34 POPULATION ECOLOGY: PATTERNS IN SPACE AND TIME

MESSAGES

1. Closely interacting biological and physical factors govern the abundance and distribution of any species in any area.

2. Ecologists can predict a population's growth with formal models.

3. The ecology of populations is closely tied to the evolution of species.

4. Most species have limited distributions.

5. The probability of survival varies with an organism's age. Different species have different survivorship curves.

6. When resources are unlimited, growth is exponential. As the population approaches carrying capacity, growth often follows a logistic curve.

7. Limiting factors may be density-dependent or density-independent, extrinsic or intrinsic, or due to competition.

8. Individuals allocate their limited energy supplies through r- and K- selected life history strategies.

9. Human population has grown exponentially and is likely to double in the near future. A demographic transition has occurred in industrialized countries.

35

THE ECOLOGY OF COMMUNITIES: POPULATIONS INTERACTING

PERSPECTIVE

This chapter continues the discussion of ecology, building on principles of population ecology presented in Chapter 34. Material in this chapter refers to predation among single-celled organisms (Chapter 16) and chemical defenses of plants (Chapter 31). Woven throughout is the recurring theme of evolution (Chapter 33). The following two chapters take the discussion of ecology to the higher levels of ecosystems (Chapter 36) and the biosphere (Chapter 37).

The themes and concepts of this chapter include:

- Interaction with other species is a major limiting factor in the abundance and distribution of organisms.
- One species can influence another's evolutionary fitness so that the two species coevolve.
- Interaction with the human species is the most powerful ecological factor in the world today.
- Interactions with other organisms often force a species into a realized niche that is more restricted than its fundamental niche.
- Interspecific competition may result in competitive exclusion, resource partitioning, or character displacement.
- Predators can be carnivores, herbivores, parasites, or pathogens. Commensal and mutualistic relationships benefit one or both species.
- Predators use strategies such as pursuit or ambush and generally have large brains. The countermeasures of prey include camouflage, warning coloration, and mimicry.
- A succession of communities colonize a newly cleared area.
- An area's species richness is dictated by latitude, isolation, and the frequency and intensity of disturbances.

CHAPTER OBJECTIVES

Students who master the material in this chapter will understand how species interact in communities. They will see how competition and predation determine a species's niche, distribution, population level, and evolutionary changes. These students will understand how communities go through succession and will appreciate the importance of environmental complexity and species diversity.

LECTURE OUTLINE

(p 710) **I. The yucca and its pollinating moth introduce community interaction. Three themes appear:**

(msg 1) **A. Interaction with other species is a major limiting factor in the abundance and distribution of organisms.**

(msg 2) (tpq 1) **B. Two species can influence each other's evolutionary fitness, so that the organisms coevolve.**

(msg 3) **C. Interaction with the human species is the most powerful biological factor in the world today.**

(p 712) **II. Organisms reside in habitats and make their living in specific niches.**

A. A niche is a species' functional role, or job, in a community.

1. Autotrophs are primary producers; animals may be herbivores or predators.

---

tpq 1. What species have coevolved with humans?

(tr 150) 2. **The potential range of all biotic and abiotic conditions under which an organism can make a living is its <u>fundamental niche</u>.**

(msg 4) 3. **Competition may force a species into a narrower <u>realized niche</u>, the part of the fundamental niche that a species actually occupies in nature.**

B. **Two species can interact in ways that increase, decrease, or leave unchanged the abundance of either or both species.**

(p 713) III. **<u>Interspecific competition</u> is characterized by a negative effect on one or both competitors.**

(msg 5) A. **Competition may have different outcomes in different circumstances.**

(tpq 2) 1. **If one species affects the other more than it affects itself, <u>competitive exclusion</u> results; one species is eliminated.**

2. **If competition affects each species' population more than it affects the other species, the two species coexist.**

B. **In natural communities, it is difficult to determine the factors critical to each competitor's success.**

---

tpq 2. Are there species that are endangered because of competition with humans?

---

1. **Exploitation competition occurs when two species exploit and have equal access to identical resources.**
2. **Interference competition is the use of aggressive behavior to keep competitors from a resource.**

C. Competition can alter a species' realized niche.

1. **The term <u>resource partitioning</u> indicates the process of dividing resources so that species with similar requirements use the same resources in different areas, times, or ways.**
2. (msg 2) **Character displacement includes evolutionary changes that reduce competition.**

(tpq 3) **D. Competition generally decreases the environment's carrying capacity for both species and limits the final density of each population.**

(p 716) **IV. Interactions between species may be beneficial for one species and harmful to the other.**

(msg 6) **A. Organisms may obtain energy from other organisms.**

1. **<u>Predators</u> kill and eat <u>prey</u> in acts of <u>predation</u>.**
2. **Herbivores eat plants or plant parts.**
3. **Parasites feed on a host without killing it immediately.**

---

tpq 3. Does competition with other species limit the human population?

---

4. **Pathogens obtain nourishment from a host and weaken or kill it.**

B. Populations of predator and prey affect each other.

1. **If the predator consumes all prey, it too will become extinct.**
2. **If the prey has a refuge, some individuals might survive to reproduce after the extinction of the predator.**
3. **If the environment is complex, the populations of predator and prey will tend to oscillate up and down.**

(msg 1)

4. **The degree of prey population control exerted by predators varies. If predators are removed, prey populations can grow beyond carrying capacity.**

(tpq 4)

5. **Integrated pest management encourages natural predators to control pests.**
6. **Populations of predator and prey may grow or shrink over the short term, but over the long term, genetic changes can influence the evolutionary balance between hunter and hunted.**

---

tpq 4. How can organic gardeners control pests?

(msg 7) **C. Predation results in a coevolutionary race between predator and prey.**

1. **Predator strategies against mobile prey include pursuit and ambush. Vertebrate predators usually have larger brains than their prey.**
2. **Countermeasures that prey organisms have evolved include camouflage as well as obvious strategies such as rapid retreat.**
   a. **Chemical defenses are common among plants and occur in some animals. Aposematic or warning coloration enables predators to recognize poisonous prey.**
   b. **Many nonpoisonous species masquerade as poisonous species, a phenomenon called mimicry.**

D. Parasites live in close physical association with their hosts.

1. **Ectoparasites live on a host's exterior.**
2. **Endoparasites inhabit internal organs or blood.**
3. (msg 2) **Parasites and hosts may coevolve in ways that reduce the harm to the host.**
4. (tpq 5) **Social parasitism involves the exploitation of the social behavior of the prey.**

---

tpq 5. Can host and parasite be of the same species?

(p 722) **V. Commensalism and mutualism involve benefit to one or both species.**

(msg 8) **A. Commensalism benefits one species without harming the other.**

B. In a mutualistic interaction, both species benefit.

1. In facultative mutualism, neither species is wholly dependent on the other.

2. In obligatory mutualism, the interacting organisms need each other to survive.

(p 724) **VI. Communities are organized in time and space.**

(msg 9) **A. Communities change over time.**

(tr 151) **1. Communities regrow at a denuded site in a process called succession.**

2. A pioneer community includes species that colonize bare ground and modify environmental conditions, leading to its replacement by a transition community.

(tpq 6) **3. Succession ends with a stable assemblage of species known as a climax community.**

4. A mosaic climax consists of patches of different climax species.

---

tpq 6. How might early-succession organisms prevent the invasion of late-succession organisms?

---

5. Facilitation involves the modification of the environment by early invaders, promoting the growth of later species.
6. Transition species may be able to invade because of their tolerance for the prevailing conditions.
7. Pioneers may inhibit other species' growth.

(msg 10) B. The total number of species found in a community is its species richness, or diversity.

1. Latitude and isolation influence the species richness of an area.
2. Competition and predation affect diversity. The activities of keystone species may determine community structure.

C. Species diversity can increase or decrease the stability of communities affected by disturbances.

1. Disturbances of intermediate frequency or intensity help to increase species diversity.
2. (msg 3) The immense increase in the human population is leading to larger and more frequent disturbances in tropical communities, threatening their destruction.

INSTRUCTIONAL AIDS

VIDEOS, FILMS, SLIDES

Camouflage in Nature Through Form and Color Matching. Pennsylvania State University AV Services, University Park, PA. Shows how mammals, birds, and fish protect themselves from natural enemies by disguise. Film.

Introduction to Ecology Collection. Carolina Biological Supply Company, Burlington, NC. This giant collection of color transparencies covers the breadth of ecological concepts.

Niches in the Environment. International Film Bureau, Chicago, IL. Compares the surface of the earth to a gigantic jigsaw puzzle of niches. Filmstrip.

REFERENCES

Clapham, W. B., Jr. 1983. *Natural Ecosystems*, Second Edition. New York: Macmillan. This compact, readable text covers a wide variety of topics.

Davies, Nicholas B., and Michael Brooke. 1991. Coevolution of the cuckoo and its hosts. *Scientific American* 264 (1):92–98. Discusses the evolutionary arms race between the parasitic cuckoo and its hosts.

Washburn, Jan O., et al. 1991. Regulatory role of parasites: Impact on host population shifts with resource availability. *Science* 253:185–188. Reveals how a ciliate parasite reduces the number of emerging mosquitoes when mosquito larvae have sufficient food but increases the number of mosquitoes when larval nutrients are scarce.

KEY TERMS

camouflage	page 719	predation	716
commensalism	722	predator	716
community	710	prey	716
competitive exclusion	713	realized niche	712
fundamental niche	712	refuge	716
habitat	712	resource partitioning	715
interspecific competition	713	species richness	725
mimicry	720	succession	724
mutualism	722	warning coloration	720
niche	712		

FUN FACTS

PLANTS THAT MIMIC

Mimicry is common in plants. The flowers of several orchid species seduce male insects into being pollinators. The flowers resemble female insects, even to their pheromones. While attempting to copulate, the male transfers pollen from one flower to another. The orchid parasitizes the insect, giving him no pollen, nectar, or sexual satisfaction for his effort.

Another orchid with less obvious mimicry is a European native, *Cephalanthera rubra,* that coexists with bellflowers of the genus *Campanula*. The bees that visit bellflowers also visit orchids although the flowers differ in color, shape, and reward for pollinators. The bellflower produces abundant nectar, orchids yield none. A scientist discovered that despite the differences obvious to human eyes the flowers look the same to a bee. In ultraviolet light, to which insect eyes are the most sensitive, the flowers reflect the same spectrum. Shape, scent, and differences in the amount of red color apparently make no difference to the bees.

Carrion plants produce the odor of rotting meat. (*Amorphophallus titanum* has a stench so overpowering that people are reported to have fainted from its smell.) Carrion plants generally do not reward their pollinators, but parasitize flies and other insects that lay their eggs in rotting meat. Female flies search for egg-laying sites from one carrion flower to another, transporting pollen and sometimes laying their eggs in a flower. The larvae that hatch from these eggs will starve, but eggs laid on carrion by a more discriminating fly will survive. Thus there is an evolutionary race as natural selection rewards flies that are better discriminators and plants that are better mimics.

Some plants have evolved camouflage. *Lithops* species look like stones and are difficult to spot against rocky backgrounds. Other plants resemble cultivated crop plants, avoiding human predators who would weed them out of farmland. Their mimicry not only includes appearance but extends to germinating, flowering, and producing seed at the same time as the crop they are imitating. In an attempt to eradicate rice-mimics, plant genticists developed a purple-leaved variety of rice so the crop could be distinguished from the weed. But as people weeded the fields, selection soon led to the appearance of purple-leaved weeds.
See: Barrett, Spencer C. H. 1987. Mimicry in plants. *Scientific American* 257 (3):76–83.

CHAPTER 35 THE ECOLOGY OF COMMUNITIES: POPULATIONS INTERACTING

MESSAGES

1. Interaction with other species is a major limiting factor in the abundance and distribution of organisms.

2. One species can influence another's evolutionary fitness so that the two species coevolve.

3. Interaction with the human species is the most powerful ecological factor in the world today.

4. Interactions with other organisms often force a species into a realized niche that is more restricted than its fundamental niche.

5. Interspecific competition may result in competitive exclusion, resource partitioning, or character displacement.

6. Predators can be carnivores, herbivores, parasites, or pathogens. Commensal and mutualistic relationships benefit one or both species.

7. Predators use strategies such as pursuit or ambush and generally have large brains. The countermeasures of prey include camouflage, warning coloration, and mimicry.

8. Commensalism and mutualism benefit one or both species.

9. A succession of communities colonizes a newly cleared area.

10. An area's species richness is dictated by latitude, isolation, and the frequency and intensity of disturbances.

36

ECOSYSTEMS: WEBS OF LIFE AND THE PHYSICAL WORLD

PERSPECTIVE

This chapter presents a global overview of the flow of energy and the cycling of materials in ecosystems. It continues the discussion of ecology that began in Chapter 34, Population Ecology, and ran through Chapter 35, The Ecology of Communities. Material in this chapter refers to the cellular importance of energy, water, and minerals, presented in the earliest chapters in the text; to animal nutrition, detailed in Part IV, How Animals Survive; and to plant physiology, the topic of Chapter 32.

The themes and concepts of this chapter include:

- Energy flows through an ecosystem in a one-way path. Materials recycle through an ecosystem.
- Organisms are dependent on each other and the physical environment for energy and materials.
- Human activities can drastically affect the health of ecosystems by altering the flow of energy and the cycling of materials.
- Producers are the base of the trophic pyramid. Consumption and energy losses occur at every trophic level.
- Higher trophic levels support fewer individuals and less biomass than do lower levels.
- Biological magnification leads to the accumulation of toxins in consumers at high trophic levels.
- The water cycle is driven by solar power. The carbon cycle is coupled to the flow of energy. The nitrogen cycle depends on bacteria. Mineral cycles that don't include a gas phase are localized.
- Acid rain and the greenhouse effect are products of the combustion of fossil fuels. Nuclear winter is the ultimate ecological threat.

CHAPTER OBJECTIVES

Students who master the material in this chapter will gain an overview of global energy and nutrient cycles. They will understand that producers are the base of every food web, that energy dissipates at every trophic level, and that the populations and biomass at higher trophic levels will be smaller than at lower levels. These students will see that human activities can have detrimental global effects on natural ecosystems.

LECTURE OUTLINE

(p 730) **I. Deep-sea vent communities introduce ecosystems. Four themes emerge:**

(msg 1) **A. Energy flows through an ecosystem in a one-way path.**

B. Materials recycle through an ecosystem.

(msg 2) **C. Because organisms need energy and materials, they are dependent on each other and on the physical environment.**

(msg 3) (tpq 1) **D. Human activities can drastically affect the health of ecosystems by altering the flow of energy and the recycling of materials.**

(p 732) **II. Energy travels in many pathways in nature.**

A. Organisms have two basic strategies for obtaining energy and materials: autotrophy and heterotrophy.

---

tpq 1. How have humans affected local ecosystems?

(tr 152) **B. Organisms in communities feed at different trophic levels.**

(msg 4) **1. Autotrophs are the primary producers. They provide the energy fixation base and the nutrient concentration base for the whole system.**

2. Heterotrophs are consumers.

a. Primary consumers eat producers.

b. Secondary consumers are carnivores that eat herbivores.

c. Tertiary consumers are carnivores that eat other carnivores.

3. Detritivores obtain energy from detritus, organic wastes and dead organisms.

C. Feeding patterns in nature include food chains and, more frequently, food webs. Producers support a grazing food web and a detritus food web.

(p 733) **III. Energy flows through ecosystems from producers to consumers.**

(tpq 2) **A. The activities of producers set a limit for the amount of energy that can be captured and channeled throughout the entire ecosystem.**

---

tpq 2. Humans consume, directly or indirectly, an estimated 40 percent of the primary productivity on earth. What does this mean for natural ecosystems?

(tr 153) **B. The energy budget of the Hubbard Brook Experimental Forest has been analyzed in a few watersheds.**

1. During the growing season, 500,000 kcal of solar energy hits each square meter.

a. About 15 percent is reflected.

b. Some 41 percent is converted to heat.

c. Evapotranspiration uses 42 percent.

(tpq 3) **d. Only 2 percent is used in photosynthesis, or gross primary productivity.**

i. Half is used by the plant's respiration.

ii. The small amount remaining is net primary productivity, the only energy that is available to consumers.

e. Most of the energy and materials fixed by producers goes to the detritus food web.

f. Energy dissipates at every trophic level.

2. Only a tiny amount of energy remains in the soil or exits from the ecosystem.

(msg 5) **C. The energy losses at each trophic level limit the number of links in the food chain and the populations at each level.**

1. The energy pyramid is a diagram with building blocks proportional in size to the amount of energy available from the level below.

---

tpq 3. Is transpiration a wasteful loss of energy?

a. The energy stored at each level is a small portion of the energy of the level below.

b. Food chains usually have no more than four links.

2. Biomass indicates energy retention.

a. A pyramid of biomass demonstrates that each level contains only about 10 percent of the biomass of the level below it.

(tpq 4) b. "Eat low on the food chain" to save resources.

(msg 6) (tr 154) 3. Biological magnification is the tendency for toxic substances to increase in concentration in progressively higher trophic levels.

(p 738) IV. Materials move through biogeochemical cycles.

A. Repositories of materials are called pools. Organismal pools are usually much smaller than environmental pools.

(msg 7) (tr 155) B. The water cycle, or hydrological cycle, is driven by solar power. Transpiration from vegetation is especially important in many parts of the world.

(tr 156) C. The nitrogen cycle depends on nitrogen-fixing bacteria.

1. During nitrogen fixation, nitrogen gas is transformed into ammonia.

---

tpq 4. Do vegetarian humans affect natural ecosystems less than meat eaters?

---

2. Plants absorb ammonium ions or nitrate ions and convert them to proteins, nucleic acids, and other compounds.

(tpq 5) 3. Consumers and detritivores assimilate plant material and release wastes.

4. Bacteria break down nitrogen-containing macromolecules to ammonium by ammonification.

5. Other bacteria convert ammonium to nitrate by nitrification.

6. Another set of bacteria change fixed nitrogen back to nitrogen gas by denitrification.

7. Nitrogen supplies in farmland may be replenished by fertilization or by crop rotation with legumes. Chemical fertilization is energy-intensive and represents a major shift in a natural cycle.

D. The nitrogen and carbon cycles are global because these elements exist as gases in the environmental reservoir.

E. Some elements, such as phosphorus and calcium, do not have a gas phase and thus are locally cycled.

---

tpq 5. Are consumers, including humans, superfluous to natural ecosystems? Could ecosystems function with only producers and detritivores?

(tr 157) 1. **The phosphorus cycle has two interlocking circuits.**

a. **On short time scales, phosphorus moves locally from the soil into organisms and back into the soil.**

b. **On long time scales, phosphorus can leave local ecosystems in streams and rivers to become locked in sediments. Only when the rock so formed weathers will the phosphorus be released.**

2. **Phosphorus is often the limiting resource in aquatic ecosystems. Fertilizers and sewage can cause oligotrophic lakes to become eutrophic.**

(tpq 6) F. **The carbon cycle is coupled to the flow of energy.**

(tr 158) 1. **Carbon moves among the atmosphere, producers, and consumers.**

2. **Carbon can accumulate in wood but is eventually returned to the atmosphere by consumption and respiration.**

3. **Sediments can bury organic litter, which may be transformed into coal or oil.**

4. **Carbon is stored when calcium carbonate shells accumulate and are transformed into limestone.**

---

tpq 6. How has human activity affected the carbon cycle?

(p 743) **V. Human intervention alters ecosystem function.**

(msg 8) (tr 159)

A. The greenhouse effect is caused in part by a buildup of carbon dioxide from burning fossil fuels and forests.

1. **The current concentration of carbon dioxide in the atmosphere is about 0.03 percent.**
2. **Other greenhouse gases include methane, chlorofluorocarbons, and nitrous oxide.**
3. **Carbon dioxide has increased by about 25 percent in the past century.**
4. (tpq 7) **Global temperatures are predicted to increase by 2° to 4°C. Rising sea level and desertification may cause disasters.**

B. Acid rain kills trees and decimates aquatic ecosystems. Fossil fuel combustion produces oxides of sulfur and nitrogen which acidify rain.

C. We must recycle to reduce waste, pollution, and energy consumption.

D. <u>Nuclear winter</u> is the ultimate ecological threat. Dust, smoke, and soot thrown up by multiple nuclear explosions (of even a fraction of current arsenals) could cause global environmental catastrophe.

---

tpq 7. What consequences might result from increased global temperatures?

INSTRUCTIONAL AIDS

VIDEOS, FILMS, SLIDES

Biological Control. The Media Guild, San Diego, CA. Examines the use of predators to control insect pests. Video.

The Ecosystem. Educational Images Ltd., Elmira, NY. Three units outline basic principles, energy flow, and material cycles in ecosystems. Transparencies, video.

The Hole in the Sky. Coronet Film & Video, Deerfield, IL. Follows a midwinter expedition to Antarctica to investigate a hole in the ozone layer. Film, video.

REFERENCES

Crutzen, Paul J., and Meinrat O. Andreae. 1990. Biomass burning in the tropics: Impact on atmospheric chemistry and biogeochemical cycles. *Science* 250:1669 – 1677. Predicts the chemical consequences of widespread burning in the tropics.

ReVelle, Penelope, and Charles ReVelle. 1988. *The Environment: Issues and Choices for Society*. Boston, MA: Jones and Bartlett. A well-written textbook that covers a wide variety of environmental problems while emphasizing ecological principles.

Soulé, Michael E. 1991. Conservation: Tactics for a constant crisis. *Science* 253:744 – 757. A review of declining biological diversity and an appraisal of strategies to combat the decline.

KEY TERMS

biological magnification	page 738
biomass	737
carbon cycle	743
consumer	732
detritivore	733
detritus	733
energy pyramid	736
eutrophic	742
food chain	733
food web	733
nitrogen cycle	739
nitrogen fixation	739
nuclear winter	748
oligotrophic	742
phosphorus cycle	741
pool	739
primary consumer	733
producer	732
pyramid of biomass	737
secondary consumer	733
tertiary consumer	733
trophic level	732
water cycle	739

FUN FACTS

THE RUIN AND RECOVERY OF A LAKE

Lake Trummen was one of Sweden's most beautiful, clear lakes until the twentieth century, when urban and industrial pollution fouled it. Its restoration offers a case study in environmental recovery.

Sewage was dumped into Lake Trummen from 1936 to 1958, increasing the concentrations of plant nutrients in the water. Eutrophication (Greek, "good feeding") resulted; algae and larger plants multiplied, reducing the water's transparency and overloading the detritus food web. Dissolved oxygen was depleted and fish died in great numbers. The once-pure lake was thoroughly unattractive to swimmers and boaters.

After 1958 the sewage was diverted but the lake remained eutrophic through the 1960s. So much sewage had been put into the lake that there was a nutrient-rich layer of black sediment, 20 to 40 centimeters deep, that acted as a reservoir of minerals.

Typically, phosphate is the limiting plant nutrient in lakes. In well-oxygenated water, phosphate combines with iron to make an insoluble compound that settles into the sediment. Phosphate acts in a positive-feedback mechanism as a lake becomes eutrophic; when the water becomes anaerobic as a consequence of eutrophication, phosphate dissolves, increasing plant and algal growth. Decomposition of the increased biomass consumes even more oxygen.

When sewage was first put into Lake Trummen, it didn't immediately become eutrophic. Phosphate was trapped by the sediment as long as the water remained aerobic. Eventually, however, a threshold was reached and decomposition depleted the dissolved oxygen. Then the phosphate dissolved and circulated freely, suddenly increasing the lake's ability to support plants and worsening the situation. Even after sewage was diverted, the sediments of Lake Trummen held so much phosphate that the lake was in a permanent positive-feedback cycle of eutrophication.

Drastic action restored Lake Trummen. During the early 1970s, half a meter of mud was dredged from the entire bottom of the lake. Phosphate levels dropped, plant growth plummeted, the water remained aerobic, and the sediments trapped the residual phosphate.

See: Goldman, Charles R., and Alexander J. Horne. 1983. *Limnology*. New York: McGraw-Hill.

SPREADING DESERTS

Changes in the global climate due to greenhouse gases may cause deserts to spread. Positive feedback systems may increase desertification. When productive grasslands become more arid, relatively uniform distributions of water and soil resources are replaced by more heterogeneous distributions, leading to invasion by desert shrubs. Wind and water remove soil from the spaces between shrubs. Overgrazing accelerates the decline in grass cover, hastening the conversion to desert. As soil is laid bare, soil and air temperatures increase, humidity drops, soil nitrogen is released, and soil erodes, making the habitat unsuitable for grasses.

See: Schlesinger, William H., et al. 1990. Biological feedbacks in global desertification. *Science* 247:1043 – 1048.

CHAPTER 36 ECOSYSTEMS: WEBS OF LIFE AND THE PHYSICAL WORLD

MESSAGES

1. Energy flows through an ecosystem in a one-way path. Materials recycle through an ecosystem.

2. Organisms are dependent on each other and the physical environment for energy and materials.

3. Human activities can drastically affect the health of ecosystems by altering the flow of energy and the cycling of materials.

4. Producers are the base of the trophic pyramid. Consumption and energy losses occur at every trophic level.

5. Higher trophic levels support fewer individuals and less biomass than do lower levels.

6. Biological magnification leads to the accumulation of toxins in consumers at high trophic levels.

7. The water cycle is driven by solar power. The carbon cycle is coupled to the flow of energy. The nitrogen cycle depends on bacteria. Mineral cycles that don't include a gas phase are localized.

8. Acid rain and the greenhouse effect are products of the combustion of fossil fuels. Nuclear winter is the ultimate ecological threat.

37

THE BIOSPHERE: EARTH'S THIN FILM OF LIFE

PERSPECTIVE

This chapter discusses global climate regions, aquatic environments, and the communities they support. It integrates and wraps up the discussion of ecology, the subject of Part VI, and refers to topics presented in Chapter 2, 6, 16, 17, 18, and 19.

The themes and concepts of this chapter include:

- Climates originate from currents of air and water on a spinning planet heated by the sun most strongly at the equator.
- Different climates promote different communities of organisms.
- Communities with similar physical environments contain organisms with similar evolutionary adaptations, even when the individual species are unrelated.
- The tilt of the earth's axis results in seasons.
- Terrestrial biomes are determined by latitude, topography, and climate.
- Standing water ecosystems are stratified because of density differences caused by temperature differences and because light cannot penetrate deeply.
- Ocean productivity varies with depth and proximity to shore.
- Human activities are causing a global biodiversity crisis.

CHAPTER OBJECTIVES

Students who master the material in this chapter will see how climate and weather affect the evolution and distribution of life. They will understand how the tilt of the earth leads to seasons, why rain and wind occur, and how currents in water and air affect climate. These students will be able to recount the characteristics of the major biomes and of the various zones in freshwater and saltwater ecosystems.

LECTURE OUTLINE

(p 750) **I. The cathedral forest introduces the subject of the biosphere. Three themes emerge:**

(msg 1) **A. Climates originate from currents of air and water on a spinning planet heated by the sun most strongly at the equator.**

(msg 2) **B. Different climates promote different communities of organisms, called biomes on land.**

(msg 3) **C. Communities with similar physical environments contain organisms with similar evolutionary adaptations, even when the individual species are unrelated.**

(p 752) **II. Climate is the major physical factor determining the abundance and distribution of living things.**

(tpq 1) **A. Weather is the condition of the atmosphere at any particular place and time; climate is the accumulation of seasonal weather events over a long period of time.**

---

tpq 1. What factors affect the local climate?

(msg 4) **B. A round, tilted earth heats unevenly.**

1. **The sun's angle of incidence depends on latitude and time of year, affecting the local heating of the earth's surface.**

(tpq 2) 2. **When the North Pole is tilted toward the sun, it is summer in the Northern Hemisphere.**

(msg 1) **C. Air temperature differences cause rain and <u>wind</u>.**

1. **Warm air rises and holds more water vapor than does cold air.**
2. **When warm, moisture-laden air rises, it cools, decreasing its capacity to hold water, which falls as precipitation.**

D. Air and water currents give rise to wind and weather.

1. **The rotation of the earth affects currents.**

(tr 160) 2. **Air currents distribute heat toward the poles.**

   a. **Six cells at different latitudes make up the overall wind pattern.**

   b. **The direction of air flow and ascent or descent determine the earth's climatic zones.**

      i. **Warm, moist air rises at the equator, dropping rain.**

      ii. **At 30° latitude, cool, dry air falls, creating deserts and trade winds.**

---

tpq 2. How would the earth's seasons and climates be different if the tilt of the axis were greater?

---

iii. At 60° (the polar front), air rises again, dropping rain and producing winds that go west to east.

iv. At the poles, air descends and occasionally carries polar weather into the U.S.

3. Wind blowing over the ocean surface causes currents, which flow as gyres that redistribute heat.

(p 754) **III. Climate determines the general character of biomes.**

(msg 5) **A. Tropical rain forests include nearly half of the world's species but only 3 percent of the land area.**

(tr 161)

1. The struggle for light has led to three main levels of plant height.
   a. The emergent layer includes treetops that rise above the surrounding vegetation.
   b. The overlapping tops of shorter trees make up the canopy.
   c. Only dim light penetrates to the understory and the forest floor.
2. Many jungle plants have evolved specializations for gathering light.
3. (tpq 3) Tropical rain forests have poor soil. Roots tend to be shallow and expanded into buttresses.

---

tpq 3. What happens to plant nutrients after a tropical rain forest is cut and burned?

---

4. **Most nutrients that are not retained by biomass are rapidly washed away.**
5. **The air in the forest is usually still, making animal pollination and seed dispersal necessary.**
6. (tpq 4) **Tropical rain forests are being rapidly cleared, threatening millions of species and perhaps leading to global climate alteration.**

B. **Tropical <u>savannas</u> occur in warm climates with extended dry seasons. The savanna's productivity is concentrated near the ground.**

C. **<u>Deserts</u> occur at dry latitudes, at the centers of continents far from marine air, where cold ocean winds blow on warm land, and in the rainshadows of mountain ranges.**

1. **Deserts experience hot day temperatures and cold night temperatures.**
2. **Desert organisms have adaptations that reduce water loss, enhance water uptake, or shorten the life cycle.**
3. **Human activities result in desertification.**

D. **Temperate <u>grasslands</u> include the prairies of North America, the pampas of South America, the steppes of Asia, and the veldt of Africa.**

---

tpq 4. How are we as individuals affected by the loss of tropical rain forests?

---

1. Grasslands are drier than forests but wetter than deserts.

(tpq 5) 2. Fire is a recurrent event, leading to the evolution of fire-tolerant species.

3. Grasslands have soil richer in organic material than soils in other biomes.

4. Soil loss due to improper farming techniques is a problem in grasslands converted to agriculture.

E. Chaparral borders grasslands and deserts in many areas with hot, dry summers and cool, wet winters. This biome experiences frequent fires.

F. Temperate forests have intermediate rainfall and fairly moderate temperatures. Seasonal temperature changes have shaped the evolution of temperate-forest organisms.

G. Coniferous forests (taiga, boreal forests) are found at latitudes with cold, snowy winters and short summers.

1. Evergreen conifers are well-adapted to short growing seasons.

2. Temperate rain forests occur along the West Coast from Canada to northern California and in parts of New Zealand.

---

tpq 5. Is fire always bad for ecosystems?

H. Tundra appears in northern regions where permafrost prevents the establishment of trees.

I. Polar cap terrain lacks major plants and is thus not considered a biome.

(tr 162) J. Life zones in mountains occur because precipitation and temperature vary with elevation.

(p 763) IV. Aquatic life includes marine ecosystems and freshwater communities.

A. The physical characteristics of water influence aquatic organisms.

B. Freshwater ecosystems may be in bodies of running or standing water.

(msg 7) 1. Communities change as one goes from cold, oxygenated, mineral-poor headwaters to lower reaches of warmer, low-oxygen, mineral-rich water.

(msg 8) 2. Lakes are vertically stratified. Different zones support different species.

a. The littoral zone includes shallow areas.

b. The limnetic zone is the top layer through which light can penetrate.

c. The profundal zone is deep and dark.

(tr 163) d. Nutrients are seasonally recycled from the sediments when the temperature of the water becomes uniform and the thermocline disappears.

i. Turnover brings oxygen-rich water from the surface to the profundal zone and mineral-rich water from the deeps to the upper layers.

ii. In temperate climates, turnover often happens in spring and fall.

(msg 9) C. Salt water supports different communities in different areas.

1. Salt and fresh water mingle in <u>estuaries</u>, some of the most productive ecosystems on earth.

(tr 164) 2. Depth, type of bottom, and light levels vary in oceans.

a. The intertidal zone, including <u>tide pools</u>, is alternately exposed to air and water.

b. The neritic zone is the ocean's most productive region. Coral reefs are the most productive ecosystems on earth.

c. The oceanic zone is nearly deserted except where upwelling occurs.

(p 766) V. Change is part of nature.

(msg 10) A. Human activities are causing rapid changes.

B. We have triggered a biodiversity crisis.

C. Economic interests conflict with preservation of natural resources.

INSTRUCTIONAL AIDS

VIDEOS, FILMS, SLIDES

Biotic Cycles Collection. Carolina Biological Supply Company, Burlington, NC. Includes descriptions of atmospheric recycling systems and cycles of nutrients on land. Transparencies.

Coniferous Forest Biome. Pennsylvania State University AV Services, University Park, PA. Discusses adaptation to extremes of climate among coniferous forests of North America. Film.

Down the Amazon. Films Incorporated, Chicago. Traces this great river from its source through the tropical rain forests. Video.

REFERENCES

Cloud, Preston. 1983. The biosphere. *Scientific American* 249 (3):176 – 189. Life is not only sustained by the lithosphere, hydrosphere, and atmosphere, but it has powerfully shaped their evolution.

Piel, Jonathan, ed. 1989. Managing planet earth. *Scientific American* 261 (3):46 – 175. This special edition includes 11 articles that address topics ranging from the changing atmosphere to strategies for sustainable economic development.

Simberloff, Daniel, et al. 1987. Special feature: Spotted owl. *Ecology* 68:766 – 779. Three papers that address the social issues surrounding the conservation of the spotted owl.

KEY TERMS

biome	page 751	marine	763
chaparral	760	rain forest	756
climate	752	savanna	757
coniferous forest	761	temperate forest	760
desert	758	tide pool	765
estuary	765	tundra	761
grassland	759	weather	752
life zone	762	wind	753

FUN FACTS

ALGAE AND ATMOSPHERES

The humble algae may affect global temperatures, moderating or aggravating climatic trends. Certain species of oceanic phytoplankton produce dimethylsulfide (DMS), which diffuses into the atmosphere. In the air, DMS oxidizes to form sulfate particles, which seed clouds by acting as condensation nuclei. Scientists predict that a negative feedback system moderates ocean temperatures. Increasing phytoplankton populations result in decreasing sunlight at the ocean's surface, suppressing phytoplankton growth.

Calculations indicate that if the ocean's cloud cover were to increase by 30 percent, the global temperature would drop by 1.3°C. During the latest ice age, the world was only 4°C colder than it is today. Antarctic ice deposited 20,000 years ago contains higher-than-average levels of particles derived from DMS, indicating that when the earth was much colder than it is today, plankton were producing many cloud nuclei. This suggests that at times the phytoplankton-temperature interaction can be an amplifying positive-feedback system that aggravates a climate shift.

See: Monastersky, Richard. 1987. The plankton-climate connection. *Science News* 132:362–365.

AMERICAN HURRICANES AND AFRICAN RAIN

The frequency and intensity of Atlantic hurricanes are strongly linked to rainfall variations in the Sahel region of Africa. When the Sahel is wet, as it was from 1947 to 1969, many intense hurricanes hammer the eastern United States. During dry periods, e.g., from 1970 to 1987, relatively few hurricanes appear. Hurricanes and rainfall are manifestations of global sea-air cycles that occur over decades. Historical records support the linkage over the past century.

Increases in Sahel precipitation during 1988 and 1989 suggest that the 1990s and the first decade of the twenty-first century may be marked by destructive hurricanes striking the southeastern coast of the U.S. Economic loss and fatalities may be greater than ever before because of the extensive development of low-lying coastal areas that has occurred in the past few decades.

See: Gray, William M. 1990. Strong association between West African rainfall and U. S. landfall of intense hurricanes. *Science* 249:1251–1256.

CHAPTER 37 THE BIOSPHERE: EARTH'S THIN FILM OF LIFE

MESSAGES

1. Climates originate from currents of air and water on a spinning planet heated by the sun most strongly at the equator.

2. Different climates promote different communities of organisms.

3. Communities with similar physical environments contain organisms with similar evolutionary adaptations, even when the individual species are unrelated.

4. The tilt of the earth's axis results in seasons.

5. Terrestrial biomes are determined by latitude, topography, and climate.

6. Biomes include rain forests, savannas, deserts, grasslands, chaparral, temperate forests, coniferous forests, and tundra.

7. Communities and conditions change along the length of a running water system.

8. Standing-water ecosystems are stratified because of density differences caused by temperature differences and because light cannot penetrate deeply.

9. Ocean productivity varies with depth and proximity to shore.

10. Human activities have triggered a biodiversity crisis.

38

ANIMAL BEHAVIOR: ADAPTATIONS FOR SURVIVAL

PERSPECTIVE

This chapter presents the evolutionary basis for animal behavior, linking the actions of animals with their anatomy, physiology, and ecology. Animal behavior is a fitting topic for the final chapter in that appropriate human behavior is the key to solving global environmental crises.

The themes and concepts of this chapter include:

- The ability to perform behaviors originates in genes, arises during development, and may be modified by the environment.
- Natural selection shapes behavior.
- Single alleles can control complex behavior.
- Behavioral patterns range from unchanging fixed-action patterns to insight learning, which allows animals to respond to novel situations.
- Animals can increase their reproductive success by oriented movement, migration, and maintenance of territory.
- The differences between male and female parental investment result in different sexual strategies and in sexual selection.
- Communication links the behavior of individuals.
- Social living increases protection from predators at the cost of increased infighting and disease.
- Altruism increases an individual's inclusive fitness.
- The principles of sociobiology may be relevant to the study of human behavior.

CHAPTER OBJECTIVES

Students who master the material in this chapter will understand how natural selection shapes behavior. They will see that genotype determines the ability to perform behaviors and that learning modifies an animal's response to a stimulus. They will gain an overview of the spectrum of behavioral patterns, from fixed-action patterns to insight learning.

These students will understand the evolutionary significance of migration, territoriality, feeding strategies, and reproductive behavior. They will be able to discuss the selective value of altruism and the application of sociobiology to human behavior.

LECTURE OUTLINE

(p 774) **I. The reproductive behavior of the black-backed gull introduces ethology.**

A. The proximate cause of a behavior is the mechanism that operates within an individual.

(tpq 1) **B. The ultimate cause of a behavior is the selective advantage that the action confers on the individual.**

C. Two major themes emerge:

(msg 1) **1. The ability to perform behaviors originates in genes, arises during development, and may be modified by the environment.**

(msg 2) **2. Natural selection shapes behavior.**

(p 776) **II. Genes direct development of neural programs that determine behavioral capabilities.**

tpq 1. What is the proximate cause of your coming to class today? The ultimate cause?

(msg 3) **A. Experiments with honeybees and garter snakes indicate that individual genes can control complex behavior.**

B. Observations of and experiments with gulls reveal the proximate and ultimate causes of eggshell-removal behavior.

(p 777) **III. Experiences in the environment can alter behavior.**

(msg 4) **A. There is a continuum of behavioral types from**
(tr 165) **instincts to learned behaviors.**

(tpq 2) **B. Innate or instinctive behaviors are automatic responses to the environment.**

1. **An innate series of precise physical movements is a fixed-action pattern.**
2. **Fixed-action patterns are characterized by five features:**
   a. **They are triggered by sign stimuli, or releasers.**
   b. **Once a releaser prompts a fixed-action pattern, the behavior must proceed from start to finish.**
   c. **A fixed-action pattern is fully formed and functional the first time it is used.**
   d. **All species members of the same sex and age perform the behavior in the same way.**

---

tpq 2. Do humans have instincts?

---

e. An animal's instinctive response to a releaser depends on its stage of maturation and physiological states of motivation, or drives.

C. Unlike fixed-action patterns, some behavior patterns have limited flexibility.

D. Imprinting is the recognition, response, and attachment of a young animal to a particular adult or object.

1. Imprinting has an inherent time limit, or critical period, after which it cannot occur.
2. Imprinting requires specific releasers and is irreversible.

E. Learning is an adaptive and enduring change in an individual's behavior based on experiences in the environment.

1. Habituation is the learned disregard of a repeated stimulus that is followed by neither reward nor punishment.
2. Trial-and-error learning, or operant conditioning, is a form of learning in which an animal associates a response, or operant, with a reinforcer, a reward or punishment.
3. (tpq 3) Classical conditioning involves the association of two separate stimuli.

---

tpq 3. How is classical conditioning especially important in advertising?

4. Insight learning, or reasoning, involves the formulation of a course of action by understanding the relationships between the parts of a problem.

(p 782) IV. Natural selection shapes behaviors.

(msg 5) A. Many animals locate and defend a home territory.

B. A taxis is an animal's movement toward light (phototaxis), specific chemicals (chemotaxis), or heat (thermotaxis).

C. Some animals have a sense of orientation and navigation that enables them to home and to migrate.

1. Orientation may be inborn, but navigation is at least partially learned.
2. Animals may rely on a circadian clock to help compute orientation relative to the sun's path.
3. Some animals have a magnetic sense that enables them to orient to the earth's magnetic field.

(tpq 4) D. Territoriality is defense of the living space from intruders.

E. Feeding behavior requires that animals find food while avoiding predators.

---

tpq 4. In what ways are humans territorial?

---

1. According to the optimality hypothesis, animals behave in ways that allow them to obtain the most food energy with the least effort and the least risk of falling prey to a predator.
2. The optimal foraging solution is neither conscious and deliberate nor universal.

F. Reproductive behavior is the central focus of natural selection.

1. Differing reproductive strategies shape the reproductive behaviors of the sexes.

(msg 6)

2. Males and females produce gametes of different sizes and numbers.
    a. Parental investment is greater for a female than for a male.
    b. Maternal investment in each young is usually greater than male investment.
    c. Male parenting behavior is likely to evolve only when females cannot support the young without help.

(tpq 5)

    d. Males typically compete among themselves to fertilize as many females as they can; females are selective about their mates.
3. Sexual selection results when members of the same species exert selective forces on one another.

---

tpq 5. Do these same sexual differences appear in human behavior?

---

a. Intrasexual selection between males results from the competition to inseminate as many females as possible, leading in some species to a dominance hierarchy.

b. A female generally picks males that provide the best genes to enhance her offspring's survival, thus engaging in intersexual selection.

(msg 7) G. Communication is a signal produced by one individual that alters the behavior of the recipient.

1. Each animal uses one or more channels of communication that fit the ecological circumstances.

2. Communication may occur by sight, sound, scent, or touch.

(tr 166) 3. Honeybees communicate by dancing for their hivemates.

a. Round dances indicate flowers that are close to the hive.

b. Waggle dances indicate the direction and distance to food sources far from the hive.

(p 788) V. Sociobiology is the study of social behavior.

(msg 8) A. Large colony size provides early detection of predators but at the costs of infighting and increased risk of communicable disease.

B. Animals will socialize only when the benefits of group living outweigh the costs.

(msg 9) C. Altruism might seem to be penalized by natural selection, but it persists by kin selection.

1. The degree of relatedness indicates the proportion of genes that are identical between two individuals.
2. Inclusive fitness is the sum of an individual's personal fitness when reproducing and its fitness from kin selection.
3. Among social insects, workers favor preservation and replication of their own alleles when they support the queen.

(msg 10) D. Debate continues on the question of whether genetic principles determine human behavior.

1. Humans do show innate behavior patterns.
2. Many behaviors are passed on as cultural traditions.
3. Studies on twins show that many human personality traits are influenced about equally by genes and environment.

INSTRUCTIONAL AIDS

VIDEOS, FILMS, SLIDES

The Discovery of Animal Behavior. Films Incorporated, Chicago. This six-part series portrays experiments in behavioral studies and includes footage shot on location. Film, video.

The Mystery of Animal Pathfinders. Coronet Film and Video, Deerfield, IL. Presents migration in organisms ranging from bacteria to trumpeter swans. Video.

Signs of the Apes, Songs of the Whales. Carolina Biological Supply Company, Burlington, NC. Explores language and the ability of animals to communicate. Video.

Sociobiology: The Human Animal. Time-Life Video, Paramus, NJ. Analyzes the arguments regarding the application of sociobiology to humans. Film, video.

REFERENCES

Maynard-Smith, John. 1978. The evolution of behavior. In: *Evolution*. A Scientific American Book. San Francisco: Freeman. Originally appearing in the September 1978 issue of *Scientific American,* this article focuses on the evolution of altruism.

Plomin, Robert. 1990. The role of inheritance in behavior. *Science* 248:183 – 188. Reviews the genetics of behavior in humans and other animals. Discusses the application of techniques of molecular biology to the study of behavior.

Scheller, Richard H., and Richard Axel. 1984. How genes control an innate behavior. *Scientific American* 250(3):54 – 62. Egg laying in a marine snail is a fixed-action pattern governed by a set of related neuropeptides defined by a family of genes.

Trivers, Robert. 1985. *Social Evolution*. Menlo Park, CA: Benjamin/Cummings. A very readable text that covers a wide range of behaviors, including natural selection, sexual behavior, altruism, parenting, and deceit.

Waterman, T. H. 1989. *Animal Navigation*. New York: Scientific American Library. An easy-to-read review of the senses and mechanisms used by migrating animals.

KEY TERMS

altruism	page 789
classical conditioning	781
communication	787
critical period	779
dominance hierarchy	786
drive	778
ethology	775
fixed-action pattern	777
habituation	780
imprinting	779
inclusive fitness	790
insight learning	781
instinct	777
kin selection	789
learning	780
navigation	783
operant	780
operant conditioning	780
optimality hypothesis	784
orientation	783
proximate cause	775
reinforcer	780
sexual selection	786
sign stimulus	778
sociobiology	789
taxis	782
territoriality	784
trial-and-error learning	780
ultimate cause	775

FUN FACTS

COMMUTER'S CLOCKS

Natural selection rewards the bird that migrates at just the right time. One that leaves too early in autumn may not have enough fat stored to make the migratory journey. One that leaves too late risks a cold death. In the spring, the bird that arrives at the breeding ground early gets a good territory; one that arrives too early risks encountering the tail end of winter. In temperate and polar regions, day length provides a reliable indicator of the seasons.

Birds have a circadian clock that tells them when autumn has arrived and it's time to migrate. In labs with constant conditions, and no clues to natural daily environmental cycles, birds maintain an activity cycle that approximates 24 hours. Over weeks and even months of isolation, they sleep, eat, and hop about in regular cycles. The cycle is often an hour or two shorter or longer than 24 hours, providing convincing evidence that the birds follow an internal clock instead of some unknown environmental cue.

Near the equator, there is little seasonal change in day length, and the circadian clock wouldn't be much help in determining seasonality. Some birds have a yearly clock as well as a daily clock. Experiments showed that birds could maintain an approximately annual rhythm for eight years or more. The birds' cycle was not precisely a year – often it was about 10 months – convincing scientists that the clock is innate.

See: Gwinner, Eberhard. 1986. Internal rhythms in bird migration. *Scientific American* 254 (4):84–92.

ALTRUISTIC VAMPIRES

Despite their creepy reputation, vampire bats (*Desmodus rotundus*) are remarkably social, flocking together while resting. Females form long-term associations with other individuals and increase their chances of survival by altruistic food-sharing.

Vampire bats will die if they go without food for more than two days. They must consume 50 to 100 percent of their weight in blood each night to satisfy their energy needs. If they must return to their roost without feeding, they solicit food from their nestmates, who then regurgitate some of the blood they had taken. Females feed not only their own offspring and relatives, as kin selection would dictate, but they also feed unrelated young and adults. They develop a buddy system; two individuals who are long-term roostmates will regurgitate almost exclusively to each other. Food sharing might cost the donor about twelve hours of survival time before starvation, but the twelve hours a recipient gains can mean the difference between life and death.

See: Wilkinson, Gerald S. 1990. Food sharing in vampire bats. *Scientific American* 262 (2):76–82.

CHAPTER 38 ANIMAL BEHAVIOR: ADAPTATIONS FOR SURVIVAL

MESSAGES

1. The ability to perform behaviors originates in genes, arises during development, and may be modified by the environment.
2. Natural selection shapes behavior.
3. Single alleles can control complex behavior.
4. Behavioral types range from unchanging fixed-action patterns to insight learning, which allows animals to respond to novel situations.
5. Animals can increase their reproductive success by oriented movement, migration, and maintenance of territory.
6. The differences between male and female parental investment result in different sexual strategies and in sexual selection.
7. Communication links the behavior of individuals. Different species use different modes of communication.
8. Social living increases protection from predators at the cost of increased infighting and disease.
9. Altruism increases an individual's inclusive fitness.
10. The principles of sociobiology may be relevant to the study of human behavior.

SECTION II
GENERAL RESOURCES

SOURCES OF VIDEOS, FILMS, AND SLIDES

The films, videos, filmstrips, and 35mm transparencies listed in each chapter of this manual represent only a small sample of the resources available. Most of the producers and distributors listed here will provide catalogs of their offerings; many will allow preview privileges. In many cases, materials may be rented or purchased.

Britannica Films & Video
425 North Michigan Avenue
Chicago, IL 60611
(800)558-6968
(312)321-7100

Bullfrog Films
Oley, PA 19547
(800)543-FROG
(215)779-8226

Central Scientific Company
11222 Melrose Avenue
Franklin Park, IL 60131
(312)451-0150

Coronet/MTI Film & Video
108 Wilmot Road
Deerfield, IL. 60015
(800)621-2131
(312)940-1260

Educational Images
P.O. Box 3456, West Side
Elmira, NY 14905
(800)527-4264
(607)732-1090

Films, Inc.
5547 N. Ravenswood
Chicago, IL 60640
(800)323-4222, ext. 43
(312)878-2600, ext. 43

Human Relations Media
175 Tompkins Avenue
Pleasantville, NY 10570
(800)431-2050
(914)769-7496

International Film Bureau
332 S. Michigan Avenue
Chicago, IL. 60604
(312)427-4545

Lab-Aids, Inc.
249 Trade Zone Drive
Ronkonkoma, NY 11779
(516)737-1133

Marty Stouffer Productions
300 S. Spring Street
Aspen, CO 81611

The Media Guild
11722 Sorrento Valley Road
San Diego, CA 92121
(619)755-9191

Miller-Fenwick, Inc.
2125 Greenspring Drive
Timonium, MD 59620
(800)638-8652
(301)252-1700

Modern Talking Picture Service
5000 Park Street North
St. Petersburg, FL 33709
(813)541-5763

The National Audiovisual Center
8700 Edgeworth Drive
Capitol Heights, MD 20743
(301)763-1896

National Geographic Society
Educational Services
17th and M Streets N.W.
Washington, D.C. 20036
(800)368-2728
(201)628-9111

Pennsylvania State University
A-V Services
Special Services Building
University Park, PA 16802
(814)865-6314

Pyramid Film & Video
Box 1048
Santa Monica, CA 90406
(800)421-2304
(800)523-0118

Sargent-Welch Scientific Co.
A-V Department
7300 N. Linder Avenue
Skokie, IL 60077
(312)677-0600

Science Videos
P.O. Box 25047
Houston, TX 77265
(713)522-1827

Time-Life Video
Distribution Center
P.O. Box 644
Paramus, NJ 07653
(800)526-4663
(201)843-4545

WGBH
Distribution
125 Western Avenue
Boston, MA 02134

Ztek Company
P.O. Box 952
Louisville, KY 40201
(502)634-0304

COMPUTER SOFTWARE FOR BIOLOGY COURSES

This list represents only a sample of the suppliers of computer software designed for biology classes. The selection of programs is expanding and changing rapidly. Most of these organizations will provide up-to-date catalogs.

A^2 Educational Support Software
P.O. Box 1828
Riverton, WY 82501
(307)856-1958

Academic Software
1415 Queen Anne Road
Teaneck, NJ 07666

Albion Software
798 North Avenue
Bridgeport, CT 06606

Bergwall Educational Software
106 Charles Lindbergh Blvd.
Uniondale, NY 11553
(800)645-1737
(516)222-1111

Berkshire Scientific Software
Concepts
Astor Square--19 U.S. Route 9
Rhinebeck, NY 12572
(914)876-7097

Biolearning Systems
Route 106
Jericho, NY 11753
(516)433-2992

Cambridge Development
Laboratory
42 Fourth Avenue
Cambridge, MA 02154
(800)637-0047
(617)890-4640

Carolina Biological Supply
2700 York Road
Burlington, NC 27215
(800)334-5551

CBS Interactive Learning
One Fawcett Place
Greenwich, CT 06830

Collamore Educational Publishing
125 Spring Street
Lexington, MA 01720

COMPress
P.O. Box 102
Wentworth, NH 03282
(800)221-0419
(603)764-5831

CONDUIT
University of Iowa
Oakdale Campus
Iowa City, IA 52242
(319)335-4100

Connecticut Valley Biological Supply
82 Valley Road, Box 326
Southampton, MA 01073
(800)628-7748

Cross Educational Software
P.O. Box 1536
Ruston, LA 71270
(318)255-8921

Datatech Software Systems
19312 E. Eldorado Drive
Aurora, CO 80013
(303)693-8982

D.C. Heath
P.O. Box 19309
Indianapolis, IN 46219
(800)428-8071
(317)359-5585

Educational Computing Network
P.O. Box 8236
Riverside, CA 92515

Educational Images
P.O. Box 3456
Elmira, NY 14905
(800)527-4264
(607)732-1090

Educational Materials & Equipment
P.O. Box 2805
Danbury, CT 06813
(800)345-2050
(203)798-2050

EduTech
1927 Culver Road
Rochester, NY 14609
(716)482-3151

Encyclopedia Britannica Educational Corporation
425 N. Michigan Avenue
Chicago, IL 60611
(800)554-9862
(404)257-1690

Exeter Publishing
100 North Country Road
Setauket, NY 11733
(516)689-7838

Fisher Scientific Company
Educational Materials Div.
4901 W. LeMoyne Street
Chicago, IL 60651
(800)621-4769
(312)378-7770

Focus Media
P.O. Box 865
Garden City, NY 11530
(800)645-8989
(516)794-8900

HRM Software
175 Tompkins Avenue
Pleasantville, NY 10570
(800)431-2050
(914)769-7496

IBM Corporation
Educational Systems
3715 Northside Parkway NW
Atlanta, GA 30327

Intellectual Software
798 North Avenue
Bridgeport, CT 06606

Life Science Associates
1 Fenimore Road
Bayport, NY 11705
(516)472-2111

Micro Learningware
Route #1, Box 162
Amboy, MN 56010
(507)674-3705

Milliken Publishing
1100 Research Blvd.
P.O. Box 21579
St. Louis, MO 63132
(314)991-4220

Mindscape Inc.
Educational Division
3444 Dundee Road
Northbrook, IL 60062
(800)221-9884

National Collegiate Software Clearinghouse
NCSU Software
North Carolina State University
School of Humanities and Social Sciences
Box 8101
Raleigh, NC 27695
(919)737-3067

Sargent-Welch Scientific
7300 N. Linder Avenue
Skokie, IL 60077
(312)677-0600

Scott, Foresman & Company
1900 E. Lake Avenue
Glenview, IL 60025
(312)729-3000

Thornton Associates
1432 Main Street
Waltham, MA 02154
(617)890-3399

University of Illinois
Computer-Based Education Research Laboratory
252 Engineering Research Laboratory
Urbana, IL 61801
(217)333-1000

Videodiscovery
1515 Dexter Avenue, #200
Seattle, WA 98109
(206)285-5400

Ward's Natural Science Establishment
P.O. Box 92912
Rochester, NY 14692
(800)962-2660
(716)359-2502

BIOLOGICAL SUPPLIES AND TEACHING MATERIALS

This list includes only a few of the suppliers of biological and teaching materials. Most of these organizations will supply current catalogs.

American Science Center/Jerryco
601 Linden Place
Evanston, IL 60202
(312)475-8440

Aquarium and Science Supply
P.O. Box 41
Dresher, PA 19025
(800)537-7979
(215)643-9696

Arbor Scientific
P.O. Box 2750
Ann Arbor, MI 48106
(800)367-6695
(313)663-3733

Bausch & Lomb Optical Division
P.O. Box 450
Rochester, NY 14692
(716)338-6000

Bergwall, Inc.
106 Charles Lindbergh Blvd.
Uniondale, NY 11553
(516)222-1111

Berkshire Scientific
Astor Square, 19 U.S. Route 9
Rhinebeck, NY 12572
(914)876-7097

Carolina Biological Supply
2700 York Road
Burlington, NC 27215
(919)584-0381

Central Scientific Co. (CENCO)
11222 Melrose Avenue
Franklin Park, IL 60131
(800)262-3626
(312)451-0150

College Biological Supply
8857 Mount Israel Road
Escondido, CA 92025
(619)745-1445

Connecticut Valley Biological
Supply
P.O. Box 326
Southampton, MA 01073
(800)628-7748
(800)282-7757 (in MA)

Delta Biologicals
P.O. Box 26666
Tucson, AZ 85726
(800)824-6778

Eagle River Media Productions
9420 Westlake Drive
Eagle River, AK 99577
(907)694-4648

Educational Dimensions
P.O. Box 126
Stamford, CT 06904
(203)327-4612

Educational Materials & Equipt.
P.O. Box 2805
Danbury, CT 06813
(203)798-2050

Frey Scientific
905 Hickory Lane
Mansfield, OH 44905
(419)589-9905

Gamco Industries
Box 1911
Big Spring, TX 79720
(915)267-6327

H & H Research
P.O. Box 5156, Station One
Wilmington, NC 28403
(919)799-4942

Hubbard Scientific
P.O. Box 104
Northbrook, IL 60062
(800)323-8368
(312)272-7810

Kemtec Educational Corp.
P.O. Box 57
Kensington, MD 20895
(301)585-0930

Kons Scientific
P.O. Box 3
Germantown, WI 53022
(800)242-5667
(414)242-3636

Lab-Aids
249 Trade Zone Drive
Ronkonkoma, NY 11779
(516)737-1133

Lab Safety Supply
P.O. Box 1368
Janesville, WI 53547
(800)356-0738
(608)754-2345

NARCO Bio-Systems
7651 Airport Blvd.
Houston, TX 77061
(713)644-7521

Nasco
901 Janesville Avenue
Fort Atkinson, WI 53538
(800)558-9595
(414)563-2446

Nasco West
P.O. Box 3837
Modesto, CA 95352
(209)529-6957

National Geographic Society
17th & M Streets, N.W.
Washington, DC 20036
(202)857-7378

National Science Programs
P.O. Box 41
Batavia, IL 60510

Programs for Learning
P.O. Box 1199
New Milford, CT 06776
(203)355-3452

Sargent-Welch Scientific
7300 N. Linder Avenue
Skokie, IL 60077
(312)677-0600

Scavengers Scientific Supply
P.O. Box 211328
Auk Bay, AK 99821

Ward's Natural Science Est.
P.O. Box 92912
Rochester, NY 14692
(800)962-2660
(716)359-2502

SECTION III

ANSWERS TO TEXT STUDY QUESTIONS

CHAPTER 1

Review What You Have Learned

1. The life characteristics of metabolism, movement, responsiveness, adaptations, and a high degree of order (structural and behavioral complexity) overcome disorganization. Reproduction, development, and the passing of inherited features via genes relate to overcoming death.

2. Life is unified in that all organisms share the same genetic code. This relatedness is a result of the evolution of life forms from a single primordial ancestral form. However, in spite of this shared ancestry and genetic code, evolution has given rise to an enormous diversity of life forms.

3. An organism is an independent individual that expresses life's characteristics.

4. All organisms must take in energy to fuel cellular processes.

5. Biosphere, ecosystem, community, population, organism, organ system, organ, tissue, cell, organelle, molecule, atom, subatomic particle.

6. Metabolism breaks down materials and, using energy, rearranges the parts into useful products.

7. The two basic types of reproduction are sexual and asexual.

8. Genes guide heredity and control the development of specific physical, chemical, or behavior traits. A mutation alters the structure of a gene.

9. Natural selection is the most important mechanism leading to evolutionary change.

10. An adaptation is an inheritable feature that increases an individual's chances of taking in energy or materials, growing, attracting a mate, or otherwise combatting disorganization or death.

11. Inductive reasoning involves generalizing from specific cases to arrive at broad principles. Deductive reasoning involves analyzing specific cases on the basis of preestablished general principles.

Apply What You Have Learned

1. Ask a question: Does a philodendron plant require weekly doses of nitrogen fertilizer? Propose a hypothesis: The plant does (or does not) require a weekly dose of fertilizer. Predict that the plant will show signs of distress if the fertilizer is discontinued. Design an experiment to test the hypothesis: Take cuttings from the plant and root them to establish two or more genetically identical plants, one or more of which will be the controls. Discontinue the fertilizer applications to one or more of the plants for six months, while maintaining the regular application of fertilizer to the controls. Based on observations during the test period and comparisons of the experimental plant with the controls, draw a conclusion about the validity of the hypothesis.

2. Predators prey less on edible butterflies resembling a brightly-colored, poisonous species than on edible butterflies without such a resemblance. Thus the edible butterflies that resemble poisonous species will have more success in

survival and reproduction than will other butterflies of the same species, and their genes will increase in frequency in the population. Among the mimic butterflies, some will have a greater similarity than others to the model poisonous species, and the more accurate the mimicry, the greater the reproductive success. Eventually natural selection will lead to very strong resemblance between the mimic and model.

CHAPTER 2

Review What You Have Learned

1. Electrons occur in energy levels around the nucleus of an atom. The number of electrons in the outer energy level dictates the atom's reactivity, the likelihood that it will combine with other atoms. In this way an atom's electrons determine its chemical behavior.

2. The atomic number is the number of protons in the nucleus. The combined number of protons and neutrons in the nucleus is the atomic mass number.

3. In a covalent bond, electrons are shared by two or more atoms. Example: Two hydrogen atoms join to form one H_2 molecule. In an ionic bond, electrons are transferred from one atom to another. Example: In a NaCl molecule, a sodium atom loses an electron, which is transferred to a chlorine atom.

4. Some atoms have isotopes, forms in which there are different numbers of neutrons in the nucleus. Thus "normal" carbon has 6 neutrons in its nucleus, while a radioactive isotope, carbon-14, has 8 neutrons in its nucleus. Also, when an element such as sodium loses an electron to form an ion (Na^+), the ion reacts differently from the normally-charged atom.

5. Water molecules may ionize into a positively charged hydrogen ion, H^+, and a negatively charged hydroxide ion, OH^-. An acid is an H^+ donor and has a high concentration of H^+ (and thus a low value on the pH scale), while a base is an H^+ acceptor and has a low concentration of H^+ (and a high pH value).

6. Monosaccharides are simple sugars such as glucose, fructose and galactose, and have the formula $C_6H_{12}O_6$. Disaccharides, such as sucrose, are composed of two-unit molecules (glucose bonded to fructose), and have the formula $C_{12}H_{22}O_{11}$. Polysaccharides are carbohydrate polymers formed by the condensation of simple sugars into long chains; examples include starch, glycogen, cellulose, and chitin.

7. Lipids include fats and oils, which have a glycerol molecule bonded to three fatty acids and are known as triglycerides; waxes, such as bees' wax and ear wax, with a different structure from triglycerides; phospholipids, which have a phosphate goup and two fatty acid chains attached to a glycerol molecule; and steroids, including vitamins A, D, and E, cholesterol, and certain hormones.

8. Proteins with a primary structure display a linear arrangement of amino acids. Secondary protein structure may include an alpha helix, a beta sheet, or a disordered loop. In the tertiary structure proteins fold into either globular shapes or fibrous shapes. The quaternary structure is based on two or more folded polypeptide chains.

9. Nucleotides are present as long polymers in DNA and RNA, which carry hereditary information; as ATP; as cAMP, which carries chemical signals; and as coenzymes, which are transport compounds in energy reactions.

10. A carbohydrate contains C, H, and O in a ratio of 1:2:1. A lipid includes these three elements in different arrangements. An amino acid contains carbon, hydrogen, oxygen, and nitrogen arranged in three functional groups bound to a central carbon atom. Amino acids may be organized into polypeptide chains.

Apply What You Have Learned
1. Water molecules exhibit the property of surface tension, the tendency to cohere to each other at the surface, but not to the air molecules above them. This results in an elastic skin on which water striders can move about.

2. A diet high in saturated fats from cheese, eggs, and meat can produce deposits in the arteries that clog the vessels. As a result blood flow may be impeded or stop, leading to a heart attack or a stroke.

3. Heat applied during pasteurization kills microorganisms by disrupting the tertiary structure of proteins, stopping metabolism as enzymes become nonfunctional.

CHAPTER 3

Review What You Have Learned
1. Robert Hooke discovered dead cells in cork; Anton van Leeuwenhoek used a simple microscope to view live euglenas and other single-celled organisms.

2. The cell theory states that all living things are composed of cells, that the cell is the basic unit of life, and that all cells arise from preexisting cells.

3. A prokaryotic cell does not have a nucleus; its DNA is in a naked, circular strand; the cell is surrounded by a cell wall. A eukaryotic cell has a nucleus and other organelles such as mitochondria, ribosomes, endoplasmic reticulum, Golgi bodies, microtubules, and microfilaments. Plant cells also have a large vacuole, chloroplasts, and a cell wall of cellulose; animal cells do not have a cell wall but are surrounded by a plasma membrane.

4. The plasma membrane is semipermeable and allows the passage of water ions and certain organic molecules into and out of the cell. The nucleus contains DNA, the genetic material that controls the cell activities, and makes and exports RNA, which relays genetic instructions to the cytoplasm. Cytoplasm acts as a pool of raw materials and contains the cytoskeleton in which the various organelles are suspended.

5. Ribosomes are the site of protein manufacture. Endoplasmic reticulum forms an interconnected set of channels throughout the cell and is the site of protein synthesis (rough ER) and lipid synthesis (smooth ER). The Golgi complex packages proteins, lipids, and cellulose. Lysosomes contain digestive enzymes within the cell. Mitochondria are the sites of aerobic respiration and produce ATP, the energy currency that fuels cellular activities.

6. False. Bacteria contain DNA in a naked, circular strand which is concentrated in the nucleoid region of the bacterial cell.

7. In a magnolia tree there are about a dozen different types of cells, such as sieve tubes in which sugars are transported throughout the tree, flattened epidermal cells which serve as an outer protective "skin" on leaves, and other structures for reproduction, food-making, and support. The spiny anteater has about 200 kinds of cells, including nerve cells for conveying information throughout the animal, muscles for movement, and other cells for digestion, reproduction and circulation.

8. Plant and animal cells possess a nucleus, cytoplasm, plasma membrane, ribosomes, ER, Golgi complex, mitochondria, and lysosomes. Plant cells also have chloroplasts and other plastids, a central vacuole, and an outer cell wall of cellulose.

9. The flagellum in *Euglena* produces movement by its whip-like lashing. Cilia in some protozoa beat in unison to propel the cell through water. An amoeba creeps along as microfilaments in its cytoskeleton rapidly extend and contract.

10. Plant cells have walls, composed largely of cellulose, immediately outside the plasma membrane. In many plant species lignin laid down inside the primary cell wall provides additional rigidity. Most animal cells are surrounded by an extracellular matrix composed of fibrous proteins such as collagen linked to polysaccharides, which establish the cell's shape.

Apply What You Have Learned
1. Water cannot pass through the lipid bilayer of a plasma membrane because water molecules are bound to one another by hydrogen bonds. The lipid bilayer does not form hydrogen bonds with water; thus the water molecules attract each other more than does the lipid bilayer.

2. No. Cell size is limited by the surface-to-volume ratio. The larger a cell's volume, the more surface area is needed for the exchange of materials through the plasma membrane. An amoeba would be unable to survive if its size increased drastically.

3. The scientist might use all three kinds of microscopes. A phase contrast microscope permits viewing a live organism's structures without the use of stains, but is limited to 1000X magnification or less. Electron microscopes can be used only with dead cells because dehydration and the application of stains is necessary. Electron microscopes have much greater magnification powers than light microscopes, and thus can be used to view smaller structures. A transmission electron microscope may be used to view internal structures; a scanning electron microscope may be used to view surface structures.

CHAPTER 4

Review What You Have Learned
1. Potential energy is energy stored and ready to do work. Kinetic energy is the energy of motion.

2. Living cells maintain a constant flow of energy for such order-producing activities as maintenance, repair, and protein synthesis.

3. Running and weight-lifting are exergonic activities because the cellular work involved liberates energy (heat). The building of proteins and the formation of new ribosomes are considered endergonic reactions because they require an input of energy.

4. ATP is an energy carrier that, like money, is "saved up" during energy-yielding exergonic reactions and "spent" during energy-costly endergonic reactions.

5. Activation energy is the minimum energy needed for biochemical reactions to take place. Enzyme activity lowers activation energy so that reactions take place much more rapidly.

6. An enzyme is a globular protein with a deep surface groove called the active site. The shape of this site fits that of a specific substrate, much like a key fits a lock.

7. In passive transport, molecules pass through the plasma membrane of a cell, moving down concentration gradients without the use of energy. In active transport, the cell expends energy to move molecules through the plasma membrane against gradients.

8. False. After an equal concentration of molecules on both sides of the membrane has been reached, molecules will continue to move through the membrane, but in equal numbers.

9. An isotonic solution outside a cell has the same concentration of solution as does the cytoplasm within the cell. A hypotonic solution outside a cell has a lower concentration of solutes than the cell's cytoplasm. A hypertonic solution outside a cell has a higher concentration of solutes than a cell's cytoplasm.

10. The sodium-potassium pump expends energy in the form of ATP to move K^+ ions into and Na^+ ions out of a red blood cell, balancing the solute concentration inside the cell with the solute concentration of the blood plasma.

Apply What You Have Learned

1. Your grandfather would be right to claim that for the most part he is less than a year old because most molecules in his body are replaced over the course of a year.

2. The greatest digestion would take place at 37°C, which is the temperature of the human body. At 17°C, the enzyme would act slowly because of the low temperature. At 57°C, the tertiary structure of the enzyme would be disrupted, preventing catalytic activity.

3. Bacteria cells are killed by the high concentration of external salt because a hypertonic solution results when the salt dissolves. Hence, water passes out of each bacterium while at the same time a great amount of salt enters the cell.

CHAPTER 5

Review What You Have Learned

1.

	Fermentation	Aerobic Respiration
Oxygen	Absent	Present
ATP Produced	2	36
Waste Products	Alcohol and CO_2 or lactic acid	H_2O and CO_2

2. A heterotroph, such as a fish, takes in preformed nutrients from the environment. An autotroph, such as a pine tree, makes its own nutrients by using the sun's energy.

3. During glycolysis, a six-carbon glucose is broken down into two molecules of the three-carbon pyruvate. Two ATPs are formed.

4. Alcoholic fermentation and lactic acid fermentation both take place in the absence of O_2; a glucose molecule is broken down to 2 pyruvate molecules, generating 2 ATPs. In alcoholic fermentation, ethyl alcohol and CO_2 are produced. In lactic acid fermentation, the waste product is lactic acid.

5. During aerobic respiration, the glucose molecule is completely broken down to waste products CO_2 and H_2O, and much of the energy in its bonds is used to regenerate 36 ATPs. In fermentation, the glucose molecule is only partially broken down to 2 pyruvate molecules, and only 2 ATPs result. The waste products still contain energy that could be extracted by aerobic respiration.

6. During the Krebs cycle, the pyruvate molecule is oxidized to CO_2. Electrons and H^+ ions are transferred to the carriers NAD^+ and FADH, and 2 ATPs form.

7. The Krebs cycle acts as a metabolic clearinghouse by breaking down lipids, proteins, or carbohydrates, and by providing acetyl-CoA or other intermediates for use in the biosynthesis of new fats, proteins, and carbohydrates.

8. The electron transport chain receives electrons from the high energy carriers, NADH and $FADH_2$, and passes them along from one electron transport protein to another in a "downward" energy flow. Oxygen acts as the final acceptor of the electrons and protons.

9. A mitochondrion is a site of aerobic respiration in the cell. Transport proteins are located within the inner membrane of the mitochondrion and pass the electrons stored in NADH and $FADH_2$ along from one to another, releasing energy in each step. Oxygen is the final acceptor in the electron transport chain and becomes reduced to H_2O. NADH and $FADH_2$ are high-energy carriers formed in the Krebs cycle that pass their electrons along to carrier proteins during the electron transport chain phase of respiration.

10. An allosteric enzyme can change shape so that its active site stops bonding to the substrate, causing a cessation in its activity. The accumulation of a metabolic product inhibits the allosteric enzyme, thus turning off its own production; this is known as feedback inhibition.

Apply What You Have Learned

1. Vintners seal the must in airless tanks to prevent oxygen from reaching the yeast. Yeast produce alcohol only if oxygen is absent. If air were bubbled through the must, the yeast would switch to aerobic respiration and consume sugars without producing alcohol.

2. When a person swims, body cells use their stores of glucose; stored lipids are then broken down and their subunits enter the Krebs cycle. Thus, calories are burned and weight is lost.

3. The muscle cells are so active that they cannot get enough O_2. During glycolysis, the pyruvate molecules cannot be broken down by aerobic respiration and instead fermentation takes place. Lactic acid forms and accumulates, causing painful leg cramps.

CHAPTER 6

Review What You Have Learned

1.

	Aerobic Respiration	Photosynthesis
Location	Mitochondrion	Chloroplast
Starting Materials	Glucose	CO_2 and H_2O
End products	CO_2 and H_2O	Glucose
Energy	Released	Stored

2. In the light reactions, the energy source is light. In the dark (light-independent) reactions, the energy sources are ATP and NADPH.

3. Sugars may be stored as starch in plastids, or used as subunits in the cellulose of plant cell walls; they may also be broken down by plant cells' mitochondria to power cellular activity.

4. Chlorophyll *a* is found in all photosynthetic autotrophs. Chlorophyll *b* occurs mostly in land plants and green algae. Carotenoid pigments are found in all autotrophs and give color to such nonphotosynthetic structures as the roots of carrots, flowers, fruits (tomatoes) and seeds (corn kernels).

5. Chlorophyll *a*, the reaction center, is surrounded by antenna complexes of chlorophylls, carotenoids, and other pigment molecules. These complexes pass the energy they absorb from light photons, in the form of electrons, to the reaction center, exciting chlorophyll *a* so that it passes electrons to an electron acceptor.

6. Noncyclic photophosphorylation involves the zigzag scheme of electron flow in a one-way direction from P680 to P700, and accounts for most of the ATP formed during photosynthesis. In cyclic photophosphorylation, an electron returns to the chlorophyll molecule that released it to the electron transport chain, producing only a small amount of ATP.

7. A glucose molecule stores in its bonds the energy equivalent of 36 ATPs. It can travel within an organism without breaking down, and can be converted to starch, glycogen, or cellulose. A molecule of glucose is thus more compact, stable, and versatile than the 36 ATPs it could be used to generate.

8. Carbon fixation occurs when a CO_2 molecule is joined to RUBP by the enzyme rubisco, forming two three-carbon molecules. Energy, electrons, and hydrogens generated during the light-dependent reactions are added to the three-carbon molecules, forming carbohydrates.

9. On a hot day when stomata are nearly closed so that little CO_2 can enter leaves, a CO_2 pump operates in C_4 plants. The CO_2 is added to a three-carbon compound, creating a four-carbon compound. When the four-carbon compound breaks down, it delivers CO_2 to the light-independent reactions.

10. Autotrophs fix the carbon of CO_2 into glucose and other organic compounds, releasing O_2. Heterotrophs feed on autotrophs directly or indirectly and, taking in O_2, release CO_2. Autotrophs then fix the released CO_2 once again and release new molecules of O_2. The cycle continues in this way.

Apply What You Have Learned
1. The direct or indirect dependence of virtually all organisms on the fixed carbon produced during photosynthesis is evidence that almost all life on earth is powered by solar energy.

2. Water serves as an electron donor and is split into oxygen and hydrogen during photosynthesis. Any artificial system would have to accomplish this step. Another goal would be to absorb light artificially and give off a flow of electrons that could be captured as chemical bond energy.

3. Possible ways to increase the rate of O_2 production include maximizing the rate of photosynthesis by increasing the exposure to light (e.g. by using electric lights at night), by increasing the amount of CO_2 in the water (e.g. by pumping it in or adding $KHCO_3$ or $NaHCO_3$ to the water), or by increasing the temperature of the water to an optimum level.

CHAPTER 7

Review What You Have Learned
1. Initially the prokaryotic cell grows, doubling in length. The circular chromosome then replicates into two side-by-side circles, each with one point of attachment to the cell membrane. These attachment points move to the opposite ends of the cell; a partition then forms in the center and splits the cell into two daughter cells, each with a complete chromosome.

2. G1 initiates interphase during which new proteins, ribosomes, mitochondria, ER, etc., are built; S phase, in which DNA is replicated, provides the cell with two sets of chromosomes; G2, in which many proteins are built; M, in which mitosis and cytokinesis occur.

3. In mitosis, the cell nucleus divides. In cytokinesis, the cytoplasm of the cell divides, yielding two new cells.

4. A chromosome is the rod-like structure that appears during the M stage. A chromatid is one of the two identical chromosomes created by replication and held together by the centromere.

5. Prophase, prometaphase, metaphase, anaphase, telophase.

6. Animal cells: A contractile ring composed of a circle of microfilaments pinches in to form a furrow in the cell surface, which deepens to eventually split the cell in two. Plant cells: Division takes place from the inside out; vesicles filled with cell wall precursors pinch off from the Golgi bodies and collect in the cell's center where they fuse, forming a cell plate which extends to the sides of the cell and divides the cell in two.

7. They do not obey the rules of contact inhibition and so form masses of cells.

8. Mitosis: The number of chromosomes in the daughter cells is identical to the number in the parent cell. Meiosis: The daughter cells have only half as many chromosomes as the parent cell.

9. The maternal and paternal chromosomes are reshuffled, leading to new chromosome combinations by independent assortment and by crossing over.

10. Sexual reproduction, because it produces various genetic combinations, some of which may be advantageous under particular environmental conditions.

Apply What You Have Learned
1. A cell unable to produce cyclins would be unable to undergo mitosis.

2. Asexual, because genetically identical offspring are produced.

3. Programmed limits on the number of cell divisions could decrease the incidence of cancer, which is uncontrolled cell proliferation. It could also prevent individuals from surviving long past their reproductive stages and monopolizing resources that could otherwise support younger individuals. The death of individuals seems to be important to the survival and evolution of species.

4. During cell division, fragments of chromosomes would no longer be attached to centromeres and would not be divided into daughter cells. The daughter cells may be dysfunctional, carcinogenic, or unable to survive, resulting in tissue damage, cancer, and an inability to heal wounds.

CHAPTER 8

Review What You Have Learned
1. They showed clear alternatives for single traits so Mendel could study one feature at a time. Peas self-fertilize; Mendel could cross-fertilize by simply removing anthers.

2. The F_1 generation would be all heterozygous, round-seeded pea plants. The F_2 generation would include round-seeded and wrinkle-seeded pea plants in an approximate 3:1 ratio.

3. There are two alleles for each gene in the parents; these separate in the gametes, which then receive only one allele of the gene.

4. The genotypic ratio and phenotypic ratio differ because two different genotypes, the homozygous recessive and heterozygous, have the same dominant phenotype.

5. LP, Lp, lP, lp.

6. The sperm may have an X or a Y chromosome. The egg may have only an X chromosome.

7. Ad, aD.

8. When white and red snapdragons are cross-fertilized, the heterozygous offspring have pink flowers, not red or white as predicted by Mendel's principle of dominance.

9. Labrador retrievers with the genotype BBEE would be pure black; with BBee, yellow; with bbEE, chocolate-colored.

10. True. During independent assortment, the genes are on different chromosomes and so are distributed into the gametes independently of each other. In linkage, genes are carried on the same chromosome and can be inherited together.

Apply What You Have Learned
1. After 10 tosses, the results may by chance be quite different from the expected 1:2:1 ratio. After 100 tosses, the ratio will be closer. According to the laws of chance, the greater the number of trials, the closer the results will be to the mathematically predicted values.

2. By means of a test cross with white homozygous recessive (bb) individuals, if only black offspring are obtained, they will be black heterozygous (Bb), indicating that the unknown parent was homozygous black (BB); if some white offspring (bb) appear, the unknown black parent was heterozygous (Bb).

3.

		mother X^+	X^+
father	X^-	X^+X^-	X^+X^-
	Y	X^+Y	X^+Y

All daughters are phenotypically normal carriers. All sons are normal.

4. D -25- A -20- B -5- C

CHAPTER 9

Review What You Have Learned
1. A gene is a series of nucleotides that specifies the sequence of amino acids making up a polypeptide, such as an enzyme.

2. A normal hemoglobin molecule contains glutamic acid in the sixth position; the sickle cell hemoglobin has valine in that position instead.

3. Each gene specifies one particular peptide, or one chain of amino acids. Genes instruct the cell to make specific enzymes.

4. Harmless bacteria were converted to a pathogenic strain after exposure to DNA extract from pathogenic bacteria. This DNA extract was incorporated into the genome of the harmless bacteria, so that progeny inherited the ability to make the protein that causes pneumonia.

5. The phage virus injects its DNA into the bacterium and induces the bacterium's genetic machinery to make many more phages. The bacterial cell then bursts and releases new phages, each of which can infect other bacteria.

6. They are all similar in containing a phosphate, a sugar and a base; they are different in that the base may be adenine (A), cytosine (C), guanine (G), or thymine (T).

7. G-A: G complements C; A complements T.

8. The two strands unwind. The now-single bases pair with complementary nucleotides present in the cytoplasm. The enzyme DNA polymerase catalyzes the formation of the covalent bonds that link these nucleotide pairs together.

9. The enzyme DNA polymerase removes any base from the end of a newly formed DNA strand that is not hydrogen-bonded to a complementary base.

10. A mutation would result which can alter protein structure.

Apply What You Have Learned

1. A possible model is that one allele of a specific gene encodes an enzyme that catalyzes the biochemical reaction leading to pigment, while the albino allele of this gene fails to make an active enzyme. Hence, no pigment is made in homozygotes.

2. RNA is the genetic material. One could test the hypothesis by inducing a mutation in the RNA and seeing if the progeny of the virus have the same mutation, then repeating the experiment with a mutation in the protein.

3. Recall that A=T and C=G, and that A + T + G + C = 100%. Therefore, T=18%; G + C = 100% − 18% − 18% = 64%. Hence G = C = 32%.

4. Radioactive phosphorus would be the best choice because it is the only element on the list that is found in DNA but not in proteins. If you used radioactive oxygen, carbon, nitrogen, and sulfur, which would be incorporated into the proteins that make up the virus's protein capsule, the radioactivity in the parts of the capsule left in the host cell would obscure or confuse the location of the radioactive DNA.

CHAPTER 10

Review What You Have Learned

1. Transcription is the formation of RNA copies of segments of DNA molecules. Translation is the formation of proteins under the direction of RNA.

2. RNA has the base uracil instead of thymine. Its sugar is ribose rather than deoxyribose. It consists of one or two strands, instead of two. It is a much shorter molecule.

3. GAUCGA

4. (a) AAA, AAG; (b) UUU, UUC; (c) GGU, GGC, GGA, GGG.

5. The near universality of the genetic code is strong evidence that all living things are descendants of an organism with the same genetic code that lived early in our planet's history.

6. GUC is the codon. Valine is the amino acid.

7. Causes of gene mutation include the replacement of one base for another; the insertion or deletion of one or more base pairs; and a rearrangement of bases, such as an inversion of base order.

8. The six components of a bacterial lac operon are: a structural gene, a promoter, a regulator gene, a repressor, an operator, and an inducer. A structural gene specifies a protein. A promoter binds RNA polymerase. A regulator encodes a regulatory protein, like a repressor. A repressor binds the operator and blocks transcription. An operator is a DNA sequence that binds the repressor or other regulatory protein to govern transcription. An inducer is a small molecule that binds to the regulator protein to control regulator protein function.

9. Liver cells are regulated by genes to produce liver proteins, even if transplanted into another organ, such as the brain. Liver cells can be removed and cultured in a petri dish, but they will still make liver proteins.

10. In both prokaryotes and eukaryotes, gene regulation occurs mainly at the level of transcription. Both use protein regulators that bind to sections of DNA. Prokaryotes turn genes on and off quickly; eukaryotic cells express the same genes throughout their lifetimes. Prokaryotes have clusters of related genes (operons), but eukaryotes do not.

Apply What You Have Learned

1. Chemicals that increase the rate of mutation in bacteria also damage human DNA. In the Ames test, bacteria would be exposed to the chemicals in hair dye. Mutation rates in the bacteria would indicate the likelihood that the chemicals would cause mutations, and hence act as carcinogens, in humans.

2. It applies to bacteria, such as *E. coli.* Operons do not occur in eukaryotes.

3. The anticodon AUU would pair with the codon UAA, a "stop codon" that normally terminates translation. If a mutant tRNA with the anticodon AUU were to carry an amino acid, translation would not stop at the appropriate time, and polypeptides would be too long and probably not function properly. The cell might die as a consequence.

4. Before a primary transcript is exported as mRNA from the nucleus, the introns are removed and the exons spliced together. A DNA copy of the resulting mRNA would not include the introns found in the original gene.

CHAPTER 11

Review What You Have Learned

1. The human gene for growth hormone was inserted into a mouse chromosome, where the extra DNA was correctly transcribed and translated into human growth hormone. This hormone caused the mouse to grow more than twice its normal size.

2. Crossing-over during meiosis brings the scarlet and rosy alleles together on the same chromosome in some of the gametes. Fusion of egg and sperm containing these recombined chromosomes results in doubly homozygous flies with pink eyes.

3. False. Recombination occurs in nearly all organisms because species with recombination have greater variation on which natural selection can act, and thus in variable environments they have an advantage over species that do not combine.

4. Mutation changes base sequences. Recombination shuffles genes and creates new combinations of alleles.

5. Restriction enzymes can cut DNA at specific points. DNA ligases can join cut DNA strands together, and, in nature, will repair damaged DNA.

6. Potential future applications of genetic engineering include gene replacements to cure genetic diseases and the production of a drug that can kill the common cold virus, an enzyme that can arrest a heart attack, bacteria that can make automobile fuel from old corn stalks, and bacteria that can protect crops from late spring freezes.

7. Improvements in animals and plants that might result from genetic engineering include the introduction of genes that increase an animal's size and milk production, that improve the efficiency of photosynthesis and the nutritional quality of seeds and vegetables, or that increase plant resistance to pests, drought, and extreme temperatures.

8. Somatic cell gene therapy may be used to introduce genes for tumor necrosis factor (TNF) into tumor-infiltrating lymphocytes (TILs). The TILs may invade tumors, produce TNF, and so destroy the cancer.

9. Ethical questions are being raised about the safety and morality of this research. Transformed bacteria having a cancer-causing gene may escape from a lab and infect people. Or there may be attempts to "improve" an already healthy child with certain desired traits. Or the transfer of genes from one species to another may influence future evolution.

Apply What You Have Learned

1. As a result of recombination and fertilization, a zygote receives a different set of genes from each parent. Mathematically, the potential combinations of thousands of genes is enormous, far exceeding the number of humans that have ever lived. Thus no two individuals (except identical twins) are likely to inherit the same combination of alleles, and each person is genetically unique.

2. Isolate human DNA and cut it with restriction enzymes. Insert the DNA fragments into plasmids, using restriction enzymes and ligases. Transfer the plasmids into bacteria and culture the bacteria. Using a radioactive probe made from pure TPA messenger RNA isolated from cells making TPA, identify the clone of bacteria bearing the TPA gene. Isolate the clone and culture the TPA-producing bacteria.

3. You should advise the circus owner that his project is not likely to be successful. Current recombinant DNA technology allows the transplantation of individual genes, such as the insertion of the gene for human growth hormone into mice. The hybridization of whole structures, such as wings onto a horse, or of whole organisms, such as humans and cattle, is far beyond today's technology.

CHAPTER 12

Review What You Have Learned

1. Relatively few offspring are produced, so statistical analysis is difficult. Humans do not mate to meet an experimenter's wishes. There is almost never a true F_2 generation. A human generation takes half a geneticist's career to follow.

2. The trait is carried in the mother's X chromosome, which when united with the father's Y chromosome, produces a male child.

3. (a) Down syndrome; (b) Burkitt's lymphoma; (c) mental retardation.

4. An oncogene is a cancer-causing gene. For example, *myc* may be translocated next to a gene that is actively transcribed in normal white blood cells. In this location, *myc* may be rapidly transcribed, causing white blood cells to grow out of control into a cancer.

5. In somatic cell fusion a human cell with its 46 chromosomes is fused with a mouse tumor cell having 40 chromosomes. As this hybrid cell divides, human chromosomes are gradually lost, resulting in clones of cells with different sets of human chromosomes. A geneticist can establish a bank of cell lines to map the remaining human genes to specific chromosomes.

6. Gene mapping by using RFLPs allows the identification of the loci of specific genes through their genetic linkage to specific RFLPs. If a person has a RFLP known to be linked to a defective gene, he or she probably also has the defective gene.

7. In amniocentesis, fetal cells collected by a needle inserted through the abdominal and uterine walls are grown in the laboratory and examined for defective chromosomes. In chorionic villus sampling, fetal cells are removed from the developing placenta and similarly examined.

8. Prospective parents are informed about the possibility of a disease the newborn baby may have and the resulting potential suffering of the child and family. Available options are considered.

9. False. Normal male cells lack Barr bodies.

Apply What You Have Learned

1. The likelihood of a homozygous recessive child being born to two heterozygotes is 25%, because there is an expected 3:1 ratio of dominant to recessive phenotypes.

2. Homozygous: None of her X chromosomes contains the gene for hemophilia, so her son would be normal. Heterozygous: One of her X chromosomes contains the gene for hemophilia; there is a 50% chance that her son may be hemophiliac.

3. Through the study of DNA "fingerprints," the allegations of paternity could be verified or disproved. Short stretches of DNA called hypervariable regions may be 37 bases long, for example, and repeated 21 times in the mother, but only 12 times on the paternal homologous chromosome. Such hypervariable sequences are inherited in a normal Mendelian pattern, allowing the fragments to be traced to one or the other parent.

4. A heterozygous woman carrying the gene for hemophilia who marries a normal male has a near zero chance of producing a hemophiliac daughter (excluding a new mutation on the father's X chromosome). There is a 50% chance that she will have a son with the disease.

5. (a) autosomal recessive; (b) sex-linked recessive; (c) autosomal dominant.

6. The argument that the man inherited the disease as a recessive allele from both parents is not likely to be correct because Huntington's disease is caused by a dominant allele. The second argument, that the disease is due to a new mutation, may be correct, as might the third argument, that the man's ancestors with the allele had not shown symptoms because they had died at relatively young ages. Additional information that would help you evaluate these possibilities might include a more detailed family history. Early deaths in a lineage might suggest that the allele had been carried through previous generations without phenotypic effect. Longevity without the disease throughout the family might be evidence against this argument. Chromosomal analysis and RFLP gene mapping of the man and his relatives might test the hypothesis that a new mutation caused the disease.

CHAPTER 13

Review What You Have Learned

1. Visual mating cues include the brilliant plumage of male birds, a lion's mane, a woman's breasts and buttocks, and a man's beard and strong muscles. Olfactory cues include the odors of mammals and insect pheromones.

2. The flagellum propels a sperm into contact with an egg's surface, where protein molecules on the sperm's surface bind to species-specific sperm receptors. Enzymes released from a sac on the sperm's tip facilitate the fusion of the plasma membranes of the egg and sperm. The sperm's nucleus enters the egg, and the egg's surface hardens, which prevents the entry of more sperm. The egg and sperm nuclei fuse.

3. The egg of a chicken is larger; it has a great deal of stored food. A whale's egg, like other mammalian eggs, has very little yolk and is microscopic.

4. Developmental determinants are chemicals in the egg's cytoplasm that act as instructions for the developing embryo by becoming localized in certain blastomeres. Here, they affect gene expression, causing cells containing the grey crescent, for example, to become the future dorsal region.

5. Ectoderm: skin and nervous tissue. Mesoderm: muscles, bones, connective tissue, blood, reproductive and excretory organs. Endoderm: inner linings of gut, lungs, liver, salivary glands.

6. Cells of the notochord induce the ectoderm cells lying above them to roll up into the neural tube; the tube then develops into the nervous system.

7. Growth factors bind to cell surface receptors and provoke a change in the cell that stimulates growth.

8. Both are clumps of cells that grow unchecked. A benign tumor remains localized in a tightly grouped mass of cells; a malignant tumor grows increasingly and metastasizes, invading other parts of the body.

9. In fertilization, a sperm unites with an egg, thus initiating development of the offspring. In parthenogenesis, the egg develops into an embryo without uniting with a sperm.

Apply What You Have Learned
1. To reduce the risk of cancer, avoid cigarette smoking; high-fat, low-fiber diet; salt-cured, nitrate-treated, or smoked foods; heavy use of alcohol; and overexposure to sunlight.

2. The mutant fish embryo will develop no nervous system because the notochord induces the neural tube to form.

3. The egg cell contains germ plasm, which appears as a region of granular cytoplasm. During cleavage, germ plasm is partitioned into a few blastula cells. Germ plasm contains cytoplasmic determinants that influence a cell to become a germ cell. The transplanted part contained such germ plasm.

4. The greatest changes during the development of a human occur during the first part of pregnancy: gastrulation, neurulation, organogenesis, and differentiation. Even minor damage during these stages may result in substantial birth defects. During the later part of a pregnancy, organs are less susceptible because they are already formed.

5. You should side with the chemical company. Birth defects such as spina bifida and cleft palate are results of incomplete closure of the neural tube and inadequate proliferation of migrating neural crest cells. Normally, these processes are completed during the first few months of life. It is therefore likely that the fetus already had spina bifida and cleft palate when the mother was exposed to the chemicals.

CHAPTER 14

Review What You Have Learned
1. Seminiferous tubules, epididymis, vas deferens, ejaculatory duct, prostate gland, urethra.

2. Male bicycle racers who train intensively spend many hours on a bicycle saddle, exercising hard. This increases the temperature of the testes, suppressing sperm production.

3. When the testosterone level is too low, the brain releases hormones that act on the testes and increase the production of sperm and testosterone. High testosterone levels block the release of brain hormones, causing the testes' production of testosterone and sperm to fall.

4. An egg is produced in the follicle of an ovary. Pathway: follicle, fallopian tubes, uterus, cervix, vagina.

5.

Female hormone	**Where formed**	**Effect**
Releasing factor	hypothalamus	stimulates pituitary to produce LH and FSH
FSH	pituitary	stimulates follicle to produce egg
Estrogen	follicle	causes uterus lining to thicken
LH	pituitary	ovulation
Progesterone	corpus luteum	with estrogen, continues buildup of uterus lining; also inhibits hypothalamus's production of releasing factors

6. One. It triggers the ovum to complete meiosis II and erect barriers to further sperm penetration.

7. Women: Oviducts or uterus are blocked or scarred; scarring resulting from pelvic diseases or use of IUDs. Men: Low sperm count; sperm lack mobility. Both sexes: venereal diseases, aging.

8. At first, the chorion absorbs nutrients from the mother's blood and passes them on to the rapidly dividing embryo. Then it develops into the placenta. Also, it produces human chorionic gonadotropin, which prevents a menstrual cycle.

9. False; they never mix. Nutrients and oxygen pass from mother's blood into embryo's blood; carbon dioxide and other wastes go in the reverse direction.

10. The Y chromosome contains the gene that causes male differentiation. The gonads develop into testes. Testes produce testosterone and Mullerian inhibiting hormone, which kills the duct cells that could develop into oviducts and uterus.

Apply What You Have Learned

1. Progesterone helps maintain the uterine lining, with its rich supply of blood vessels that nourish the developing embryo. When the hormone is missing, the lining sloughs off. If an embryo has recently implanted, it will thus be miscarried.

2. The finding helps identify Alzheimer's as a genetic disease and may possibly aid in identification of potential victims.

3. Anabolic steroids may cause women to develop male secondary sex characteristics, including facial hair, muscle growth, deeper voice, and aggressive

behavior. They may also interfere with the normal female hormonal cycles, causing infertility.

4. The arguments of the technicians are more likely to be correct. In a negative feedback loop, high levels of testosterone or the anabolic steroid can act on the brain and block the secretion of LH and FSH. Lowered amounts of these gonad-stimulating hormones would result in less production of natural testosterone by the testes.

CHAPTER 15

Review What You Have Learned

1. The inorganic chemicals they metabolize (H_2 and CO_2) were common in the Earth's first atmosphere; they cannot tolerate free O_2; their cell walls, cell membranes, and electron transport chains contain molecules different from those of the true bacteria; they fix carbon in a unique pathway; their RNA has a different base sequence from that of other bacteria.

2. The explosion of an enormous ball of searing gases about 18 billion years ago that created all matter in the universe.

3. Organic compounds such as formaldehyde and hydrogen cyanide may have been present in the clouds of organic molecules that were present as planets coalesced. Meteorites called carbonaceous chondrites which contained the five nucleotide bases may have fallen to earth. Natural energy sources, such as lightning, ultraviolet light, and heat from volcanic activity could have driven endergonic reactions that converted atmospheric gases into monomers.

4. Polymers form; self-replication; molecular interactions; cell-like containers take shape; coordinated cell-like activities.

5. (a) Formation of the earth; (b) appearance of photosynthesis, with oxygen being given off into the seas and into the atmosphere for the first time; (c) the emergence of eukaryotic cells.

6. Mitochondria are similar in size and shape to modern aerobic bacteria; they have their own DNA which replicates independently. The genes in their DNA are more similar to bacterial genes than to the genes of eukaryotes.

7. Plate tectonics, the building and movement of crustal plates, causes earthquakes, volcanoes, the uplifting of mountain ranges and other changes in the earth's surface.

8. Linnaeus invented the binomial system of nomenclature; he also classified living things in increasingly specific groups, depending on their structural traits.

9. A species consists of organisms that share the same set of structural traits and can interbreed.

10. Species, genus, family, order, class, phylum, kingdom.

Apply What You Have Learned

1. Scientists conclude from photographs of Mars that it once had flowing rivers, lakes, and seas, and that life may have evolved there. Traces of that evolution

may be present in rocks, and there may be evidence of the emergence of life from organic building blocks.

2. At that great depth, light could not penetrate and thus photosynthesis could not take place.

3. The prokaryotes that inhabit harsh environments, such the extremely hot, acidic, or sulfurladen habitats that might have existed on early earth, differ from other organisms, including other bacteria, in the structure of cell walls, RNA sequences, and other characteristics. Thus the sixth kingdom would be the Kingdom Archaea.

4. You should answer that conditions on earth today are different from the conditions when life first arose. The extraterrestrial sources of organic compounds, such as the solar nebula and carbonaceous chondrites, have largely been consumed by the solar system, and thus, few of life's monomers now reach earth from space. The planet's atmosphere no longer includes much carbon monoxide, hydrogen sulfide, ammonia, and methane, and thus, few monomers are created in the atmosphere. The oxygen now present in the atmosphere would rapidly oxidize and break down any monomers or polymers that might form spontaneously. Today, there are hordes of organisms that consume organic compounds available in the environment. Thus, it is highly unlikely that conditions would now allow the spontaneous generation of life.

CHAPTER 16

Review What You Have Learned

1. Protists are larger, contain a true nucleus and have other membrane-bound organelles.

2. They fix carbon, nitrogen, and other elements into biologically useful compounds; they release massive amounts of O_2; they decompose organic matter; they are parasites and cause disease; they are useful in scientific research and genetic engineering.

3. It has an outer cell wall made of peptidoglycans, a plasma membrane, cytoplasm dotted with ribosomes but without membrane-enclosed organelles, and a circular strand of DNA coiled in one region of the cell.

4. Gram positive bacteria have an outer single layer of peptidoglycans (sugar-protein complexes) that allow the bacteria to take a bright purple stain. Gram negative bacteria have an outer layer containing proteins and fat-sugar complexes (lipopolysaccharides) that prevent uptake of the purple stain but allow uptake of a red stain.

5. (a) rod; (b) sphere; (c) spiral; (d) curved.

6. A saprobe lives on dead organisms. A parasite lives on living organisms. A symbiont lives together with other organisms.

7. False. Prokaryotes recombine genetically through conjugation, transduction, or transformation.

8. Archaebacteria may be found in a cow's intestines, in salt lakes, in hot sulfur springs, or in volcanic mud pots.

9. Heterocysts are special, thick-walled cells that carry on nitrogen fixation in cyanobacteria, which have the advantage that they can live in environments where fixed nitrogen is rare.

10.

Type of bacteria	**Characteristics**
archaebacteria	anaerobes; characteristic RNA, carbon fixation, biochemistry
halophiles	salt-loving archaebacteria
thermophiles	sulfur metabolism, thrive in hot spots
eubacteria	characteristic RNA, cell walls and membranes, electron transport chains; many aerobic, photosynthetic
gram-positive	stain purple
gram-negative	stain red
cocci	spherical shape
bacilli	rod shape
spirilla	spiral shape
vibrios	curved rods
green and purple	autotrophs, contain chlorophyll
cyanobacteria	photosynthetic; some fix N
actinomycetes	excrete antibiotics
rickettsias	tiny, rod-shaped, cause disease
mycoplasmas	simplified parasites lacking cell walls
spirochetes	spiral shape

11. Bacteria that live within the human gut produce useful vitamins (K, B_{12}, riboflavin, biotin) and inhibit pathogens from gaining access into blood. Cellulose-decomposing bacteria help herbivores digest plant material. Bacteria also produce food (e.g., cheese), generate chemical reagents (e.g., ethanol), decompose dead organisms in the soil, and are useful in studies of biochemistry and molecular biology.

12. The virus attaches part of the capsid to the outside of the cell, injects its DNA or RNA into the cell, and causes the cell to produce new viral particles. When the cell bursts, thousands of viruses are liberated which can infect other cells.

13. Viroid: intracellular parasite lacking a protein coat and consisting only of small RNA molecules. Prion: smallest of intracellular transmissible disease-causing agents, which lacks genetic material and consists only of proteins.

14. Mastigophora: whip-like flagella. Sarcodina: amoeboid movement. Apicomplexa: spore-like and non-motile as adults. Ciliates: propelled by cilia.

15. Phytoplankton include euglenoids, dinoflagellates, golden brown algae and diatoms, cyanobacteria, and simple algae. These organisms form the basis of the aquatic food chain by fixing carbon and by releasing oxygen.

16. Golden brown algae and diatoms are the most common of the phytoplankton species; they contain a carotenoid pigment and chlorophylls. Cell walls contain silicon instead of cellulose. They store oil rather than starch.

17. True slime mold: mass called a plasmodium; consists of continuous cytoplasm with many diploid nuclei, surrounded by a plasma membrane. Cellular slime mold: single-celled, amoeba-like.

Apply What You Have Learned

1. Bacterial endospores may be present and can germinate into bacteria that produce deadly toxins. Instruments must be sterilized with high heat and pressure to kill these spores.

2. No. Phytoplankton, including golden brown algae and diatoms, produce more oxygen than all land plants combined.

3. The cysts of *Giardia* excreted in the feces of humans, dogs, beavers, and deer may enter the water, survive, and infect someone drinking the water many weeks later.

4. No, his argument is not likely to be correct. Viruses are too simple to be capable of independent life. They must rely on host cells for their reproduction. The first organisms to arise must have had the ability to replicate themselves without relying on other organisms.

CHAPTER 17

Review What You Have Learned

1. Fungi secrete into their environments enzymes that break down organic matter. They absorb the digested nutrients through their cell membranes.

2. They are fungi that live in symbiotic relationship with roots of perhaps 90% of all land plants. They expand the surface area for the uptake of water and minerals.

3. Zygomycotes erect stalks bearing spore-producing structures that split open and spread haploid spores. Ascomycetes produce spores on an ascus; rows of these are borne in an ascocarp. Basidiomycetes produce basidiospores in club-shaped structures called basidia. Deuteromycetes reproduce with asexual spores called conidia that are pinched off from the tips of spore-forming hyphae called conidiophores.

4. Basidiospores are formed in the gills of mushrooms, disperse, and grow into primary mycelia in the ground. When opposite mating types touch, a cell from each fuses, thus forming a single cell with two nuclei. This divides into a secondary mycelium, which spreads underground and produces new basidiocarps, or mushrooms.

5. The haploid stage is the gametophyte, which produces male and/or female gametes. After the gametes fuse, the diploid phase, the sporophyte, forms. Meiosis takes place in the sporophyte and produces haploid spores, which develop into gametophytes.

6. Geological changes created new environments and climates that at first stressed aquatic plants. As land was raised, plants that could resist desiccation developed by mutation and random genetic combinations, leading to the

evolution of plants with a vascular system, a dominant sporophyte generation, and that formed seeds.

7. Rhodophytes contain red, bluish, or purple pigments and can carry on photosynthesis at great depths; the haploid generation dominates. Phaeophytes, the brown algae, include the largest algae; the diploid generation is dominant in many species. Chlorophytes contain carotene and chlorophylls *a* and *b*; many have conspicuous haploid and diploid generations.

8. False. They are important because they produce oxygen and contribute carbohydrates to aquatic food chains.

9. Their leaves have waterproof coatings and tiny holes that allow gas exchange. They also have rigid tissues that keep them upright and rhizoids that act as anchors.

10.

Type of plant	Gametophyte	Sporophyte
Bryophyte	conspicuous	small; dependent on gametophyte
Low vascular	small free-living	conspicuous green plant
Seed plants	very small, non-photosynthetic, and depend on sporophyte	dominant

11. Fern: Haploid spores produced by the large sporophyte fronds germinate into tiny heartshaped gametophytes. They produce eggs and motile sperm which unite in fertilization, leading to the development of new adult sporophytes. Pine: Meiosis occurs in male cones and female cones. Male cones release pollen grains that are carried by wind to female cones, where they become fixed to a sticky fluid secreted by the ovule; when the fluid dries, it draws the pollen inside. During the year, the pollen tube, which contains two sperm nuclei, grows slowly toward the ovule. In the ovule a female gametophyte develops containing haploid eggs. One of the sperm nuclei fertilizes an egg to form an embryo inside the ovule, surrounded by nutrient-bearing tissue of the female gametophyte. Seeds now form, and when the scales of the woody female cone open in late summer of the second year, the seeds fall to the gound where they germinate the following spring.

12. (a) The ovary contains ovules; it has a style with sticky stigma on top. (b) A stamen includes an anther, which produces pollen grains, on a stalk-like filament.

13. During double fertilization, one sperm nucleus fuses with the egg and forms a diploid zygote that will become the embryo of the seed; the second sperm nucleus fuses with the two polar nuclei which forms the triploid cells of the endosperm.

Apply What Have You Learned

1. Apple trees depend on honeybees for pollination. If the bees are killed by the insecticide, the apples will not be pollinated and no fruit will be produced.

2. *Chlamydomonas* is a single-celled organism, and hence might be considered a protist. However, it bears a striking resemblance to a spore from a multicellular green alga, leading some scientists to classify it as a plant.

3. No, the homeowner is probably injuring or killing his trees by spraying fungicides on the ground. Trees depend on soil fungi to recycle nutrients from dead material and on mycorrhizal fungi for water and mineral uptake.

4. The forester suggested hiring a mycologist because he recognized that mycorrhizal fungi are essential to the health of trees. He thought that the seedlings were not being colonized by mycorrhizal fungi in time to sustain them through drought. A mycologist might be able to select the right strain of mycorrhizal fungus for the situation and might be able to inoculate the seedlings before they were transplanted.

CHAPTER 18

Review What You Have Learned

1. A trend away from circular symmetry toward bilateral symmetry. Cephalization. Development of a digestive tube with openings at both ends. Suspension of the digestive tube in a coelom. Segmentation.

2. Water containing food particles is swept by the beating of the choanocytes' flagella through pores into the osculum where embedded amoebocytes digest the particles.

3. False. The sponge cells are organized into primitive tissues and function as a single individual, and whereas each cell in a protozoan colony can reproduce independently, in sponges reproduction requires a cellular division of labor.

4. Cnidarians have three-layered bodies, nematocysts, and a digestive cavity, the coelenteron. They have a loose network of nerve cells and can regenerate lost parts.

5. The jellyfish develops male and female gonads that release gametes into the water. After fertilization, the egg develops into a planula larva, which differentiates into a many-sectioned strobila. Each section develops tentacles and becomes a medusa.

6. Digestive: branched intestine. Excretory: flame cells. Muscular: contractile muscles. Nervous: nerve cord. Reproductive: testes and ovaries.

7. A pseudocoelom and a digestive tube with openings at both ends.

8. A mollusk's gills exchange oxygen and water. There is an open circulatory system with a heart that sends blood vessels to the gills where carbon dioxide is released and oxygen picked up. Some of these animals have large brains and acute senses. They produce highly mobile, fringed larvae called trochophores.

9. Gastropods: snails, slugs. Bivalves: oysters, clams. Cephalopods: squids, octopuses.

10. Polychaeta: fire worms. Oligochaeta: earthworms. Hirudinea: leeches.

11. Bilateral symmetry, cephalization, gut tube, coelom, segmentation.

12. Each segment is separated from the next by septa. It contains two nephridia that collect wastes via a ciliated funnel and excrete them through a pore. Each segment includes a fluid-filled compartment of the coelom and is surrounded by circular and longitudinal muscles in the body wall. Setae are attached to each segment. A segment also contains clusters of nerve cells connected to the brain by nerve cords.

13. An exoskeleton composed of chitin; jointed appendages; specialized segments; tracheas for breathing (insects) or book lungs (spiders); acute senses.

14. Chilopoda: centipedes; Diplopoda: millipedes; Crustacea: crabs, lobsters; Arachnida: spiders, ticks; Insecta: mosquitoes, grasshoppers.

15. An endoskeleton; water vascular system for locomotion; radial symmetry; larvae with bilateral symmetry; regeneration ability.

16. A sea star has a set of canals with many short branches that end in tube feet. A bulb-like portion of each tube foot can contract and force water from the canals into the foot. This pressure extends the foot and a terminal sucker grips the surface. The animal moves forward by the extension, adhesion, and release of countless tube feet.

Apply What You Have Learned

1. Tapeworms shed thousands of embryos that bore into the fish's muscles where they become encased in protective cysts. Uncooked pieces of infested fish can infect a person who eats them; the embryos can grow into new tapeworms inside the person's intestines.

2. Undercooked pork may contain tiny nematode worms that cause trichinosis.

3. A single starfish arm with a small piece of the central part still attached can regenerate an entire body.

4. This animal is an echinoderm. The features that define it as an echinoderm include tube feet, lack of an excretory system, absence of a brain, and deuterostomic larval development.

CHAPTER 19

Review What You Have Learned

1. The larval tunicate resembles other chordates more than the adult does. The larva has all five chordate traits, but during the transformation to the adult form, it loses the tail, notochord, and nerve cord.

2. Notochord, hollow spinal cord, gill slits, tail, muscle blocks.

3. Chondrichthyes have skulls, vertebrae, and skeletons made entirely of cartilage. Osteichthyes have skeletons composed of bone.

4. Lobe-finned fish have large, muscular lobed fins and lungs. Example: lungfish, coelacanth.

5. Lungs; heart with partial separation of oxygenated and deoxygenated blood; thin, moist, smooth skin; eggs that must remain moist; tadpole stage.

6. Dry scaly skin; expandable rib cage; heart and circulatory systems that separate oxygenated and deoxygenated blood better than amphibians' hearts; internal fertilization; amniotic eggs.

7. It had characteristics of reptiles (scaly skin, claws, long jointed tail, teeth) and birds (feathers on forelimbs and tail).

8. Birds: feathers, hollow bones, large sternum, homeothermic, series of connected air sacs and lungs, four-chambered heart, hard-shelled eggs. Mammals: milk, body hair, four-chambered heart, homeothermic; most have internal development of young.

9. Monotremes: duck-billed platypus, spiny anteater. Marsupials: kangaroo, opossum. Placentals: human, whale.

10. Stereoscopic vision, opposable thumb, large brain, extended infant care, upright gait, omnivore-type teeth.

11. Gorillas probably branched off the evolutionary line leading to humans and chimpanzees about ten million years ago; chimpanzees and the human lineage diverged about six million years ago.

12. *Homo habilis* had a mixture of ape and human characteristics, with ape-like skull and teeth, brain (700cc) larger than a chimpanzee's, head located on top of backbone like a human's, human-like hand bones, upright gait. Its fossils date from 2 million years ago. It used tools and butchered animals. *Homo erectus* had a larger brain than *H. habilis* and differently shaped teeth and skull. It appeared about 1.5 million years ago and spread widely, showing finer stone tool workmanship and greater ability to deal with varied food resources and extremes of climate. It hunted large game in bands and used fire.

13. Neandertals lived from about 400,000 years ago to between 100,000 and 30,000 years ago. They were short, stocky people with projecting brow ridges and large brains (1,400cc), who made spears, built shelters, cared for the elderly, and buried the dead ceremoniously.

14. Fossil and genetic evidence support the single-origin model for *H. sapiens*. Studies of mitochondrial DNA indicate that all modern human lineages can be traced back to a single female who lived about 200,000 years ago, most likely in Africa.

Apply What You Have Learned

1. No. Dinosaurs died out at the end of the Mesozoic era, long before humans appeared on the scene.

2. No, placing *H. sapiens* at the top of the evolutionary tree is no more correct than placing any other living species at the top of the tree. All existing species are descended from equally long lineages. One might argue that cockroaches, for example, have greater claim to the top of the tree because they have survived much longer than the recently-evolved humans have.

3. They are similar to fish that arose about 400 million years ago and were believed to have died out long ago.

4. An example of an adaptive trait that has become maladaptive might be humans' intertribal aggression. When weapons were limited to stones, spears, and arrows, aggression kept tribes dispersed enough that resources were adequate for all. But now that we have nuclear weapons, this same trait of aggression imperils the global population.

CHAPTER 20

Review What You Have Learned

1. Anatomy is the study of biological structure; physiology is the study of how such structures work.

2. Homeostasis is the maintenance of a constant internal environment despite external fluctuation. Examples include the maintenance within a narrow range of salt level in body fluids and of body temperature.

3. Each system has a tube leading from the outside into the body; each tube has special regions where substances can be exchanged with body fluids; body fluids are circulated to every cell in the body where substances diffuse into and out of the cell.

4. A tissue is a group of similar cells performing the same function (epithelial, connective, nervous, muscle). An organ consists of two or more tissues that perform a certain function (liver, stomach, intestines, pancreas).

5. No. Each organ system depends on the others: the gut requires hormones, blood supply, and nerve connections.

6. Streamlined body, baleen plates, nostril position, and reductions of flaps and appendages that would add drag and increase heat loss.

7. By natural selection, alleles that led to greater survival, food intake, and reproduction replaced those that led to lower fitness and reproduction.

8. In cold water, sensors in the skin relay a signal to the hypothalamus which causes a message to be sent to muscles to contract, giving off heat to warm the animal or restricting blood flow to the skin; as internal temperature rises, sensors stop signalling the brain.

9. An example of a positive feedback loop is the birth of a baby whale. Oxytocin causes uterine contractions, which cause more oxytocin to be secreted, leading in turn to more contractions, which are spaced closer and closer. When the infant whale is expelled into the sea, the loop stops and the uterus no longer contracts; this restores homeostasis.

10. A lizard instinctively moves in or out of the sun or does heat generating push-ups, keeping its body temperature between 35°C and 40°C.

11. During an infection, white blood cells attack the invaders and release pyrogens, which cause the release of prostaglandins, hormonelike molecules that act on the hypothalamus to turn up its thermostatic setting.

Apply What You Have Learned
1. The whale is so heavy that it requires the buoyant support of sea water; otherwise the weight of its body will cause the lungs to collapse and it will suffocate.

2. False. Bacteria, like all other organisms, must maintain internal homeostasis despite internal changes and fluctuations in the environment. An example is the cellular concentration of ATP, which must be maintained at a level high enough to power the cell's activities but not at a level so high that energy is wasted.

3. A healthy person has 0.1 to 0.2 mg of bilirubin in 100 ml of blood; a hepatitis patient may have 3.6 mg, as well as higher levels of enzymes ALK PHOS, SGPT(ALT), GGT, SGOT(AST), and LDH. These differences imply that the liver is important in maintaining the homeostasis of many blood components.

4. No, the giant flatworm is scientifically unbelievable. Its cells would be unable to survive because their surface-to-volume ratio would be too low for diffusion to transport materials in and out of the cells rapidly enough to accommodate the need for oxygen and materials and the disposal of wastes. The giant flatworm would be unable to transport materials throughout its body and exchange substances with the environment without complex distribution systems such as a vascular system and gills or lungs.

CHAPTER 21

Review What You Have Learned
1. High blood pressure; high pressure fluid outside the blood vessels due to tight skin on legs; rete mirabile lowers and regulates the blood pressure reaching the brain and eyes.

2. Salts, sugars, fats, and dissolved proteins (globulins, albumins, fibrinogen).

3. Red blood cells contain hemoglobin and carbonic anhydrase enzyme. They have a biconcave disk stage.

4. When the red blood cell content decreases, and less oxygen circulates to the liver and kidneys, these organs produce the hormone erythropoietin and secrete it into the blood. Erythropoetin stimulates stem cells in the bone marrow to divide more rapidly and form more red blood cells, which carry more oxygen throughout the body, including the liver and kidneys. These organs then produce less erythropoietin, and the division of the marrow stem cells slows.

5. The atrium of the two-chambered heart receives oxygen-depleted blood from veins and pumps it into the ventricle, which pumps it to the gills. In the gills, oxygen is absorbed and carbon dioxide is released. Oxygenated blood passes from the gills into arteries that deliver it to capillaries throughout the fish's body. Oxygen leaves the blood in the capillaries and enters tissue cells, and the remaining deoxygenated blood collects in veins that lead back to the atrium.

6. Arteries, arterioles, capillaries, venules, veins.

7. Beginning with deoxygenated blood from body tissues: superior and inferior vena cavae, right atrium, right ventricle, pulmonary artery, lungs, pulmonary vein, left atrium, left ventricle, aorta, capillary beds (via systemic circulation).

8. Pacemaker cells are located in the SA node of the upper right atrium. An electrical impulse spreads from the SA node across the walls of both atria, causing them to contract in unison. When the electrical signal from the SA node reaches the AV node located at the junction of the atria and the ventricles, there is a brief delay before the signal reaches the ventricle. The ventricles then both contract and send blood to the pulmonary artery and aorta.

9. Systole phase: the atria or ventricles contract. Diastole: atria or ventricles relax.

10. In the arteries, blood flows from regions of high pressure near the heart to regions of lower pressure near the capillaries. In the veins, valves cause a one-way flow.

11. Smooth muscles in the walls of the damaged vessel contract; circulating platelets release the hormone serotonin, which keeps the muscles contracting; platelets collect in the wound and form a plug; a series of changes in blood proteins transform soluble fibrinogen into insoluble threads of fibrin that trap red blood cells, forming a blood clot.

12. Lymph nodes prevent dead cells and debris in lymph from mixing into the blood and contain large numbers of lymphocytes that attack bacteria and other foreign materials.

13. Thymus: active in the immune system, specifically in the maturation of white blood cells. Spleen: filters impurities from the blood, including worn-out red blood cells; also stores white blood cells.

14. No. Cholesterol is manufactured naturally by the liver and is an important component of cell membranes.

Apply What You Have Learned

1. The rate of heart contractions increases with more vigorous physical activity.

2. The higher diastolic reading probably resulted from stress the patient experienced while waiting for the doctor.

3. She should swim, because exercise that uses massive amounts of oxygen will stimulate her body to produce more red blood cells, thus preparing her for the lower concentration of oxygen molecules at high elevations. Weight lifting would not stimulate red blood cell production although it would strengthen her muscles for the trek.

4. If dinosaurs had three-chambered hearts, they were probably cold blooded, as are modern reptiles. Three-chambered hearts do not completely separate oxygenated blood from deoxygenated blood and thus may not support the high metabolic rate that characterizes warm-blooded animals. If the dinosaurs had four-chambered hearts, as do modern mammals and birds, it would be consistent with the hypothesis that they, too, were warm blooded.

CHAPTER 22

Review What You Have Learned

1. Skin and blood cells release chemicals that raise the temperature. Blood flow to the area increases. Phagocytes enter the area and engulf bacteria and cell debris.

2. B cells make and secrete antibody proteins that bind to antigens, like the surface molecules of bacteria, and mark them for destruction. T cells kill bacteria directly and help regulate the activities of other lymphocytes. Macrophages stimulate lymphocytes to attack bacteria and consume resulting debris.

3. They trap foreign cells and molecules and house white blood cells.

4. Antigens bind to antibodies on the surface of B lymphocytes. These B cells move to a nearby lymph node where helper T lymphocytes stimulate the B cells to divide into hundreds of plasma B cells, each of which secretes thousands of antibody molecules. These antibodies are carried in the blood and combine with bacterial proteins, causing the bacteria to clump. Macrophages are attracted and devour the bacteria. There are also complement proteins that bind to antibody/bacterial complexes and kill the bacteria by perforating their cell membranes.

5. (See chart, Fig. 22.9.)

6. Yes. The arms of the Y recognize and bind to invading bacteria, while the stem stimulates macrophages or complement proteins to destroy the antigen-antibody complex.

7. Primary response: Only a small number of B cells respond to the first exposure to an antigen, but they divide into many memory cells. Secondary response: The second time the same antigen enters the body, a large pool of memory cells quickly divides and produces more memory cells as well as plasma cells which secrete specific antibodies that bind to the invader's antigenic surface.

8. Antibody molecules are encoded by three or four separate groups of genes on a chromosome. During the development of B cells, one gene from each of the groups is randomly selected, and the chosen genes are brought together, making a single combined gene. This shuffling can create an almost infinite number of antibody protein shapes.

9. Humoral tissue response: Antibodies produced by B cells travel through blood and lymph to the location of antigens. Cell mediated response: T cells directly kill microbes.

10. Self-tolerance is the lack of immune response to the body's own components. As macrophages digest foreign antigens, they insert the fragments in their membranes and present them to T cells. However, macrophages do not insert self-antigens in their membranes. During development of the fetus, immature B cells with antibodies to selfantigens are eliminated. Self tolerance may also be due to inhibition of the immune system by suppressor T-cells. Lack of self-tolerance can result in autoimmune conditions, like rheumatoid arthritis.

11. In passive immunization, antibodies made by one individual (the donor) are transferred to another individual (the host), where they can protect the host from disease. In active immunization, an individual's own body cells are stimulated to generate antibodies. Active immunity provides long-term protection.

Apply What You Have Learned
1. The proteins on the flu virus's outer coat change, so that the vaccine is no longer effective against the new antigens.

2. She had measles when she was young, and thus generated many memory cells. When re-exposed to the virus, her memory cells will be stimulated to produce enough antibodies to eliminate the virus.

3. The AIDS virus attacks the lymphocytes that confer protection. The proteins in the virus's outer coat change, thus requiring that a new set of antibodies be built up. The virus can remain latent inside cells for many years.

4. Diagnosis 1, that the patient is unable to produce histamines, is unlikely to explain his symptoms. His skin's redness and swelling indicate that histamines are being produced. Diagnosis 2, that his helper T cells' production of interleukins is inadequate, is a likely explanation. Without interleukins, B cells would not be stimulated to divide into hundreds of antibody-secreting cells and hence his production of antibodies would be inadequate to fight infections. Diagnosis 3, that his immune system lacks antibody variety, is a possibility if his immune system is generally suppressed, but given the immune system's ability to generate antibody diversity by shuffling antibody genes, a healthy immune system is capable of generating antibodies to combat almost any antigen. Diagnosis 4, that he has lost self-tolerance, is unlikely to be correct. Loss of self-tolerance would be manifested as autoimmune diseases such as rheumatoid arthritis, not as an inability to fight off disease-causing organisms.

CHAPTER 23

Review What You Have Learned
1. The seal has twice as much blood as a human per kilogram of body weight, thus storing a higher level of oxygen. During a dive reflex, blood is shunted to organs involved in navigation and controlling swimming, the heart rate slows, and the muscles switch to anaerobic respiration, thus requiring less oxygen.

2. Water passes over the gills in the direction opposite to the direction of blood flow within the capillaries. Thus the blood leaving the gills, with higher oxygen content than blood entering the gills, is exposed to water entering the gills, which has the highest oxygen content. If the water and blood flows were parallel, the oxygen content of the water would decline as the oxygen content of the blood increased, reducing the diffusion of oxygen from water to blood.

3. Spiracles, holes on the sides of the abdomen, lead into air tubes called tracheae, which branch into tiny air capillaries. At the inner tip of an air capillary, the incoming oxygen dissolves in a droplet of fluid and diffuses into nearby cells.

4. Air follows a one-way path in a bird's respiratory system. As the bird takes two breaths of air, the first draws air into the posterior air sacs and then the lungs while the second pushes air from the lungs into the anterior air sacs and then out of the body. Thus deoxygenated air is not mixed with incoming air.

5. Nasal cavities, larynx, trachea, bronchi, bronchioles, alveoli, blood capillaries.

6. Smoke damages the cells lining the alveoli and paralyzes the cilia, often leading to a hacking cough due to the accumulation of dust and mucus. Chemicals in the smoke may cause cancer or emphysema. The blood vessel walls are also damaged, leading to possible heart attack or stroke.

7. No. Groups of neurons in the medulla and elsewhere in the brain regulate the breathing rate.

8. Oxyhemoglobin results when the four iron atoms in a hemoglobin molecule bind with oxygen in lung capillaries, giving it a bright red color. Deoxyhemoglobin is hemoglobin depleted of oxygen and has a darker, almost bluish color.

9. A yak has three or four times as many red blood cells/cc of blood as does a human. Yak hemoglobin may have a special affinity for oxygen.

10. Some carbon dioxide dissolves in the plasma or combines with deoxygenated hemoglobin. Most of the carbon dioxide dissolves in the fluid of the blood, forming carbonic acid, which dissociates into bicarbonate and hydrogen ions. When the plasma reaches the lungs, the bicarbonate and hydrogen ions are reconverted to carbon dioxide and water. The carbon dioxide diffuses through the capillary wall into an alveolus.

11. During heavy exercise the muscles produce much carbon dioxide, resulting in increased acidity as it combines with water and forms carbonic acid. Additional acidity results from the production of lactic acid during anaerobic muscle activity. In this increasingly acidic environment, hemoglobin gives up oxygen more readily to the oxygen-starved cells.

Apply What You Have Learned

1. Smoke paralyzes the cilia that would normally sweep mucus and dust up toward the pharynx to be swallowed or expelled. Cough drops will not expel mucus and dust and thus will not remedy the cause of the smoker's cough.

2. The worker could not breathe because the breathing mechanism involves rib cage expansion, and the weight of the debris prevented the necessary movement of his ribs and diaphragm.

3. The study is likely to reveal an increased breathing rate due to the increase in the amount of carbon dioxide in the air, which would stimulate the medulla to trigger an increase in intercostal muscle contraction and diaphragm movement.

4. The respiratory centers in the brain sense the level of carbon dioxide in the blood. When carbon dioxide levels rise, the brain signals the body to breathe. In normal breathing, a certain amount of carbon dioxide remains in the blood, but hyperventilation reduces the carbon dioxide level below the normal resting level. As the diver consumes oxygen while submerged, the oxygen level in the blood

may drop below the level required to support consciousness, and this may occur before the carbon dioxide level rises high enough to signal the body to resurface.

CHAPTER 24

Review What You Have Learned

1. Herbivore: an animal that eats only plant matter; horse, rabbit, grasshopper. Carnivore: eats animal flesh; crocodile, wolf, leech. Omnivore: eats both plant and animal matter; human, pig, bear.

2. Monosaccharides: glucose, fructose. Disaccharides: sucrose, maltose. Polysaccharides: starch, glycogen.

3. Carbohydrates: glucose. Lipids: fatty acids and glycerol. Proteins: amino acids.

4. Fats can be oxidized during aerobic respiration, thereby yielding usable energy. They serve as energy storage products. They cushion internal organs and provide thermal insulation. They are necessary for the absorption of the fat-soluble vitamins A, D, E, and K. They make cell membranes.

5. Proteins make up connective tissue, muscle tissue, hair, cornea of the eye, enzymes, antibodies, hemoglobin, and many hormones.

6. Most vitamins serve as precursors for coenzymes or activators for enzymes that participate in cell metabolism.

7. Calcium is used in building bones and teeth. Phosphorus is found in bones and teeth and is a basic constituent of nucleic acids and ATP. Iron is essential for hemoglobin. Iodine is a constituent of thyroid hormones. Zinc is a common coenzyme. Copper is needed for hemoglobin production. Fluorine is needed for healthy teeth. Sulfur is found in many proteins. Magnesium is a constituent of enzymes and bones.

8. The teeth cut, tear, and grind food. The tongue moves food to the molars for grinding and shapes the bolus. The stomach mixes, churns, and stores food.

9. Gastric juices contain hydrochloric acid and pepsinogen, which is converted into the enzyme pepsin by the action of hydrochloric acid; pepsin partially digests proteins into short polypeptide segments.

10. The pancreas secretes enzymes into the small intestine (proteases which digest proteins, lipases which digest fat, amylase which digests carbohydrates) and bicarbonate ions, which help neutralize acid entering the small intestine from the stomach.

11. The liver destroys worn-out red blood cells, stores glycogen, sends glucose into the bloodstream, produces bile salts, and detoxifies many poisons.

12. Bile salts emulsify fat globules into tiny fat droplets, thus increasing their surface-to-volume ratio and making it easier for lipases to digest fat. Bile salts also help transport fully digested fat molecules to the lining of the small intestine for absorption into the lymph.

13. Peristalsis moves the chyme arriving from the stomach, while its pH becomes less acidic and the intestinal and pancreatic enzymes complete the digestion of proteins, carbohydrates, and lipids into their end products. The nutrients are absorbed from the small intestine into the blood.

14. The colon absorbs water, ions, and vitamins from the food mass and stores undigested wastes until they are excreted.

15. Nerves provide communication along the digestive tract thus speeding or slowing the propulsion of food and the digestive functions of secretory glands. Gastrin is a hormone secreted by the stomach lining that stimulates the production of hydrochloric acid. The hormone secretin stimulates the pancreas to secrete bicarbonate ions into the small intestine, neutralizing the incoming stomach acid. The hormone cholecystokinin is released by the small intestine when partially-digested protein in the chyme enters; it stimulates the pancreas to secrete digestive juices; it also causes the regulatory centers in the brain to produce the sensation of being "full."

Apply What You Have Learned

1. Vegetarians need to eat food combinations that provide adequate and balanced amino acid intake. Thus, rice, which has methionine but little lysine, is combined with beans, which have little methionine but sufficient lysine.

2. No. They both contain an identical form of vitamin C, ascorbic acid.

3. It binds to bile salts and passes through the intestinal tract without being absorbed. This prevents the recycling of cholesterol from bile salts, thereby lowering the amount of cholesterol in the blood.

4. No, you should not follow the coach's advice to consume excess animal protein and to skip the vegetables. One reason is that vegetables contain vitamins and other nutrients essential to health. Another reason is that excess amino acids will not be incorporated into muscle tissue but will be consumed for energy. A third reason is that a regular diet of red meat contributes to a higher risk of diseases such as cancer. However, an adequate supply of all 20 amino acids, as found in animal proteins and balanced plant proteins, is necessary to build muscle mass.

CHAPTER 25

Review What You Have Learned

1. Excretion is the removal of cellular metabolic by-products (especially nitrogen-containing wastes), excess water, and salts filtered from the blood by the kidneys. Elimination is the expulsion of undigested food materials that were not absorbed by the digestive tract.

2. Ammonia: aquatic invertebrates. Uric acid: birds and many insects. Urea: mammals.

3. Glomerulus: a bundle of capillaries enclosed within Bowman's capsule. Podocytes: cells lining Bowman's capsule that have finger-like projections that enclose the capillaries. Peritubular capillaries: a network of capillaries that twists around the nephron tubule's looped portion.

4. False. It is responsible mainly for reabsorption of water.

5. Water, sugars, amino acids, urea, and sodium, potassium, and chloride ions.

6. Active transport in the tubule returns most of the sodium and chloride ions, sugars, and amino acids into the peritubular capillaries. Most of the water follows the ions and solutes passively by osmosis back into the capillaries.

7. Sweating or evaporation of water from the lungs causes an increase in the concentration of blood solutes which is detected by nerve cells in the hypothalamus. These stimulated nerve cells cause you to become aware of thirst and you drink water, which distends the stomach walls. Expansion of the stomach stimulates nerve impulses that inhibit the thirst center. Water entering the bloodstream restores the osmotic balance.

8. An increase in solute concentration stimulates the hypothalamus to send nerve impulses to the pituitary, causing it to release stored ADH into the blood. In the kidneys, ADH makes the walls of the distal tubule and collecting ducts more permeable to water, so more water is reabsorbed back into the blood stream, diluting the concentration of solutes. This blood dilution then acts on the hypothalamus and decreases the secretion of ADH, and the negative feedback loop is completed.

9. If Na^+ concentration in the blood is reduced, the adrenals secrete more aldosterone, causing the tubules to reabsorb more sodium from the filtrate into the blood. The increased level of sodium feeds back to the adrenals, reducing aldosterone production, and then the tubules reabsorb less sodium. As a result, the sodium level in the blood remains within a narrow range. The system works in the opposite way on potassium ions.

10. When blood flow to kidney cells falls below a certain volume the cells release renin, which leads to the formation of the hormone angiotensin; this causes blood vessels to constrict, raising blood pressure. This hormone also stimulates the adrenal to secrete aldosterone, which causes sodium and water to be reabsorbed, leading to even higher blood volume and hence higher blood pressure.

11. When blood volume and pressure increase, cells in the heart's atria secrete ANF (atrial natriuretic factor, a hormone), which decreases sodium reabsorption by the distal tubule. With less sodium reabsorbed into the blood, the less water will be reabsorbed, and the net result is an increase in the quantity of urine and a corresponding lowering of blood volume and blood pressure.

12. In fresh water or dilute sea water, a special excretory gland expels excess water and the gills actively transport salt back in. In briny water, the brine shrimp swallows salt water and the gills expel excess salt.

13. In freshwater fish, the blood is hypertonic (its salt concentration is higher than that of the environment), so much water flows in through the skin; their kidneys produce large amounts of dilute urine and expel the excess water. But salt leaves with the water, so the gills actively absorb salt from the outside. In oceanic sharks, the animal's body fluids are isotonic to (have the same solute concentration as) the sea water, so there is little net flow of water in either direction; high levels of urea accumulate in the blood, raising the solute concentration to nearly that of the environment; salts build up due to

consumption of salty foods, and excess salts are excreted by special rectal glands. In saltwater bony fish, the body fluids are hypotonic to (have lower solute concentrations than) sea water, and the fish tend to lose water and gain salt. To compensate, they drink seawater, and produce little urine, while their gills actively pump out excess salt.

14. The kidney exchanges water and solutes with the circulatory system. The nervous system initiates thirst and drinking. The digestive system delivers water to the blood. The excretory system works with the respiratory system in conserving water. It is regulated by hormones from the brain and other body parts that coordinate urination and blood volume and pressure.

Apply What You Have Learned

1. Nitrogen released during the breakdown of nucleic acids is incorporated into uric acid. The diets of poor people included little food that contained great amounts of nucleic acids, while the diets of the wealthy included many foods, such as organ meats, that were high in nucleic acids. Thus, rich people were more likely to get gout.

2. Tubular secretion in the proximal and distal tubules removes unwanted materials from the blood in the peritubular capillaries and secretes them into the urine. Drug testing of urine reveals even minute traces of residues of drugs such as cocaine, marijuana, heroin, and tranquilizers.

3. Alcohol inhibits ADH secretion, leading to the excretion of large quantities of urine. The resulting dehydration of the body causes the "hangover" feeling and requires that the lost fluid be replaced.

4. The doctor suspects that cancer of an adrenal gland is causing the increase in blood pressure because the adrenals secrete aldosterone. If cancer caused an adrenal gland to grow uncontrollably, it would produce greater amounts of aldosterone. Aldosterone would cause the distal tubules of the kidney to reabsorb sodium and water, increasing the volume of blood and hence increasing blood pressure. A blood test of sodium and potassium levels might verify the diagnosis if it revealed elevated sodium levels and depressed potassium levels. High levels of aldosterone might also appear in the test.

CHAPTER 26

Review What You Have Learned

1. An endocrine gland is a ductless gland that secretes hormones directly into the bloodstream. Exocrine glands have ducts that carry substances such as milk and sweat out of the body.

2. Paracrine hormones are primitive hormones that act on cells immediately adjacent to the ones that secreted them. Neurotransmitters are nerve signal compounds. Neurohormones are hormones secreted by nerve cells. True hormones are secreted by glands such as the pituitary or thyroid. Pheromones are compounds that act on other organisms at a distance.

3. A hydrophobic messenger passes into a cell and binds to a receptor protein. The receptor/hormone complex binds to DNA, allowing certain genes to be transcribed to mRNA and new proteins to be formed as a result. Hydrophilic messengers, such as proteins, cannot pass through the hydrophobic cell

membrane and instead bind to cell surface receptors; this leads to the activity of second messengers which cause a change in the cell.

4. A second messenger is a small molecule that appears inside a cell as a result of an external signal; it acts by changing the activity of specific enzymes or other proteins; these, in turn, alter the activity of the entire cell. Second messengers such as cyclic AMP can affect various functions, including muscle movement, reproduction, fat oxidation, and others.

5. Nerve cells in the hypothalamus make hormones that are transported down long cell extensions (axons) to the posterior pituitary where they are stored at the tips of the extensions. Cells in the anterior pituitary make hormones. Nerve cells in the hypothalamus secrete releasing hormones or inhibiting hormones into special blood vessels that lead to the anterior pituitary and cause cells there to release or not to release their hormones. Both lobes of the pituitary release their hormones when stimulated by the hypothalamus, which receives information from other parts of the brain. The hypothalamus and the pituitary together influence the activities of the ovaries, testes, thyroid, and adrenals.

6. A child with an underactive thyroid develops cretinism – low rates of protein synthesis and carbohydrate breakdown, leading to low body temperature and sluggishness; growth is stunted and mental development is retarded. An affected adult gains weight easily and is mentally slow.

7. If the blood level of calcium ions rises, the thyroid releases the hormone calcitonin, which causes excess calcium to be deposited in bones. If calcium in the blood falls, the parathyroid glands release parathyroid hormone, which causes bones to release calcium and the body to absorb more of it from food.

8. During stress, epinephrine is released from the adrenal medulla, causing the heart to beat faster, sugar levels to rise, breathing rate to speed up, and blood to be shunted away from the stomach and intestines to the muscles, resulting in more blood volume to the muscles and enhanced ability to fight or flee. Cortisol is secreted by the adrenal cortex and speeds up the metabolism of sugars, proteins, and fats; it also decreases inflammation in stressed tissues.

9. When blood sugar level rises, beta cells in the pancreas secrete insulin, which causes liver cells and cells in other tissues to remove glucose from the blood and store it as glycogen. If the blood sugar level falls, alpha cells in the pancreas release glucagon, which causes the liver to release stored glycogen as glucose.

Apply What You Have Learned

1. No. Without the hormone, the caterpillar never metamorphoses into a moth.

2. Suckling stimulates the production of milk by causing nerves in the nipple to send signals to the hypothalamus, which in turn causes oxytocin to be released from the posterior pituitary. This hormone causes muscle-like cells lining the breast's milk ducts to contract, forcing milk into the nipple.

3. Insulin causes cells to take glucose from the blood. High-carbohydrate snacks put glucose into the blood so that the sugar is available to enter cells responding to insulin.

4. The best choice is (2); the thyroid gland concentrates iodine, and radioactivity can cause cancer. Reject (1) because the residents are unlikely to be related closely enough to each other to share a genetic tendency to developing cancer. Reject (3) because a lack of iodine would cause goiter but not cancer, and the diet of most people includes enough iodine that its absence from the water supply would be unimportant.

CHAPTER 27

Review What You Have Learned

1. Dendrites receive signals from other nerve cells or from the environment. The soma is the main cell body, consisting of a nucleus, ribosomes, mitochondria, endoplasmic reticulum, Golgi apparatus; it produces proteins and enzymes. An axon is a long, tubular process. The axon terminal is a knob-like structure at the end of the axon where communication with another neuron takes place across a synapse.

2. The potassium channel is a protein-lined pore allowing potassium ions to pass through the cell membrane. The sodium channel is a protein-lined pore, usually tightly closed and preventing outside sodium ions from leaking inward. The sodium/potassium pump uses ATP energy to transport potassium ions into the cell and sodium ions out of the cell.

3. A neurotransmitter signal stimulates the resting cell so that the sodium channels open, allowing sodium ions to enter; the inside of the cell becomes positively charged and the action potential rapidly passes to other parts of the neuron.

4. False. A neuron cannot fire again until the sodium-potassium pump restores its resting potential and the refractory period, the time during which a neuron cannot have another action potential, passes.

5. A stimulus induces an action potential and sodium ions rush in, causing the adjacent membrane to lose its polarity. Depolarization causes nearby sodium gates to open, propagating the impulse. Thus, sodium influx, depolarization, and the action potential take place sequentially along the entire length of an axon.

6. The myelin sheath is composed of Schwann cells, forming a lipid-rich layer that insulates the axon. Ions can flow rapidly across the axon membrane only at the nodes of Ranvier, and the nerve impulse leaps rapidly from one node to the next.

7. At electrical synapses, membranes of the presynaptic cell and the postsynaptic cell are fastened together by gap junctions, which allow ions to flow directly through pores from one cell to the next, propagating the impulse. At boutons of chemical synapses, an action potential causes calcium ion channels in the cell membrane to open, allowing calcium ions to rush in; the synaptic vesicles fuse with the membrane of the presynaptic cell, liberating thousands of neurotransmitter molecules into the synaptic cleft. They bind to receptor proteins in the membrane of the postsynaptic cell, causing this cell to react.

8. Acetylcholine transmits nerve impulses within the brain and relays messages from neurons to skeletal muscles. Norepinephrine is found in synapses

throughout the brain and helps to keep mood and behavior on an even keel; outside of the brain, it works with epinephrine in a "flight or fight" response.

9. Excitatory synapse: a molecular messenger brings the postsynaptic cell closer to firing an impulse by causing a few channels for sodium ions to open, thus slightly depolarizing the cell. Inhibitory synapse: the neurotransmitter opens channels to potassium or chloride ions or both, making the receiving cells' interior more electrically negative, and thus more polarized. Many more excitatory impulses than usual would be needed to trigger an impulse in the postsynaptic neuron.

10. Sensory neuron: receives information and transmits it to the brain or spinal cord. Interneuron: relays messages between nerve cells. Motor neuron: sends messages from the brain or spinal cord to muscles or glands.

11. Habituation is a progressive decrease in the strength of a response to a constant weak stimulus that is repeated over and over again with no undesirable consequences. Habituation occurs at synapses, decreasing the likelihood that post-synaptic cells will fire.

12. Sensitization is a heightened response to the same level of stimulation. A neuromodulator released at the synapse can cause more neurotransmitter to be released, leading to a stronger and faster response.

Apply What You Have Learned

1. Neurotoxins are likely to be better because if the nerves of the prey cease to function, the animal will become paralyzed and cannot escape from the snake. Poisons that cause generalized tissue damage are not likely to immobilize the prey as rapidly.

2. In Parkinson's disease there is a decrease in the production of the neurotransmitter dopamine. The drug L-dopa is modified by the brain into dopamine, which helps alleviate the symptoms.

3. Amphetamine increases the release of norepinephrine and dopamine in brain synapses. This results in increased alertness, but also can cause negative side effects such as blurred vision, insomnia, anxiety, and a rise in blood pressure.

4. One diagnostic clue would be the resting potential of a motor neuron. If it has a potential of about -70 mV, the sodium-potassium pumps are working properly. If it fails to propagate an action potential when stimulated, the sodium channels may be blocked and prevented from opening. If the neuron is permanently depolarized, the venom may be disrupting the sodium-potassium pumps or blocking the sodium channels open. Another diagnostic clue would be the concentration of neurotransmitters at a synapse. If it fails to rise after the presynaptic neuron is stimulated to produce an action potential, the venom may be blocking neurotransmitter release. If it rises then stays high, you would suspect that the venom is blocking the neurotransmitter-degrading enzymes. If the concentration of neurotransmitters rises and falls normally after stimulation of the presynaptic neuron, but the postsynaptic neuron does not respond, you might deduce that the venom is blocking the receptors.

CHAPTER 28

Review What You Have Learned

1. Dendrites in the eye's light receptor cells make a light-sensitive protein, while nearby support cells focus light on these receptors.

2. Sound waves vibrate the eardrum, which transmits the vibrations to three bones in the middle ear and on to the cochlea. In the cochlea, the vibrations move the basilar membrane, bending stereocilia on hair cells; this mechanical movement is converted into an electric signal, which crosses a synapse and stimulates a sensory neuron leading to the brain.

3. The semicircular canals are filled with fluid. As the position of a person (or bird) changes, the stereocilia on the hair cells bend, releasing nerve impulses to the brain, which interprets the change of body position or balance.

4. Light passes through the cornea, enters the pupil and passes through the lens, which focuses the image on the retina. There, photoreceptor cells – the rods that perceive light and cones that detect color – synapse with sensory neurons that connect with the optic nerve leading to the brain.

5. The rear part of the rods detects light. It contains stacks of membranous disks in which millions of rhodopsin molecules are embedded. Each of these molecules contains retinal, a molecule that changes shape upon exposure to light, leading to the perception of light.

6. The hydra has the simplest nervous organization, the nerve net. The flatworm has concentrations of nerve cells called ganglia in the head region that act like a primitive brain. There are nerve trunks on each side of the body that connect the periphery with the head ganglia.

7. The dendrites of pain receptors in the skin send action potentials to the cell body in the dorsal root ganglion; the action potentials continue along the sensory neuron's axon into the spinal cord, where the neuron synapses with interneurons. Interneurons not in the reflex arc send a message to the brain where it is interpreted as pain; interneurons in the reflex arc send messages to motor neurons that leave the spinal cord through the ventral root of the spinal nerve and stimulate muscles in the arm, hand, and finger that contract and jerk the finger from the flame.

8. The autonomic nervous system, including the parasympathetic and sympathetic nerves, controls glands, the heart muscle, and smooth muscles in the digestive and circulatory systems. Parasympathetic nerves leave the spinal cord in the neck and tail bone; they secrete acetylcholine, which slows the heart beat. Sympathetic nerves leave the spinal cord in the central region of the back; they secrete norepinephrine, which speeds up the heart beat. The two sets of autonomic nerves function in opposition.

9. The lowest part, the medulla oblongata, regulates respiratory rate, heart rate, and blood pressure. The pons and midbrain relay sensory information from the eyes, ears, and tongue to other parts of the brain. The hypothalamus regulates the pituitary and helps maintain homeostasis, regulating body temperature, water balance, hunger, and the digestive system. The thalamus is a relay area to

the cerebrum. The reticular formation, a network of tracts reaching into the cerebellum and cerebrum, controls sleep and wakefulness.

10. Cerebellum: coordinates movement and awareness of the position of the body. Cerebrum: controls speech, emotions, sensory perception, motor action, and memory.

11. The surface regions of the sensory and motor cortex are responsible for similar but not identical parts of the body; motor nerves go to muscles from the motor cortex; information from sensory nerves and the sense organs arrives in the sensory cortex.

12. Right hemisphere: responsible for intuitive thought, musical aptitude, the recognition of complex visual patterns, the emotions; controls left side of the body. Left hemisphere: responsible for language ability, analytical thought, fine motor control; controls the right side of the body.

Apply What You Have Learned

1. Loud sounds can cause the stereocilia to break off hair cells in the cochlea, thus damaging the sense of hearing that depends on them.

2. Olfactory nerves in the nose and back of the throat have receptors to which volatile aroma molecules of food bind. This binding stimulates sensory nerves that stimulate olfactory regions of the brain. Some of this sensation is interpreted as taste as well as smell. During a bad head cold, these receptors are prevented from functioning, leading to a decrease in taste perception.

3. There are ethical questions about whether fetuses should be sacrificed to help diseased adults. Also, who should have the right to make decisions about the fate of tissues from aborted fetuses?

4. You correctly argue that the patient's brain stem is damaged; specifically, his hypothalamus has been injured. The hypothalamus controls body temperature, hunger, and mood. A damaged gland is not likely to be at fault because the symptoms come from organ systems controlled by several different glands. The cerebral cortex is probably not injured because the patient's memory and motor control are normal. The cerebellum is undamaged, as demonstrated by the patient's normal motor control.

CHAPTER 29

Review What You Have Learned

1. In a typical bone such as the femur, an outer layer of compact bone surrounds an inner region of bone marrow tissues; a looser spongy bone layer with cross-girders for support occurs at bone ends. Embedded within the bone are cells that secrete the proteins and minerals of the bone and reabsorb calcium and phosphate from bone for use in other parts of the body.

2. In a homeostatic mechanism, the stress of weight-bearing exercise stimulates bone cells to lay down more material, making stronger bones. The hormone estrogen helps women absorb calcium from their diets, strengthening bone.

3. Joints with limited mobility, such as the joints between vertebrae, have pads of cartilage that absorb shock. Freely moveable joints, such as the knee, have a bursa filled with synovial fluid in addition to pads of cartilage.

4. Antagonistic muscle groups act in opposition to each other. For example, the shin muscle (anterior tibialis) flexes the foot toward the body while its antagonist, the calf muscles (gastrocnemius and soleus), extends the foot to point away from the body.

5. A myofibril contains repeating units of dark and light bands. Each unit is a sarcomere; it contains actin (light) and myosin (dark) filaments. When sarcomeres in the myofibril shorten, the muscle contracts.

6. Myosin and actin filaments lie side by side in the sarcomere. The head of the myosin molecule binds to the actin filament, swings like an oar, and slides the actin past it, causing the sarcomere to shorten a bit. At the end of the stroke, a molecule of ATP joins the myosin head, causing it to disengage from actin. As ATP is cleaved to form ADP and phosphate, the released energy causes the myosin head to move forward and bind once more to actin. The actin sliding past the myosin causes the sarcomere to shorten and the muscle to contract.

7. A motor neuron releases acetylcholine into the synaptic cleft and onto a muscle cell's plasma membrane, generating an action potential that enters the transverse tubules. A signal passes from here to the sarcoplasmic reticulum, causing the membranous sac to release stored calcium into the cell's cytoplasm. As a result the regulatory proteins (troponin and tropomyosin) move aside, allowing the myosin heads to slide actin filaments past myosin. Hence, the muscle cell contracts.

8. Heart muscle contains networks of cells that are electrically connected by intercalated discs, while skeletal muscle fibers do not communicate with each other. Both types of muscle are striated. Smooth muscle lacks striations; the cells have a single nucleus and communicate with each other via gap junctions.

9. The oxidative energy system, which is based on aerobic breakdown of glucose and other energy sources, is more effective at powering sustained muscle contractions. Slow oxidative muscle fibers (slow twitch muscle fibers) obtain most of their ATP from the oxidative system; they are rich in mitochondria, blood supply, and the red protein myoglobin, which stores oxygen. As a result, they resist fatigue and can contract for long periods of time.

10. Slow-oxidative muscle fibers contain many mitochondria, are highly vascularized, have large quantities of myoglobin, are red in color, and are capable of sustained contraction. Fast glycolytic fibers contain much actin and myosin, little myoglobin, are white in color, and provide greater power than red fibers but rapidly fatigue. Fast oxidative-glycolytic fibers have characteristics midway between first two fiber types.

11. Health benefits of regular exercise include lower body weight, decreased desire to smoke, increased relative amount of HDLP (high density lipoprotein), slightly lowered blood pressure, and increased blood flow to the heart.

Apply What You Have Learned

1. Bone processes are projections that serve as attachment sites for ligaments and tendons. On a skull these can be measured to determine the relative sizes of the muscles and bands of cartilage that once attached to them. From this information, an artist can recreate the shape of the face.

2. ATP is necessary for the myosin head to release from actin. At death, the supply of ATP ends and myosin becomes permanently attached to actin; the muscles remain stiff, resulting in rigor mortis.

3. Exercise stimulates the supply of endorphins from the pituitary and other brain regions. These substances act like morphine, reducing pain and stimulating a sense of well-being, resulting in a feeling of relaxation and contentment after the exercise.

4. The suggestion that the patient is paralyzed because of inadequate ATP production is not likely to be correct, because his heart and central nervous system are functioning normally. If ATP were in short supply these systems too would cease to function. A calcium deficiency is also unlikely, because the heart, like the skeletal muscles, depends on the movement of calcium ions to control contraction. If the enzymes that degrade acetylcholine were poisoned, his muscles would probably stay contracted rather than flaccid. Of these suggested causes, the most likely explanation for his paralysis is that acetylcholine release is inhibited, which would lead to flaccidity because skeletal muscles would not be stimulated by neurons. Because the heart generates its own signal, which is propagated by gap junctions, it would not be affected by a lack of acetylcholine.

CHAPTER 30

Review What You Have Learned

1. Taproot: carrot. Fibrous: grasses. Adventitious: corn.

2.

Ground Tissue	Description	Example
Parenchyma	loosely packed, thin-walled rounded storage cells	potato
Collenchyma	cylindrical cells, thick walls, long stringy fibers	celery
Sclerenchyma	die at maturity, leaving cell wall hardened with lignin	walnut shells

3. Xylem is composed of tracheids and vessel elements that become hollow at maturity. Tracheids transport water and minerals in all vascular plants; they are long narrow cells with pointed ends that overlap; there are pits in the cell walls through which fluid can move from cell to cell. Vessel elements are found mainly in flowering plants; their cells are stacked directly end-to-end and also have perforation plates separating them. They are dead when functional. Phloem vessels consist of cells, called sieve tubes, arranged end-to-end; they contain cytoplasm, but are without a functioning nucleus; each sieve tube is next to a companion cell whose nucleus directs the activities of both types of cells. Phloem conducts sugars and other substances from source to sink.

4. Phloem cells are alive and thin-walled. They transport sugars and amino acids rather than water and minerals. They move sugars from regions of high concentration to areas that require them for metabolism or growth.

5. A meristem is a region of parenchyma tissue that remains undifferentiated and gives rise to new cells and cell types during the plant's life.

6. Stamen: anther containing pollen, filament. Carpel: ovary, ovule containing egg, stigma, style.

7. The zone of cell division includes apical meristem protected by root cap cells. In the zone of elongation individual cells lengthen. In the zone of maturation, cells differentiate and begin to carry out specialized roles of dermal, ground, and vascular systems.

8. Endodermis borders the innermost part of the cortex and helps regulate the flow of water and minerals into the vascular system. The pericycle lies just inside the endodermis and consists of undifferentiated parenchyma cells that give rise to lateral roots, vascular and cork cambium.

9. The vascular cambium produces new phloem and xylem. The meristematic tissue of cork cambium forms the waterproof, protective layer, the periderm.

10. No. Most photosynthesis takes place in the palisade parenchyma.

Apply What You Have Learned
1. The root hairs, where most of a plant's water absorption occurs, may be damaged if the plant is roughly handled during transplanting.

2. The rings reflect the growth activity of the tree. When light, water, and nutrients are plentiful, as in the spring, the cambium produces large cells and wide rings. Because of the volcanic eruption on the Greek island of Santorini, the sun was obscured by volcanic dust, resulting in reduced photosynthesis and very narrow growth rings in the long-lived bristlecone pines that were in existence at the time.

3. Removing the bark would eliminate the phloem that normally transports sugars and amino acids downward from the leaves to the roots, eventually resulting in the death of the tree.

4. Among the population of trees bearing hard, unappetizing seeds, a few had seeds surrounded by a more fleshy, nutritious ovary, which attracted seed-eating birds. The birds spread the seeds widely, increasing the population of fleshy-fruited trees. Among this population, birds were most attracted to trees bearing the most tasty, fleshy fruit. This continued selection would result in attractive, tasty fruit such as pears and apples.

CHAPTER 31

Review What You Have Learned
1. Gibberellins regulate plant height and induce germination of seeds of rice, barley, and other grasses; they also play a role in flowering, fertilization, and

growth of fruits, new leaves, and young branches. Auxins simulate cell elongation in stems, leaves, and ovary wall; auxin prevents leaves, fruits, and flowers from falling off and causes overgowth and death in high concentration. Cytokinins stimulate cell division and the growth of lateral buds and prevent leaf aging.

2. Abscisic acid acts as a growth inhibitor by inhibiting translation of RNA, thus blocking protein synthesis. Abscisic acid induces and maintains dormancy and the closing of stomata.

3. Moisture leaches away abscisic acid from the seed, removing the hormone's growth inhibiting influence. Water may cause the embryo to secrete gibberellin. Increasing warmth in the spring may trigger hormonal activities that lead to germination. Ground cover may block light from reaching the seed, preventing the germination of small seeds such as lettuce seeds.

4. Gibberellins cause cells in a corn embryo to synthesize enzymes that break down starch to sugar, which can be used by the embryo as fuel for germination.

5. Tropism is directional growth in response to a stimulus, such as light (phototropism), gravity (gravitropism), or water (hydrotropism).

6. Some plants go through "sleep movements" when kept in total darkness, drooping their leaves during the hours of night and raising them during the hours of day.

7. Yes. Phytochrome in the P_{fr} form reflects the fact that the plant is receiving sufficient light at wavelengths needed to maintain a healthy level of chlorophyll. By contrast, the phytochrome of a plant that does not receive sufficient light will convert spontaneously to the P_r form.

8. Auxin promotes growth in the stem but inhibits development of the lateral buds into side branches. As it moves down the stem, its concentration is reduced and branches near the bottom of the tree are the least inhibited, allowing them to grow. The result is a conical shape. Cytokinin rising from the roots counteracts auxin's effect and allows the lowest branches to grow more strongly. In a plant with a more spherical shape, the apical bud probably produces less auxin and thus allows more lateral growth. If the apical bud is removed, there is more lateral growth, resulting in a bushy appearance.

9. Leaves of a chrysanthemum, a short-day plant, were exposed to short days and the flower buds to long days. The buds bloomed into flowers, indicating that the leaves sent a substance, florigen, to the bud, stimulating it to bloom. When the leaves were exposed to a long day, the plant did not flower.

10. Fruit growth is stimulated by auxin secreted by cells of the developing seed. Ethylene secreted by the cells of the fruit stimulates ripening. When auxin levels drop in a ripe fruit, enzymes break down cell walls in the abscission zone and the fruit drops.

11. Abscisic acid induces and maintains dormancy. Cytokinin keeps leaves green. Auxin inhibits the formation of an abscission zone. During the shorter days of autumn, there is a changing ratio of the hormones, leading to enzyme action that causes the cell walls in the abscission zone to break down and the leaf falls from the tree.

12. Repellents include camphor and tannins. Nicotine and cyanide are poisons. Plants make antibiotics that kill attacking fungi and bacteria. Some plants produce compounds that mimic insect hormones, disrupting normal insect metamorphosis.

Apply What You Have Learned

1. The 2,4-D stimulates such rapid cell division and tissue proliferation that the phloem becomes plugged, leaves and stems droop, twist, and curl, chlorophyll degrades, and the plant dries up and dies.

2. To increase the bushiness of Christmas trees, you should decrease the rate of stem elongation, reduce the growth of apical meristems, and promote the growth of lateral buds. You should decrease the production of gibberellins and auxins because these hormones promote stem elongation. You should increase the production of cytokinins, which stimulate the growth of lateral buds. To keep the plants growing vigorously and to prevent needle abscission, you should reduce the production of abscisic acid and ethylene.

3. Since 14 ¼ hours is less than 14 ½ hours, ragweed will bloom, and since 14 ¼ hours is more than 14 hours, the spinach will also bloom. This proves that the absolute amount of daylight is not the critical factor for designating a plant a short-day or long-day plant, but rather whether the photoperiod is longer or shorter than a specific time.

4. In a short day plant, flowering will be induced if there is a reduction in the number of hours of daylight. By watering it only during the day, the long night will be uninterrupted, allowing the accumulation of P_r during the long dark period, which promotes flowering. If you watered it at night, any stray light would convert P_r to P_{fr}, and this would inhibit flowering.

CHAPTER 32

Review What You Have Learned

1. Root hairs take up water across the plasma membrane via simple diffusion. The endodermis, the cell layer that separates the root cortex from the central vascular cylinder, regulates the flow of water into and out of the vascular tissues. Each endodermal cell is surrounded by a waxy belt (the Casparian strip) that prevents water from flowing through the cell walls and forces water to move through the cells' cytoplasm.

2. When membrane proteins in parenchyma cells actively pump ions into the xylem, water follows osmotically causing the volume of water in the vascular system to increase and hence causing root pressure to increase in the xylem.

3. The hydrogen bonds between water molecules cause them to cohere to each other in a long unbroken chain; water adheres to cellulose inside xylem vessels by adhesion. As a water molecule evaporates from a stoma, the entire liquid column moves up in the xylem tube, and at the same time a new water molecule moves into the roots.

4. Factors affecting the opening and closing of stomata include hormone activity and levels of carbon dioxide, light, temperature, moisture, and ions.

5. A legume gets nitrogen when nitrogen-fixing bacteria in the nodules convert nitrogen from the air into ammonia, which combines with a hydrogen ion to form the ammonium ion. A non-legume takes up ammonium ions from the soil where they were released from legume nodules, generated by free-living nitrogen-fixing bacteria in the soil, or released by ammonifying bacteria in the soil acting upon decaying organic matter. Nitrifying bacteria in the soil convert ammonium ions to nitrate ions, which are also taken up by the plant.

6. Calcium controls cell membrane permeability and is a component of pectin, which contributes to the elasticity of cell walls. Potassium regulates osmosis in plant cells and helps activate enzymes. Magnesium occurs in chlorophyll molecules and in cofactors of enzymes. Phosphorus is a part of DNA molecules and membrane phospholipids. Sulfur is a component of amino acids needed to build proteins and a component of some coenzymes.

7. Iron is involved in the synthesis of chlorophyll and is part of certain electron acceptors. Copper is present in chlorophyll and some enzymes. A deficiency of chlorine will stunt roots and fruit and wilt the entire plant. Zinc plays a role in protein synthesis.

8. Weathering by water, wind, heat, and cold disintegrates bedrock to smaller particles. Bacteria, fungi, algae, and lichens convert minerals in the particles to organic material, which becomes part of the soil when organisms excrete wastes and when they die.

9. After minerals are dissolved in soil water, they pass through the plasma membrane of the root hairs to the cells of the root cortex, are transported through the cytoplasm of endodermal cells, and are secreted into the xylem. They are lifted up by transpiration through the xylem and are distributed to all parts of the plant.

10. Sources include regions where sugars and other high-energy compounds are produced, such as leaves, or occur in high concentration, such as storage regions in the roots. Sinks are regions where cells are actively using the compounds, for example, in growing tips of roots and shoots.

11. A bit of plant tissue is placed in a culture medium. The cells then form an undifferentiated callus. Bits of callus are placed in a new culture medium with hormones that promote differentiation. Each bit of callus can differentiate into a tiny plant or a tiny embryo. Since all the tiny plants are derived from the parent's somatic cells, they have identical genes and grow into identical adults.

12. Somaclonal variations are new heritable traits that arise by mutation in a tissue-cultured callus. A breeder can select desired types for further growth of new useful varieties.

13. Researchers have inserted genes from *Bacillus thuringiensis* (whose toxin kills certain insect larvae) into tobacco plants, rendering these plants immune to the tobacco hornworm. Tobacco and petunia plants have received a gene transfer for conferring resistance to the herbicide glyphosate, which kills off other plants when sprayed on a field.

Apply What You Have Learned

1. By this type of crop rotation, clover plants, which have nitrogen-fixing bacteria in nodules on their roots, replenish the soil with ammonium ions that are needed by plants for the biosynthesis of amino acids and other nitrogen-containing biological molecules. The farmer is thus fertilizing his corn crop by growing clover and plowing it into the soil.

2. In early spring, the roots of maple trees break down stored starch into sugar, which passes into the phloem sieve tubes. Water from the xylem follows, along with water from the soil, creating enough root pressure to push the sap up into the trunk of the tree. By carefully tapping into the phloem of the tree, the sweet maple sap can be drained off and later boiled down into maple syrup.

3. The technique requires that some tomato cells already have genes for salt-tolerance. If such genes are lacking, it is not possible to develop salt-tolerant tomato plants from tissue culture. It may be necessary instead to resort to gene transfer techniques.

4. His vegetables died when he over-fertilized them because the chemical fertilizer increased the solute concentration in the soil to a level greater than the solute concentration in the plant roots. Water then left the roots by osmosis instead of entering the roots, and the plants died of dehydration.

5. No, you should not accept the job of developing a nitrogen-fixing strain of tomato. Your lab specializes in plant tissue culture, and as such can develop only the traits that already appear in plant tissues. The client needs to ask a lab specializing in genetic engineering to create the tomato because genes would have to be transferred from nitrogen-fixing bacteria into the tomato.

CHAPTER 33

Review What You Have Learned

1. A gene pool is the sum total of all alleles carried in all members of a population.

2. Mutation, gene duplication, exon shuffling, and recombination are causes of genetic variation.

3. (a) The frequency of genotype *AA* is p^2; (b) The frequency of *Aa* is 2pq; (c) The frequency of *aa* is q^2.

4. The Hardy-Weinberg principle assumes random mating, large population size, no natural selection, no mutation, and no gene flow.

5. Evolution is a change in allele frequencies over a period of time.

6. Directional selection favors one extreme form of a trait over all others. For example, the lightest and fastest-running cheetahs reproduced most successfully over time, favoring those with alleles for light weight. Stabilizing selection favors intermediate individuals. For example, more human babies are born weighing about 7 pounds than any other weight; heavier and lighter babies have lower chances of survival. Disruptive selection favors the extremes of variation, as in females of the African swallowtail butterfly that exactly resemble one or another butterfly species that is bad-tasting to predatory birds.

7. Mutation is the change of one allele into a different allele and provides new raw material for natural selection to act upon. When gene flow occurs, as individuals from one population migrate to an adjacent population, they may introduce alleles that are new in the second group. Genetic drift includes unpredictable changes in allele frequency caused by drastic reduction of population size, thus changing the composition of the gene pool. Non-random mating behavior can alter the frequency of alleles through sexual selection and can change the frequencies of heterozygotes and homozygotes from what random mating would predict.

8. Species are populations of organisms that can interbreed with each other in nature to produce fertile offspring.

9. Allopatric speciation is based on geographical isolation in which a physical barrier separates populations of a species and prevents gene flow between the two groups. Genetic drift and selection cause the two populations to diverge. Parapatric speciation occurs in populations that overlap and is common among organisms that can move only a short distance. Sympatric speciation occurs when a population becomes subdivided into reproductively isolated groups even without any spatial separation between the groups.

10. Microevolution is small changes in allele frequencies within species. Macroevolution is large changes above the species level.

11. Phyletic gradualism: Evolutionary change is slow and even, based on the process of microevolution. Punctuated equilibrium: Most species remain stable for long periods and then undergo periods of rapid change; during speciation, changes occur almost exclusively in small populations.

Apply What You Have Learned

1. Cheetahs are nearly identical genetically and the genome contains a significant number of genes for maladaptive traits. Through inbreeding, genetic variability was greatly reduced, leading to homozygosity of alleles for negative characters, such as susceptibility to disease and defective sperm. Elimination of habitat has contributed to the current small population of genetically inferior individuals that now face extinction.

2. Scientists may differ in their theories about how evolution takes place, but they generally accept the doctrine of evolution itself. Unlike Darwin, who emphasized phyletic gradualism, many scientists now argue that major evolutionary changes occur during brief periods interspersed by long periods that show little change.

3. The Hardy-Weinberg principle is a standard against which one can measure changes in allele frequency in populations over time, or, in other words, to measure evolution.

4. The argument that a minimum of several hundred owl pairs must be maintained is based on the assumption that genetic variability is essential to the survival of any species in nature. The argument that a few owl pairs in zoos represent a viable population assumes that all individuals are equal and identical to all others and that genetic variability is unimportant. A population of several hundred pairs is more likely to remain viable under natural conditions because

the gene pool may include enough variability to allow adaptation to a changing environment. A population reduced to a few breeding pairs of owls is likely to suffer many of the same problems as the cheetah, including reduced fertility and fitness because of homozygosity of maladaptive alleles, increased susceptibility to disease because of genetic uniformity, and reduced ability to accommodate environmental changes because of lack of variability in the gene pool.

CHAPTER 34

Review What You Have Learned

1. The biotic environment includes an environment's living organisms. The abiotic environment is the non-living environment, for example, the soil, air, rocks, weather, and so on.

2. A population is a group of interacting individuals of the same species in a particular geographic area, e.g., saguaro cactus of Arizona. A community includes two or more populations of different species occupying the same geographical area, e.g., Hohokam Indians and saguaro cactus. An ecosystem consists of interacting communities together with the physical factors of the environment, e.g., Hohokam Indians, saguaro cactus, corn, rainfall, and temperature. The biosphere comprises all ecosystems on Earth, together with the atmosphere, oceans, and soil.

3. Crude density indicates the number of organisms in an area of a certain size. Ecological density is related to the number of organisms only where they actually live, in their habitats.

4. An S-shaped curve represents the growth of a real population in a constant but limited environment. A J- shaped curve represents the innate capacity for growth given limitless resources.

5. Yes. High amounts of unused resources allow growth rates to reach their maximum; e.g., the population of Tasmanian sheep grew exponentially when they were first introduced to the island.

6. Delta N = change in the number of individuals; K = carrying capacity; r = rate of population growth per individual; r_m = innate capacity for increase; N = number already present.

7. Density dependent mechanisms, such as communicable diseases, become more influential as the population density increases. Density-independent mechanisms, such as hurricanes, exert their effects regardless of population density.

8. *Daphnia*, the water fleas, reproduce rapidly in a J-shaped curve until they exceed the carrying capacity of their environment; then, with the resources depleted, the population crashes.

9. *Daphnia* show r-selection, characterized by high rates of reproduction, an adaptation to colonized habitats that are generally unoccupied and unpredictable. Tasmanian sheep show K-selection, characterized by slow rates of reproduction and maintenance of the population near the carrying capacity of the environment.

10. The agricultural revolution and industrial revolution are factors that have contributed to the exponential growth of the human population.

11. Reduced growth rates have occurred in industrialized countries because both birth and death rates have dropped. In Third World countries there has been rapid population growth because death rates have declined while birth rates have remained high.

12. No. In 1900 there were more boys than girls up to the age of four.

Apply What You Have Learned
1. Postponing the decision to have a child until after the age of 30 probably will shift the curve slightly to the right.

2. The hunting of rhinos is reducing their population greatly, to the point of extinction, because they are K-selected, have low reproductive rates, and are highly specialized to compete for resources in their environments.

3. With limited food and other needed resources in the future, the human population will probably level out to an S-shaped curve. However, the possibility of a crash does exist if the population continues to grow at the present rate and if, like reindeer newly introduced to islands, we continue to consume resources at a rate that reduces the carrying capacity of the environment.

4. The maximum sustainable catch is likely to be gained from a population of 500,000 fish. The net rate of reproduction per individual is greatest at a population level of 100,000, because there is the greatest surplus of resources available to the fish, but the total rate of reproduction of the population is greater at a population level that is one-half the carrying capacity because N is much larger. At ninety percent of the carrying capacity, or a population of 900,000 fish, the net rate of reproduction per individual is so low that the total rate of reproduction for the population is lower than that of a population of 500,000. A simple way to visualize the relationship is to look at the slope of the logistic curve. The slope represents the rate of change per unit of time and is steepest at $K/2$.

CHAPTER 35

Review What You Have Learned
1. Habitat is the physical home of a species. Niche is the functional role of a species in the community.

2. Gause grew a large and a small species of *Paramecium* in the same container. Interspecific competition occurred and the smaller drove the larger to extinction. The smaller species was more resistant to bacterial waste products and so reproduced faster than the larger species.

3. Yes. On the island of St. Martin, there are two lizard species that eat the same kinds of insects but differ in size. In an experimental plot containing the two species, researchers found that individuals of the larger species had less food in their stomachs, grew more slowly, laid fewer eggs, and were forced to perch higher in the bushes than when that species lived alone in an enclosure. In other words, the presence of one species can limit another's niche and eventually may force it into extinction by competitive exclusion.

4. In North America 25,000 years ago, directional selection led to the evolution of saber-toothed tigers that specialized in elephants and giant ground sloths, lions that went after horses, and cheetahs that pursued small antelopes.

5. Coevolution is the simultaneous evolution of one or more species with another species, based on their interaction in the community. For example, running speed increased in cheetahs and in their antelope prey.

6. Predators may regulate prey populations by killing their prey and keeping the prey population below the carrying capacity of the enviroment. An example is dingos and red kangaroos. When dingos were excluded from range lands, red kangaroo populations shot up 170 times higher than before.

7. Predator strategies include pursuit of the prey, such as porpoises going after fish; ambushing prey, as when a frog ambushes an insect by snapping out its sticky tongue; or luring prey in, as when an alligator snapping turtle lures fish with its worm-shaped tongue.

8. Defensive adaptations include camouflage, as among insects that resemble a leaf; chemical warfare, as among eucalyptus and creosote plants that produce distasteful or toxic substances that deter herbivores; aposematic coloration, as in the bright colors of the poisonous frogs of Costa Rica that warn away predators; mimicry, in which the viceroy butterfly resembles the monarch butterfly, which is foul-tasting to predators; and weapons, such as sharp spines on the porcupine.

9. Not true. Parasites usually debilitate their hosts but often do not kill them outright. If they were more dangerous, they would destroy their own habitat.

10. In commensalism, one species benefits from the relationship while the other is neither harmed nor helped (e.g., epiphytes growing on trees). In mutualism, both species benefit from their relationship (e.g., the yucca plant and the yucca moth).

11. A pioneer community includes the first organisms to grow in an area. A climax community appears after a process of succession of different types of organisms; the species composition of the community becomes stable.

12. By eating prey, a predator reduces competition for space or other resources and thus preserves species richness. For example, a starfish species along the coast of Washington state preys on 15 species of barnacles, snails, clams, and mussels. When the starfish were removed, the number of prey species dropped to eight because the mussel population increased and crowded out some of the other invertebrates.

13. There are a great many species in a tropical rain forest. The more species that are interwoven into complex networks of competition, predation, and mutualism, the more fragile the entire community becomes. The extinction of any species may affect many other organisms.

Apply What You Have Learned

1. When the predators were removed, the deer multiplied to such an extent that there was not enough food for them all.

2. Darwin had observed a flower in the rain forest with a 12-inch floral tube and foresaw that there would be a moth with a 12-inch proboscis that was adapted to pollinate this flower.

3. The barren area was at first colonized by pioneer plants such as lichens and mosses, followed by horsetails and *Dryas*. Then alders added nitrogen to the soil, leading to abundant growths of willows and cottonwoods. Sitka spruce formed dense forests that shaded these trees out. After that, shade-tolerant hemlock grew below the spruce canopy and became the dominant tree. In low places, sphagnum moss soaked up large amounts of water and killed trees by choking off their roots' oxygen supply. So a mosaic community now exists of patches of spruce-hemlock forest and sphagnum bog.

4. At first, the insecticides killed most of the bollworms, reducing the damage to the cotton crop and increasing the productivity. Repeated exposure to the insecticide selected insecticide-resistant strains of the rapidly reproducing bollworm, and after the fourth year of pesticide application, the pesticide-resistant bollworm population increased greatly. The slower-reproducing predators of the bollworm were largely exterminated by the spray and could not control the population growth of the insecticide-resistant bollworm.

CHAPTER 36

Review What You Have Learned

1. A producer is an autotroph, which can produce needed biological molecules from nonliving substances. A consumer is a heterotroph, which obtains its biological molecules by consuming autotrophs, other heterotrophs, or detritus.

2. In chemosynthesis, chemoautotrophic bacteria in deep-sea vents or other habitats harvest the stored energy of hydrogen sulfide or other inorganic molecules to fix carbon dioxide into sugar molecules. In photosynthesis, green plants absorb solar energy to fix carbon dioxide into sugar molecules.

3. Fungi, bacteria, worms, vultures, and many kinds of insects obtain energy from organic wastes and dead organisms.

4. The trees serve as producers; grazing herbivores such as chipmunks and jays consume fruits and seeds, while other herbivores such as mice, deer, and caterpillars graze on leaves. These primary consumers are then eaten by hawks, owls, skunks, foxes, and snakes. In the detritus food web, fungi and bacteria break down dead tree parts, and then themselves serve as food for grubs and earthworms. Spiders and beetles consume soil invertebrates and, in turn, serve as food for salamanders, shrews, mice, and jays.

5. True. Photosynthesis converts only about 2% of the solar energy to chemical energy. Of this, only about half is actually stored in leaves, stems, roots, or fruits. This supports practically all other life in the ecosystem. Plants use the other half (about 1% of the solar energy) for their own cellular respiration and metabolism.

6. In an energy pyramid, the energy stored by the organisms at each trophic level is greater (often about ten times greater) than the energy stored by the organisms in the next higher trophic level.

7. The levels in a pyramid of biomass represent different trophic levels. The lowest level in the pyramid represents the autotrophs; the next higher level, primary consumers; the third level, secondary consumers, and so forth. The width of the bar represents the amount of energy stored at that level.

8. Water circulates from the atmosphere to the earth's surface as rain or snow. It then circulates back again to the atmosphere by evaporation from soil, puddles, ponds, rivers, and oceans, and by transpiration from plants.

9. During nitrogen fixation, bacteria in soil or in root nodules of legumes convert nitrogen gas from the air into ammonia, which dissolves in water to form ammonium ions; these are absorbed by plants and assimilated into proteins, nucleic acids, and other molecules. Bacteria produce ammonium from plant and animal remains and wastes, a process known as ammonification. In nitrification, certain bacteria change ammonium to nitrate, which can be taken up by plants. In denitrification, other bacteria change nitrate to nitrogen gas.

10. Plants assimilate phosphate from soil and water. Animals obtain phosphorus from plants or by eating other animals. When plants and animals die, bacteria convert the organic phosphorus in their tissues into phosphate, which enters the soil again.

11. Light energy from the sun passes through the atmosphere and warms the earth, which reradiates the energy as infrared radiation. This infrared radiation is trapped by the atmospheric blanket of carbon dioxide, thus heating the earth.

12. Oxides of nitrogen and sulfur are emitted by the burning of coal and oil by power plants, factories, and automobiles. These oxides in the air combine with water molecules and form nitric and sulfuric acids, which are carried down to the earth's surface by rain.

13. A nuclear winter would set in as clouds of dust block out sunlight, lowering the earth's temperature and causing the death of many plants; monsoon rains would be disrupted; massive flooding and erosion would result from rainfall over burned-out forests; radiation levels would be lethal to animal and plant life; poisonous substances released by urban fires would impede the regrowth of vegetation. The net result would be the extinction of many species and the permanent disruption of many ecosystems.

14. Human activity can disrupt the carbon cycle by increasing the rate at which carbon dioxide is released into the atmosphere (burning fossil fuels, kilning lime, and destroying forests) and decreasing the rate at which it is removed from the atmosphere (by suppressing vegetation).

Apply What You Have Learned

1. DDT resists degradation in the environment and tends to be stored in body fat. When used as an insecticide, it washes into streams and lakes, where it enters water plants that are later eaten by herbivorous fish. DDT then accumulates in fish tissues. Likewise, an osprey that eats the fish retains this DDT in its tissues. DDT interferes with calcium metabolism, resulting in thin-shelled, easily broken eggs.

2. Ammonifying soil bacteria produced ammonia from the fish proteins and nucleic acids. The growing corn assimilated this available nitrogen and yielded a fast-growing, productive crop.

3. Phosphates from the detergent would run off into rivers and lakes, stimulating algae and aquatic plants to grow so fast that eutrophication would set in. As algae and aquatic plants die, their tissues fuel the metabolism of decomposers. As these processes go forward, so much dissolved oxygen is consumed that fish suffocate.

4. The greenhouse effect might change the global temperature so rapidly that climatic zones would shift toward the poles at a pace faster than plants can colonize new ground. The maturation and seed dispersal rates of many plants do not permit rapid expansion of their ranges. For example, a process of succession usually must precede the growth of a mature forest. Succession and establishment of the forest community may take two centuries or longer before the climax species appear and set seed. Then the seeds may be dispersed only a few kilometers, limiting the rate of spread of the species. The retreat of the glaciers ten thousand years ago was slow enough to accommodate this process. Another factor is that human-caused disturbance limits the spread of natural ecosystems. Cultivated farmland, for example, is a barrier to the spread of forests. And the animal species that are dependent on certain plant species, directly as consumers or indirectly as predators on those consumers, cannot survive outside of the range of those plant species.

CHAPTER 37

Review What You Have Learned

1. Biosphere: the portion of planet earth which includes the water, air, and soil that supports life. Biome: a major community of organisms, defined mainly by their vegetation, and characterized by specific adaptations to a particular climate.

2. Coils I north and south: Warm, moist air rises at the equator and drops its moisture in tropical rains as it gains altitude and cools; the air then descends at latitude 30 degrees, north and south, and since it is dry, creates great deserts. Trade winds are generated as the air flows near the earth's surface and toward the equator. Coils 2 north and south: Dry air descends at latitude 30 degrees, moves toward the poles along the earth's surface and ascends at latitude 60 degrees, producing winds from west to east. Coil 3: The mild air moving toward the poles in Coil 2 meets cold air flowing from the poles in Coil 3. This polar front causes the air to rise again at latitudes 60 degrees north and south and to give up its moisture, supporting the great temperate forests. Most of the air moves toward the poles, where the frigid air descends, bringing polar weather into the United States.

3. Factors influencing ocean currents include the earth's rotation, the prevailing winds blowing over ocean surfaces, and the positions of the continents.

4. The trade winds reversed and the warm ocean currents from the Tahiti area flowed back to South America, warming the ocean around the Galapagos Islands, Ecuador, and Peru. This surface layer of warm water prevented much of the normal upwelling of nutrients, reducing the ocean's productivity and starving fishes, squids, and seals.

5. Tropical rain forests are characterized by heavy rainfall, temperatures above average, light and minerals in short supply, and dense vegetation that blocks winds. Tropical savannas are characterized by year-long warmth, an extended dry season, and stunted trees interspersed with tall grasses.

6. Some deserts arise as dehydrated air descends back toward earth. Others, like the Gobi desert, lie at the centers of continents, far from moist sea air. Some deserts form because they are located in the rainshadows of tall mountain ranges.

7. Grasslands lack trees and have extremes of hot and cold weather. The soils are richer in organic matter than those of other biomes and the grasses have dense roots. Chaparral is found in areas with hot, dry summers and cool, wet winters. The plants are generally less than 2 meters tall and have hairy, leathery leaves. The roots and underground stems are fire-resistant and some species have seeds that must be seared by fire before they can germinate.

8. Temperate forests are dominated by broad-leaved, deciduous trees, receive intermediate amounts of rainfall, and experience moderate temperatures between summer highs and winter lows. Coniferous forest consist primarily of evergreen trees with needles. The southern border of the coniferous forest tends to fall at the southern limit of the polar front.

9. The permafrost zone, a meter below the surface, is (as its name suggests) permanently frozen, even in summer. It inhibits the growth of trees due to problems with root penetration and survival.

10. Polar bears, seals, penguins, a few insects, and microbes are found on the polar ice caps; plants are generally not.

11.

Zone	**Amount of Light**	**Organisms**
Littoral	ample; penetrates to the bottom	water lilies, algae, cattails, protozoa, insects, snails, amphibians, fish, birds
Limnetic	penetrates through the top layer	phytoplankton, zooplankton
Profundal	dark	decomposers, scavenger fish, insect larvae

12. Fall turnover occurs when surface water cools to 4°C and becomes heavy and sinks to the bottom, carrying oxygen with it. The lower water is stirred and rises, carrying nutrients up to the photosynthetic layer. After the winter's ice melts, spring turnover takes place. The upper water warms to 4°C, sinks to the bottom through the colder, less dense water, and churns up nutrients from the bottom.

13. An estuary is an area where salt water from the ocean mingles with fresh water from a river. The daily change of tides stirs up nutrients, making an estuary one of the most productive of ecosystems. There is rapid growth of plants and algae, supporting fish, shellfish, worms, birds, and a host of other organisms.

14. The intertidal zone extends along the coastline between high and low tides; it includes tide pools. The neritic zone extends from the intertidal zone to the edge of the continental shelf. The pelagic zone is the open ocean beyond the continental shelf.

Apply What You Have Learned

1. Less than 2% of light reaches the floor of a tropical rain forest because of the thick canopy above. Philodendrons can exist in this setting because their leaves are capable of capturing the greatest amount of the available pale light. It is thus useful as a house plant because it can survive on indirect, indoor light.

2. Yes, one would expect tundra type vegetation. The top of the mountain is snow-covered; conditions there resemble those of the tundra in polar regions. This is where small plants grow, similar to those of the arctic zone.

3. No. Trout thrive only where there is plentiful oxygen and the water is cold and clear, such as in a rapidly-flowing stream. The calmer water downstream moves sluggishly and is cloudy, with less dissolved oxygen; conditions here are not conducive to trout.

4. Once a tropical rainforest is clearcut, it is gone forever (for all practical purposes). Most of the nutrients are held in living plant tissue; when the forest is cut and burned over large areas, the nutrients are rapidly washed away, leaving the soil infertile. Organic matter in the soil is quickly lost due to rapid decomposition, leaving the soil with a texture unsuitable for seed germination. Transpiration is greatly reduced when the forest is removed, reducing local rainfall and changing the climate so that rain forests are unlikely to return. The species that are driven to extinction by habitat destruction will not reappear.

CHAPTER 38

Review What You Have Learned

1. Proximate causes include the physiological mechanisms that direct an animal to perform a behavior, such as the neuronal pathways that cause a gull to remove an eggshell from the nest. The ultimate cause is the selective advantage the action confers on the active individual; e.g., gulls possessing genes that allow the bird to recognize and remove eggshells will produce more surviving chicks than those that don't.

2. Instincts are actions that are inborn and unchangeable. Learned behaviors can be modified by experience.

3. A fixed-action pattern is triggered by a specific sign stimulus or releaser. The behavior must proceed in a particular sequence, in an all-or-none fashion. It is innate, fully formed, and functional the first time it is used. All members of a species of the same age and sex perform the behavior the same way under similar environmental conditions. It depends on the animal's maturation and drive, which is influenced by hormones or other physiological states.

4. One-trial learning. Habituation. Trial-and-error learning. Classical conditioning. Insight learning.

5. A taxis is an animal's movement toward or away from a specific stimulus, such as light, a chemical, or heat.

6. Factors that influence animal migration include the sun's position, the circadian clock, and the earth's magnetic field.

7. When a male gull claims a territory on a beach, he assumes stylized body positions and issues specific calls to warn away other males.

8. Females generally invest more resources in their offspring than do males. Female birds allocate much nutrition to form the eggs; female mammals nourish and shelter their offspring in their uterus and then feed them with milk long after birth.

9. Sight: a firefly's blinking light. Sound: the alarm call of a prairie dog as a warning of an approaching predator. Scent: sex pheromones released into the air by female moths, attracting male moths from miles away. Touch: a returning bee performs a "dance" that is sensed by other bees via touch and interpreted to communicate the whereabouts of nearby food.

10. Kin selection is a type of natural selection in which an individual increases its genetic contribution to the next generation by aiding and increasing the survival and reproductive success of relatives, with whom the individual shares alleles.

11. (a) half; (b) half; (c) one quarter; (d) one quarter; (e) one eighth.

Apply What You Have Learned

1. Since the chicken egg does not have jagged edges, the gull's neural programs are set up by its genes to accept it.

2. No, the goslings will not imprint on you. Imprinting is impossible from the third day on. It lasts from the time of hatching to the second day.

3. This type of advertising relies on classical conditioning, involving the association of the cereal with a desirable role model.

4. The newly dominant male monkey increases his reproductive success by killing all nursing infants and newborns for several months after the takeover. These infants were sired by the previously dominant male and thus they do not contribute to the newly dominant male's reproductive success. Their mothers ovulate sooner than they would if the infants survived, thus allowing the newly dominant male to reproduce sooner, increasing his reproductive success. The genes for infanticide thus increase in the population.